Amyloid Proteins

METHODS IN MOLECULAR BIOLOGY™

John M. Walker, SERIES EDITOR

309. **RNA Silencing:** *Methods and Protocols,* edited by *Gordon Carmichael, 2005*

308. **Therapeutic Proteins:** *Methods and Protocols,* edited by *C. Mark Smales and David C. James, 2005*

307. **Phosphodiesterase Methods and Protocols,** edited by *Claire Lugnier, 2005*

306. **Receptor Binding Techniques:** *Second Edition,* edited by *Anthony P. Davenport, 2005*

305. **Protein–Ligand Interactions:** *Methods and Protocols,* edited by *G. Ulrich Nienhaus, 2005*

304. **Human Retrovirus Protocols:** *Virology and Molecular Biology,* edited by *Tuofu Zhu, 2005*

303. **NanoBiotechnology Protocols,** edited by *Sandra J. Rosenthal and David W. Wright, 2005*

302. **Handbook of ELISPOT: Methods and Protocols,** edited by *Alexander E. Kalyuzhny, 2005*

301. **Ubiquitin–Proteasome Protocols,** edited by *Cam Patterson and Douglas M. Cyr, 2005*

300. **Protein Nanotechnology:** *Protocols, Instrumentation, and Applications,* edited by *Tuan Vo-Dinh, 2005*

299. **Amyloid Proteins:** *Methods and Protocols,* edited by *Einar M. Sigurdsson, 2005*

298. **Peptide Synthesis and Application,** edited by *John Howl, 2005*

297. **Forensic DNA Typing Protocols,** edited by *Angel Carracedo, 2005*

296. **Cell Cycle Protocols:** *Mechanisms and Protocols,* edited by *Tim Humphrey and Gavin Brooks, 2005*

295. **Immunochemical Protocols,** *Third Edition,* edited by *Robert Burns, 2005*

294. **Cell Migration:** *Developmental Methods and Protocols,* edited by *Jun-Lin Guan, 2005*

293. **Laser Capture Microdissection:** *Methods and Protocols,* edited by *Graeme I. Murray and Stephanie Curran, 2005*

292. **DNA Viruses:** *Methods and Protocols,* edited by *Paul M. Lieberman, 2005*

291. **Molecular Toxicology Protocols,** edited by *Phouthone Keohavong and Stephen G. Grant, 2005*

290. **Basic Cell Culture Protocols,** *Third Edition,* edited by *Cheryl D. Helgason and Cindy L. Miller, 2005*

289. **Epidermal Cells,** *Methods and Applications,* edited by *Kursad Turksen, 2005*

288. **Oligonucleotide Synthesis,** *Methods and Applications,* edited by *Piet Herdewijn, 2005*

287. **Epigenetics Protocols,** edited by *Trygve O. Tollefsbol, 2004*

286. **Transgenic Plants:** *Methods and Protocols,* edited by *Leandro Peña, 2005*

285. **Cell Cycle Control and Dysregulation Protocols:** *Cyclins, Cyclin-Dependent Kinases, and Other Factors,* edited by *Antonio Giordano and Gaetano Romano, 2004*

284. **Signal Transduction Protocols,** *Second Edition,* edited by *Robert C. Dickson and Michael D. Mendenhall, 2004*

283. **Bioconjugation Protocols,** edited by *Christof M. Niemeyer, 2004*

282. **Apoptosis Methods and Protocols,** edited by *Hugh J. M. Brady, 2004*

281. **Checkpoint Controls and Cancer, Volume 2:** *Activation and Regulation Protocols,* edited by *Axel H. Schönthal, 2004*

280. **Checkpoint Controls and Cancer, Volume 1:** *Reviews and Model Systems,* edited by *Axel H. Schönthal, 2004*

279. **Nitric Oxide Protocols,** *Second Edition,* edited by *Aviv Hassid, 2004*

278. **Protein NMR Techniques,** *Second Edition,* edited by *A. Kristina Downing, 2004*

277. **Trinucleotide Repeat Protocols,** edited by *Yoshinori Kohwi, 2004*

276. **Capillary Electrophoresis of Proteins and Peptides,** edited by *Mark A. Strege and Avinash L. Lagu, 2004*

275. **Chemoinformatics,** edited by *Jürgen Bajorath, 2004*

274. **Photosynthesis Research Protocols,** edited by *Robert Carpentier, 2004*

273. **Platelets and Megakaryocytes, Volume 2:** *Perspectives and Techniques,* edited by *Jonathan M. Gibbins and Martyn P. Mahaut-Smith, 2004*

272. **Platelets and Megakaryocytes, Volume 1:** *Functional Assays,* edited by *Jonathan M. Gibbins and Martyn P. Mahaut-Smith, 2004*

271. **B Cell Protocols,** edited by *Hua Gu and Klaus Rajewsky, 2004*

270. **Parasite Genomics Protocols,** edited by *Sara E. Melville, 2004*

269. **Vaccina Virus and Poxvirology:** *Methods and Protocols,* edited by *Stuart N. Isaacs, 2004*

268. **Public Health Microbiology:** *Methods and Protocols,* edited by *John F. T. Spencer and Alicia L. Ragout de Spencer, 2004*

267. **Recombinant Gene Expression:** *Reviews and Protocols, Second Edition,* edited by *Paulina Balbas and Argelia Johnson, 2004*

266. **Genomics, Proteomics, and Clinical Bacteriology:** *Methods and Reviews,* edited by *Neil Woodford and Alan Johnson, 2004*

265. **RNA Interference, Editing, and Modification:** *Methods and Protocols,* edited by *Jonatha M. Gott, 2004*

METHODS IN MOLECULAR BIOLOGY™

Amyloid Proteins

Methods and Protocols

Edited by

Einar M. Sigurdsson

Departments of Psychiatry and Pathology,
New York University School of Medicine, New York, NY

HUMANA PRESS ✳ TOTOWA, NEW JERSEY

© 2005 Humana Press Inc.
999 Riverview Drive, Suite 208
Totowa, New Jersey 07512

www.humanapress.com

This publication is printed on acid-free paper. ∞
ANSI Z39.48-1984 (American Standards Institute)

Permanence of Paper for Printed Library Materials.

Production Editor: Nicole E. Furia
Cover design by Patricia F. Cleary
Cover art provided by Einar M. Sigurdsson.

For additional copies, pricing for bulk purchases, and/or information about other Humana titles, contact Humana at the above address or at any of the following numbers: Tel.: 973-256-1699; Fax: 973-256-8341; E-mail: humana@humanapr.com; or visit our Website: www.humanapress.com

Printed in the United States of America. 10 9 8 7 6 5 4 3 2 1

eISBN 1-59259-874-9

ISSN 1064-3745

Library of Congress Cataloging in Publication Data

Amyloid proteins : methods and protocols / edited by Einar M. Sigurdsson.
    p. ; cm. — (Methods in molecular biology ; v. 299)
  Includes bibliographical references and index.
  ISBN 1-58829-337-8 (hardcover : alk. paper)
  1. Amyloid—Laboratory manuals. 2. Amyloid beta-protein—Laboratory manuals.
  [DNLM: 1. Amyloid—biosynthesis—Laboratory Manuals. 2. Amyloid—isolation & purification—Laboratory Manuals. 3. Biological Assay—methods—Laboratory Manuals. 4. Cells, Cultured—cytology—Laboratory Manuals. QU 25 A531 2005] I. Sigurdsson, Einar M. II. Series: Methods in molecular biology (Clifton, N.J.) ; v. 299.
  QP552.A45A495 2005
  616.3'995—dc22
                            2004012023

Preface

Amyloid diseases are characterized by the deposition of insoluble fibrous amyloid proteins. The word "amyloid" indicates a starch-like compound, and though a misnomer, continues to be the accepted term for this group of protein conformational disorders. Approximately 30 different proteins can form amyloid and although there is usually no homology in their amino acid sequence, all share a β-pleated secondary structure. Historically, these β-pleated deposits were detected by histological dyes, and the characteristic fibril structure confirmed with electron microscopy. As these amyloids were purified and sequenced, various in vitro techniques were developed, often using synthetic peptides and/or highly purified amyloid derived from diseased tissue. Development of animal models occurred concurrently and some of these diseases can now be passed on to animals by injecting them with amyloid-rich tissue fractions, suggesting an infectious nature for these proteins. For other amyloidoses, transgenic technology has been necessary for recapitulating the disease. Together, these in vitro and in vivo models have been used to understand the etiology and pathogenesis of amyloid diseases as well as to screen for drugs that can block amyloid formation and/or disassemble the fibrils.

Several of these methods and protocols are detailed in *Amyloid Proteins: Methods and Protocols,* using examples from various amyloids. The volume is divided into three parts. Part I contains in vitro assays, starting with a few chapters that focus on preparation of amyloid and its precursors (oligomers and protofibrils). These are followed by chapters detailing specific analytical methods for studying these proteins. Part II describes cell culture models and assays for production of amyloid proteins, and Part III consists of protocols for amyloid extraction from tissue, its detection in vitro and in vivo, as well as nontransgenic methods for developing amyloid mouse models. Most of the chapters follow a similar format and are detailed protocols for performing a particular procedure. However, certain chapters focus more on the general principles and theoretical issues of a particular method.

It is my hope that these articles will be useful both for students and scientists new to the amyloid field as well as for seasoned investigators learning new techniques to further their research.

I would like to thank the authors for their contribution and the series editor, Dr. John M. Walker, for the opportunity to edit this book.

Einar M. Sigurdsson

Contents

Preface v

Contributors xi

PART I. IN VITRO ASSAYS

1 Preparation of Aggregate-Free, Low Molecular Weight Amyloid-β
    for Assembly and Toxicity Assays
    Gal Bitan and David B. Teplow 3

2 Determination of Peptide Oligomerization State Using Rapid
    Photochemical Crosslinking
    Sabrina S. Vollers, David B. Teplow, and Gal Bitan 11

3 In Vitro Preparation of Prefibrillar Intermediates of Amyloid-β
    and α-Synuclein
    Hilal A. Lashuel and Dolors Grillo-Bosch 19

4 Purification of Recombinant Tau Protein and Preparation
    of Alzheimer-Paired Helical Filaments In Vitro
    Stefan Barghorn, Jacek Biernat, and Eckhard Mandelkow 35

5 Cyclic Amplification of Protein Misfolding and Aggregation
    Paula Saá, Joaquín Castilla, and Claudio Soto 53

6 X-Ray Diffraction Studies of Amyloid Structure
    O. Sumner Makin and Louise C. Serpell 67

7 Molecular Electron Microscopy Approaches to Elucidating
    the Mechanisms of Protein Fibrillogenesis
    Hilal A. Lashuel and Joseph S. Wall 81

8 Time-Lapse Atomic Force Microscopy in the Characterization
    of Amyloid-Like Fibril Assembly and Oligomeric Intermediates
    Claire Goldsbury and Janelle Green 103

9 Fourier Transform Infrared and Circular Dichroism Spectroscopies
    for Amyloid Studies
    Miguel Calero and María Gasset 129

10 Quasielastic Light Scattering for Protein Assembly Studies
    Aleksey Lomakin, David B. Teplow, and George B. Benedek 153

11 Intrinsic Fluorescent Detection of Tau Conformation
    and Aggregation
    Martin von Bergen, Li Li, and Eckhard Mandelkow 175

vii

12 Quantitative Measurement of Fibrillogenesis by Mass Spectrometry
    Andrew D. Miranker *185*

PART II. CELL CULTURE ASSAYS

13 Isolation and Culturing of Human Vascular Smooth Muscle Cells
    Finnbogi R. Thormodsson and Ingvar H. Olafsson *197*

14 Murine Cerebrovascular Cells as a Cell Culture Model
    for Cerebral Amyloid Angiopathy: *Isolation of Smooth Muscle
    and Endothelial Cells From Mouse Brain*
    Sonia S. Jung and Efrat Levy *211*

15 Purification of Human Wild-Type or Variant Cystatin C
    From Conditioned Media of Transfected Cells
    *Frances Prelli, Monika Pawlik, Blas Frangione,
    and Efrat Levy* *221*

16 Prion Propagation in Cell Culture
    Sylvain Lehmann *227*

PART III. IN VIVO-RELATED ASSAYS

17 Preparation and Propagation of Amyloid-Enhancing Factor
    Robert Kisilevsky *237*

18 Purification of Amyloid Protein AA Subspecies From Amyloid-Rich
    Human Tissues
    Gunilla T. Westermark and Per Westermark *243*

19 Purification of Transthyretin and Transthyretin Fragments
    From Amyloid-Rich Human Tissues
    Per Westermark and Gunilla T. Westermark *255*

20 Extraction and Chemical Characterization of Tissue-Deposited
    Proteins From Minute Diagnostic Biopsy Specimens
    Fernando Goñi and Gloria Gallo *261*

21 Tissue Processing Prior to Protein Analysis and Amyloid-β Quantitation
    *Stephen D. Schmidt, Ying Jiang, Ralph A. Nixon,
    and Paul M. Mathews* *267*

22 ELISA Method for Measurement of Amyloid-β Levels
    Stephen D. Schmidt, Ralph A. Nixon, and Paul M. Mathews *279*

23 Histological Staining of Amyloid-β in Mouse Brains
    Einar M. Sigurdsson *299*

24 The Mouse Model for Scrapie: *Inoculation, Clinical Scoring,
    and Histopathological Techniques*
    Harry C. Meeker, Xuemin Ye, and Richard I. Carp *309*

25 Radiolabeling of Amyloid-β Peptides
Miguel Calero and Jorge Ghiso 325

26 In Vivo Imaging of Amyloid-β Deposits in Mouse Brain
With Multiphoton Microscopy
Jesse Skoch, Gregory A. Hickey, Stephen T. Kajdasz,
Bradley T. Hyman, and Brian J. Bacskai 349

27 Magnetic Resonance Imaging of Amyloid Plaques in Transgenic Mice
Youssef Zaim Wadghiri, Einar M. Sigurdsson,
Thomas Wisniewski, and Daniel H. Turnbull 365

Index 381

Contributors

BRIAN J. BACSKAI • *Department of Neurology, Massachusetts General Hospital, Charlestown, MA*

STEFAN BARGHORN • *Max-Planck-Unit for Structural Molecular Biology, Hamburg, Germany*

GEORGE B. BENEDEK • *Department of Physics and Center for Materials Science and Engineering, Massachusetts Institute of Technology, Cambridge, MA*

JACEK BIERNAT • *Max-Planck-Unit for Structural Molecular Biology, Hamburg, Germany*

GAL BITAN • *Center for Neurologic Diseases, Brigham and Women's Hospital and Department of Neurology, Harvard Medical School, Boston, MA*

DOLORS GRILLO-BOSCH • *Departament de Quimica Organica, Universitat de Barcelona, Barcelona, Spain*

MIGUEL CALERO • *Centro Nacional de Microbiología, Instituto de Salud Carlos III, Majadahonda, Madrid, Spain*

RICHARD I. CARP • *Department of Virology, New York State Institute for Basic Research in Developmental Disabilities, Staten Island, NY*

JOAQUÍN CASTILLA • *Department of Neurology, University of Texas Medical Branch, Galveston, TX*

BLAS FRANGIONE • *Departments of Pathology and Psychiatry, New York University School of Medicine, New York, NY*

GLORIA GALLO • *Department of Pathology, New York University School of Medicine, New York, NY*

MARÍA GASSET • *Instituto Química-Física Rocasolano, CSIC, Madrid, Spain*

JORGE GHISO • *Departments of Pathology and Psychiatry, New York University School of Medicine, New York, NY*

CLAIRE GOLDSBURY • *Max-Planck-Unit for Structural Molecular Biology, Hamburg, Germany*

FERNANDO GOÑI • *Departments of Neurology and Psychiatry, New York University School of Medicine, New York, NY; Department of Immunology, University of Uruguay School of Chemistry, Montevideo, Uruguay*

JANELLE GREEN • *M.E. Müller Institute for Structural Biology at the Biozentrum, University of Basel, Basel, Switzerland*

GREGORY A. HICKEY • *Department of Neurology, Massachusetts General Hospital, Charlestown, MA*

BRADLEY T. HYMAN • *Department of Neurology, Massachusetts General Hospital, Charlestown, MA*

YING JIANG • *Center for Dementia Research, Nathan Kline Institute, Orangeburg, NY*

SONIA S. JUNG • *Center for Dementia Research, Nathan S. Kline Institute for Psychiatric Research, Orangeburg, NY*

STEPHEN T. KAJDASZ • *Department of Neurology, Massachusetts General Hospital, Charlestown, MA*

ROBERT KISILEVSKY • *Department of Pathology and Molecular Medicine, Queen's University and The Syl and Molly Apps Research Center, Kingston General Hospital, Kingston, Ontario Canada*

HILAL A. LASHUEL • *Center for Neurologic Diseases, Brigham and Women's Hospital and Department of Neurology, Harvard Medical School, Cambridge, MA*

SYLVAIN LEHMANN • *Institut de Génétique Humaine du CNRS Montpellier, France and Laboratoire de Biochimie, Hôpital St. Eloi, Montpellier, France*

EFRAT LEVY • *Departments of Psychiatry and Pharmacology, New York University School of Medicine, New York, NY and Center for Dementia Research, Nathan S. Kline Institute for Psychiatric Research, Orangeburg, NY*

LI LI • *Max-Planck-Unit for Structural Molecular Biology, Hamburg, Germany*

ALEKSEY LOMAKIN • *Department of Physics and Center for Materials Science and Engineering, Massachusetts Institute of Technology, Cambridge, MA*

O. SUMNER MAKIN • *Structural Medicine Unit, Department of Haematology, Cambridge Institute for Medical Research, University of Cambridge, Cambridge, United Kingdom*

ECKHARD MANDELKOW • *Max-Planck-Unit for Structural Molecular Biology, Hamburg, Germany*

PAUL M. MATHEWS • *Center for Dementia Research, Nathan Kline Institute, Orangeburg, NY; Department of Psychiatry, New York University School of Medicine, New York, NY*

HARRY C. MEEKER • *Department of Virology, New York State Institute for Basic Research in Developmental Disabilities, Staten Island, NY*

ANDREW D. MIRANKER • *Department of Molecular Biophysics and Biochemistry, Yale University, New Haven, CT*

RALPH A. NIXON • *Center for Dementia Research, Nathan Kline Institute, Orangeburg, NY; Departments of Psychiatry and Cell Biology, New York University School of Medicine, New York, NY*

INGVAR H. OLAFSSON • *Karolinska Institute and University Hospital, Stockholm, Sweden*

MONIKA PAWLIK • *Department of Pharmacology, New York University School of Medicine, New York, NY; Nathan Kline Institute Orangeburg, NY*

FRANCES PRELLI • *Department of Pathology, New York University School of Medicine, New York, NY*

PAULA SAÁ • *Department of Neurology, University of Texas Medical Branch, Galveston, TX*

STEPHEN D. SCHMIDT • *Center for Dementia Research, Nathan Kline Institute, Orangeburg, NY*

LOUISE C. SERPELL • *Structural Medicine Unit, Department of Haematology, Cambridge Institute for Medical Research, University of Cambridge, Cambridge, United Kingdom*

JESSE SKOCH • *Department of Neurology, Massachusetts General Hospital, Charlestown, MA*

EINAR M. SIGURDSSON • *Departments of Psychiatry and Pathology, New York University School of Medicine, New York, NY*

CLAUDIO SOTO • *Department of Neurology, University of Texas Medical Branch, Galveston, TX*

DAVID B. TEPLOW • *Center for Neurologic Diseases, Brigham and Women's Hospital and Department of Neurology, Harvard Medical School, Boston, MA*

FINNBOGI R. THORMODSSON • *Faculty of Medicine, University of Iceland, Reykjavik, Iceland*

DANIEL H. TURNBULL • *Skirball Institute of Biomolecular Medicine and Departments of Radiology and Pathology, New York University School of Medicine, New York, NY*

SABRINA S. VOLLERS • *Center for Neurologic Diseases, Brigham and Women's Hospital and Department of Neurology, Harvard Medical School, Boston, MA*

MARTIN VON BERGEN • *Max-Planck-Unit for Structural Molecular Biology, Hamburg, Germany*

YOUSSEF ZAIM WADGHIRI • *Skirball Institute of Biomolecular Medicine and Department of Radiology, New York University School of Medicine, New York, NY*

JOSEPH S. WALL • *Department of Biology, Brookhaven National Laboratory, Upton, NY*

GUNILLA T. WESTERMARK • *Department of Biomedicine and Surgery, Division of Cell Biology, Linköping University, Linköping, Sweden*

PER WESTERMARK • *Department of Genetics and Pathology, Rudbeck Laboratory, Uppsala University, Uppsala, Sweden*

THOMAS WISNIEWSKI • *Departments of Neurology, Psychiatry, and Pathology, New York University School of Medicine, New York, NY*

XUEMIN YE • *Department of Virology, New York State Institute for Basic Research in Developmental Disabilities, Staten Island, NY*

I

In Vitro Assays

1

Preparation of Aggregate-Free, Low Molecular Weight Amyloid-β for Assembly and Toxicity Assays

Gal Bitan and David B. Teplow

Summary

More than 20 diseases have been identified which are caused by the deposition of amyloid. Natural and chemically synthesized amyloidogenic proteins are used widely to study the structure, assembly, and physiologic effects of both oligomeric and fibrillar forms of these proteins. In many cases, conflicting results arise in these studies, in part owing to difficulties in reproducibly preparing amyloidogenic proteins in a well-defined assembly state. To avoid these problems, several methods have been devised that provide reliable means of preparing amyloid-forming proteins for experimental use. Here, we discuss methods that have been used successfully to prepare one such protein, the amyloid β-protein (Aβ), involved in Alzheimer's disease.

Methods for reproducible preparation of Aβ in a well-defined assembly state include isolation of low molecular weight (LMW) Aβ by size exclusion chromatography, filtration through LMW cut-off filters, and solubilization/lyophilization in the presence of reagents which facilitate disassembly of Aβ. These reagents include strong bases and acids, and fluorinated alcohols. These methods, which were originally developed for Aβ, are generally applicable to amyloidogenic peptides and proteins. In this chapter, we describe the preparation of LMW Aβ using size exclusion chromatography and filtration. The advantages and disadvantages of each method are discussed.

Key Words: Aggregation; solubility; size exclusion chromatography; filtration; amyloid β-protein.

1. Introduction

In vitro studies using synthetic amyloidogenic proteins have been highly important for understanding the mechanisms by which these proteins assemble into neurotoxic species and thus may be involved in aberrant protein folding diseases. A significant difficulty in these studies has been obtaining reproducible data. For example, in studies of the amyloid β-protein (Aβ), significant differences in assembly kinetics and neurotoxicity have been observed using

From: *Methods in Molecular Biology, vol. 299: Amyloid Proteins: Methods and Protocols*
Edited by: E. M. Sigurdsson © Humana Press Inc., Totowa, NJ

synthetic Aβ from different manufacturers or even using different lots from the same manufacturer *(1–4)*. This irreproducibility likely resulted from the presence of preexisting aggregates in the peptide stocks. These seeds must be removed or dissociated in order to improve reproducibility. An "aggregate-free" protein solution contains only monomer and small oligomers *(5)* and is termed low molecular weight (LMW) protein solution. Removal of preexisting aggregates is performed based on the size difference between the aggregates and the LMW protein solution. The methods for Aβ illustrated here are generally applicable to other amyloidogenic proteins. LMW Aβ has been isolated by using size exclusion chromatography (SEC) *(6)*, or filtration through a low molecular weight cutoff (MWCO) filter *(7)*. Alternately, dissociation of aggregates can be accomplished by treatment of the lyophilized Aβ stock with strong acids *(8)*, bases *(7)*, or polyfluorinated alcohols *(8,9)*. An important advantage of mechanical, rather than chemical, elimination of the aggregates is that heterogeneous seeds, which have been shown to induce rapid aggregation *(10)*, are also eliminated. In addition, side reactions such as oxidation or racemization may result upon treatment of the protein with strong acids or bases, respectively. An additional advantage of SEC, relative to the other methods, is that the separation of the LMW protein from preexisting aggregates is monitored online, providing a means to assess the quality of the preparation. Here we describe methods for preparation of LMW Aβ by either SEC or filtration through a 10 kDa MWCO filter.

2. Materials

1. Waters (Milford, MA) FPLC system, including a Rheodyne (Ronnert Park, CA) 9725i injector, 1 mL injection loop (*see* **Note 1**), Waters 650 system controller or Waters 515 pump, and Waters 486 tunable absorbance UV detector set to 254 nm (*see* **Note 2**).
2. Branson (Danbury, CT) 1200 bath sonicator.
3. Eppendorf (Westbury, NY) 5415 C microcentrifuge.
4. Amersham Biosciences (Piscataway, NJ) 30/10 Superdex 75 HR column.
5. 10 mM Sodium phosphate buffer, pH 7.4 (*see* **Note 3**).
6. 20 mM Sodium phosphate buffer, pH 7.4 (*see* **Note 4**).
7. Dimethyl sulfoxide (DMSO).
8. 2 mM NaOH.
9. 1 N NaOH.
10. Millipore (Bedford, MA) Microcon-10 filters.

3. Methods

The methods described here outline preparation of LMW Aβ by using SEC and preparation of LMW Aβ by filtration through a 10 kDa MWCO filter. Both methods give comparable results for Aβ40 *(5)*, whereas the oligomer size distributions obtained for Aβ42 differ. Both filtration and SEC yield similar low order Aβ42 oligomers (up to octamer) but high order oligomers (~30–60 kDa) are detected only when LMW Aβ42 is isolated by SEC *(11)*.

3.1. Preparation of LMW Aβ by SEC

1. Prepare 10 m*M* sodium phosphate buffer and adjust the pH to 7.4 using solutions of phosphoric acid or NaOH.
2. Filter the resulting solution through a 0.22 μm filter (Corning [Corning, NY] sterilizing filter system, part no. 430517) to eliminate contamination by bacteria or dust.
3. Using this buffer, equilibrate the Superdex 75 HR column at a flow rate of 0.5 mL/min. Do not prepare the peptide solution until the column is equilibrated and the system is ready for injection (*see* **Note 5**).
4. Dissolve 350–400 μg of lyophilized Aβ in DMSO at 2 mg/mL (*see* **Note 6**).
5. Sonicate this solution for 1 min in a bath sonicator (*see* **Note 7**).
6. Centrifuge for 10 min at 16,000*g* to pellet large aggregates.
7. Inject 160–180 μL of the solution onto the Superdex 75 column.

A typical chromatogram would display 2 to 3 peaks (**Fig. 1**). A small void volume peak is observed in all samples, as is a major peak in the included volume that corresponds to LMW Aβ. A protofibril peak is consistently observed in Aβ42 samples but not in all Aβ40 samples. The top third of the LMW peak is collected for further experiments. The LMW fraction contains small to medium oligomers (**Fig. 2**). It should be used immediately after its isolation to ensure minimal aggregation.

3.2. Preparation of LMW Aβ by Filtration

For preparation of LMW Aβ by filtration, the peptide may be prepared using either of two methods. In both methods, the peptide is dissolved in the presence of a strong base. Using one method, the peptide is re-lyophilized. With the other method, the peptide is diluted into buffer and filtered immediately.

3.2.1. Preparation of LMW Aβ for Filtration With Re-Lyophilization

1. Dissolve lyophilized Aβ (normally, a TFA salt) at nominal concentration 1 mg/mL in 2 m*M* NaOH. The final pH of the solution should be ≥10.5 (*see* **Note 8**).
2. Sonicate this solution for 1 min in a bath sonicator and lyophilize.
3. Dissolve this lyophilizate at 2 mg/mL and filter (*see* **Subheading 3.2.3.**).

3.2.2. Preparation of LMW Aβ for Filtration Without Re-Lyophilization

1. Dissolve lyophilized Aβ in deionized water (*see* **Note 9**) at 4 mg/mL (*see* **Note 10**).
2. Add 1 *M* NaOH to adjust the pH to ≥10.5 (*see* **Note 11**).
3. Add 20 m*M* sodium phosphate, pH 7.4, to dilute the Aβ concentration to 2 mg/mL.
4. Sonicate this solution for 1 min in a bath sonicator and continue directly to filtration (*see* **Subheading 3.2.3.**).

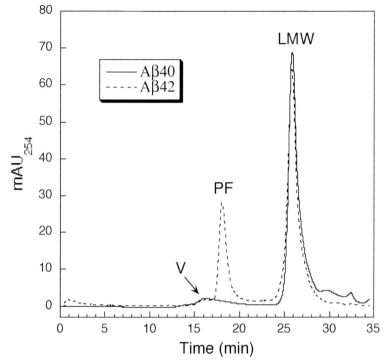

Fig. 1. Preparation of LMW Aβ by SEC (adapted with permission from *[11]*). Aβ40 (solid line) and Aβ42 (dotted line) fractionated by SEC using a Superdex 75 matrix and 10 m*M* sodium phosphate, pH 7.4, as the mobile phase. A small void volume peak (V) is observed in both samples, as is a major peak (LMW) in the included volume which corresponds to LMW Aβ. A protofibril peak (PF) is consistently observed in Aβ42 samples but not in all Aβ40 samples. This peak is always larger for Aβ42 than for Aβ40, likely reflecting the faster rate of fibrillogenesis of the longer Aβ alloform.

3.2.3. Filtration

1. Prepare the appropriate number of Microcon-10 filters (MWCO 10 kDa) (*see* **Note 12**) by washing each filter twice with 200 µL of 10 m*M* sodium phosphate buffer. The washing is done by centrifugation at room temperature for 20–25 min at 16,000*g*.
2. Transfer the filters into new collecting tubes.
3. Apply the Aβ solution prepared by either method in **Subheadings 3.2.1.** or **3.2.2.** and centrifuge at room temperature for 30 min at 16,000*g*.

The filtrate contains the LMW fraction. The LMW fraction contains small oligomers (**Fig. 2**). It should be used immediately after its isolation to minimize aggregation.

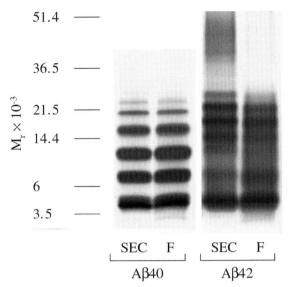

Fig. 2. Oligomer size distribution of LMW Aβ40 and Aβ42 isolated by SEC or filtration (adapted with permission from *[5]* and *[11]*). LMW Aβ40 and Aβ42 were isolated either by SEC or by filtration through a 10,000 MWCO filter. The peptides were photochemically crosslinked to produce a quantitative "snapshot" of the oligomer size distribution *(5)* and the products were analyzed by SDS-PAGE. The mobilities of molecular mass markers are shown on the left. The oligomer size distributions obtained using the two methods are identical for Aβ40 but differ for Aβ42 *(11)*.

4. Notes

1. A PEEK injector and loop, or equivalent non-metallic fluid path, should be used because stainless steel injectors and loops facilitate Aβ aggregation and tend to get clogged.
2. Comparable, isocratic FPLC systems equipped with all nonmetallic tubing are suitable for this procedure. Other wavelengths may be used for detection. For example, tryptophan- or tyrosine-containing proteins may be detected at 279 nm. Proteins without chromophores can be detected at 214 nm.
3. Other buffers may be used. However, buffers containing a high amount of salt (such as PBS) accelerate Aβ aggregation.
4. In case a different buffer is desired, this is a 2X buffer corresponding to the buffer in item 5.
5. A flat UV trace indicates that the column is equilibrated.
6. The peptide content (% peptide by weight) should be taken into account when calculating the volume of DMSO used for preparing this solution.
7. Sonication is an efficient way to break apart loosely attached aggregates. However, sonication for longer than 1 min may induce aggregation and should be avoided.

8. The pH may be adjusted using 0.1 M NaOH if necessary.
9. Deionized water of the highest quality should be used at all steps, including preparation of all buffers. Water of conductivity 18.2 MΩ is considered good quality.
10. Dissolution may take 1–2 min. Tap the tube gently to dislodge air bubbles but do not vortex or mix by pipetting up and down. Aβ40 normally gives a clear solution, whereas Aβ42 preparations tend to be turbid.
11. Approximately 1.5 µL of 1 M NaOH per 1 mL of Aβ solution is required.
12. Each filter will hold up to 550 µL. Since some of the peptide adsorbs to the filter membrane, higher final concentrations of LMW Aβ are obtained when higher volumes are used.

Acknowledgments

This work was supported by National Institutes of Health Grants AG14366, AG18921, and NS38328 (to DBT), by the Foundation for Neurologic Diseases (to DBT), and by the Massachusetts Alzheimer's Disease Research Center (1042312909A1, to GB).

References

1. Simmons, L. K., May, P. C., Tomaselli, K. J., et al. (1994) Secondary structure of amyloid β peptide correlates with neurotoxic activity *in vitro. Mol. Pharmacol.* **45,** 373–379.
2. Howlett, D. R., Jennings, K. H., Lee, D. C., et al. (1995) Aggregation state and neurotoxic properties of Alzheimer β-amyloid peptide. *Neurodegeneration* **4,** 23–32.
3. Soto, C., Castaño, E. M., Kumar, R. A., Beavis, R. C., and Frangione, B. (1995) Fibrillogenesis of synthetic amyloid-β peptides is dependent on their initial secondary structure. *Neurosci. Lett.* **200,** 105–108.
4. Rabanal, F., Tusell, J. M., Sastre, L., et al. (2002) Structural, kinetic and cytotoxicity aspects of 12–28 β-amyloid protein fragment: A reappraisal. *J. Pept. Sci.* **8,** 578–588.
5. Bitan, G., Lomakin, A., and Teplow, D. B. (2001) Amyloid β-protein oligomerization: prenucleation interactions revealed by photo-induced cross-linking of unmodified proteins. *J. Biol. Chem.* **276,** 35176–35184.
6. Walsh, D. M., Lomakin, A., Benedek, G. B., Condron, M. M., and Teplow, D. B. (1997) Amyloid β-protein fibrillogenesis—Detection of a protofibrillar intermediate. *J. Biol. Chem.* **272,** 22364–22372.
7. Fezoui, Y., Hartley, D. M., Harper, J. D., et al. (2000) An improved method of preparing the amyloid β-protein for fibrillogenesis and neurotoxicity experiments. *Amyloid: Int. J. Exp. Clin. Invest.* **7,** 166–178.
8. Jao, S. C., Ma, K., Talafous, J., Orlando, R., and Zagorski, M. G. (1997) Trifluoroacetic acid pretreatment reproducibly disaggregates the amyloid β-peptide. *Amyloid: Int. J. Exp. Clin. Invest.* **4,** 240–252.
9. Nilsson, M. R., Nguyen, L. L., and Raleigh, D. P. (2001) Synthesis and purification of amyloidogenic peptides. *Anal. Biochem.* **288,** 76–82.

10. Harper, J. D. and Lansbury, P. T. Jr. (1997) Models of amyloid seeding in Alzheimer's disease and scrapie: mechanistic truths and physiological consequences of the time-dependent solubility of amyloid proteins. *Annu. Rev. Biochem.* **66,** 385–407.
11. Bitan, G., Kirkitadze, M. D., Lomakin, A., Vollers, S. S., Benedek, G. B., and Teplow, D. B. (2003) Amyloid β-protein (Aβ) assembly: Aβ40 and Aβ42 oligomerize through distinct pathways. *Proc. Natl. Acad. Sci. USA* **100,** 330–335.

2

Determination of Peptide Oligomerization State Using Rapid Photochemical Crosslinking

Sabrina S. Vollers, David B. Teplow, and Gal Bitan

Summary

The assembly of the amyloid β-protein (Aβ) into neurotoxic oligomers and fibrils is a seminal pathogenic process in Alzheimer's disease (AD). Understanding the mechanisms of Aβ assembly could prove useful in the identification of therapeutic targets. Owing to the metastable nature of Aβ oligomers, it is difficult to obtain interpretable data through application of classical methods, such as electrophoresis, chromotography, fluorescence, and light scattering. Here, we apply the method Photo-Induced Crosslinking of Unmodified Proteins (PICUP) to the study of Aβ oligomerization. This method directly produces covalent bonds among unmodified polypeptide chains through *in situ* generation of peptide free radicals. PICUP provides a snapshot of the native oligomerization state of proteins and can be used for assembly state analysis of a wide variety of peptides and proteins.

Key Words: Crosslinking; oligomer; oligomerization state; protein assembly; amyloid β-protein.

1. Introduction

Parkinson's disease, amyotrophic lateral sclerosis, Huntington's disease, prion diseases, and Alzheimer's disease (AD) are all associated with pathologic protein folding and amyloid formation *(1,2)*. For example, the amyloid β-protein (Aβ) plays a key role in the etiology of Alzheimer's disease *(3)*. Recent findings support the hypothesis that oligomeric forms of Aβ are the proximate effectors of neurotoxicity *(4,5)*. Oligomerization also appears to be involved in the assembly and cytotoxicity of a variety of other amyloidogenic proteins, including PrP^{Sc} *(6)*, α-synuclein *(7)*, and huntingtin *(8)*. Therefore, elucidating pathways of oligomerization and assembly of Aβ should not only facilitate the development of therapeutic strategies for AD, but also should be of value for understanding and treating other diseases.

From: *Methods in Molecular Biology, vol. 299: Amyloid Proteins: Methods and Protocols*
Edited by: E. M. Sigurdsson © Humana Press Inc., Totowa, NJ

Biophysical studies of Aβ oligomerization have been difficult owing to the metastable nature of the oligomers and have not yielded a consensus regarding the oligomer size distribution. Size exclusion chromatography (SEC) *(9)*, polyacrylamide gel electrophoresis (PAGE) *(10)*, dynamic light scattering (DLS) *(11)*, fluorescence resonance energy transfer (FRET) *(12)*, analytical ultracentrifugation *(13)*, and nuclear magnetic resonance (NMR) spectroscopy *(14)* have been used to study Aβ oligomerization. Low resolution, peptide denaturation, and induced aggregation often have complicated the interpretation of the data. We have shown that stabilization of Aβ oligomers through covalent chemical crosslinking allows quantitative analysis of the Aβ oligomer size distribution *(15)*. This approach is a useful, general tool for studying metastable protein oligomers, and in addition, produces results which readily distinguish natively associated protein oligomers from oligomers formed through random collisions in solution (**Fig. 1**).

Photo-Induced Crosslinking of Unmodified Proteins (PICUP) *(16)* is a powerful method for forming covalent bonds between polypeptides utilizing photolysis of a light-harvesting catalyst (Ru(Bpy)) in the presence of an electron acceptor (**Fig. 2**). PICUP offers several advantages relative to other crosslinking methods. It requires a very short reaction time (≤1 s), no *pre facto* peptide modifications or insertions, and produces high yields (approx 80%) of crosslinked material. The method is applicable across a wide range of pH and minimizes potential damage to proteins and other biological molecules by employing visible rather than UV light. In this chapter, we discuss the use of PICUP to study peptide and protein oligomerization, both for analytical and preparative purposes.

2. Materials

1. Light source: Dolan-Jenner (Lawrence, MA) 200 W incandescent lamp (*see* **Note 1**).
2. Reaction apparatus allowing controlled exposure and positioning of samples a fixed distance from the light source. An inexpensive, yet highly reliable and flexible apparatus, may be constructed using any 35 mm SLR (single lens reflex) camera body and an attached bellows (**Fig. 3**).
3. 0.2 mL Clear, thin-walled plastic polymerase chain reaction (PCR) tubes (Eppendorf, Westbury, NY).
4. 1.8 mL glass vial (Kimble Chromatography, Vineland, NJ).
5. 35 × 10 mm Plastic petri dishes (Falcon, Franklin Lakes, NJ).
6. Tris(2,2'-bipyridyl)dichlororuthenium(II) hexahydrate (Ru(bpy)) (Sigma, St. Louis, MO), 1 m*M*, in 10 m*M* sodium phosphate, pH 7.4 (*see* **Notes 2** and **3**).
7. Ammonium persulfate (APS) (Sigma), 20 m*M*, in 10 m*M* sodium phosphate, pH 7.4.
8. Low molecular weight Aβ (*see* **Note 4**).

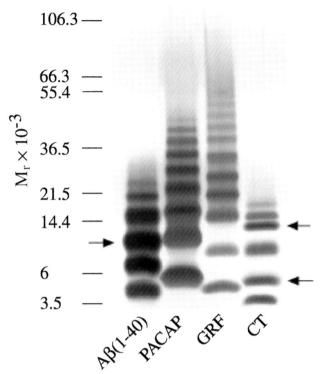

Fig. 1. PICUP results for amyloidogenic and non-amyloidogenic proteins analyzed by SDS-PAGE (adapted with permission from **ref. *15***). The non-amyloidogenic proteins pituitary adenylate cyclase-activating peptide (PACAP) and growth hormone-releasing factor (GRF) show continuous oligomer distributions whose elements decrease in intensity as molecular weight increases. The amyloidogenic proteins Aβ and calcitonin (CT) display restricted oligomer distributions with non-monotonic intensity changes. Arrows indicate nodes in the intensity distributions of these latter proteins.

9. Quenching reagent: 5% (v/v) β-mercaptoethanol (Sigma) in 2X SDS-tricine Sample Buffer (Invitrogen, Carlsbad, CA), or 1 *M* dithiothreitol (DTT) (Fisher, Fair Lawn, NJ) in water (*see* **Note 5**).

3. Methods

The methods described in the following detail crosslinking for analytical purposes and for preparative purposes. Factors such as the amount of sample available and the intended purpose of crosslinking determine which method is most appropriate. The method can be customized to satisfy special conditions or parameters (*see* **Note 6**).

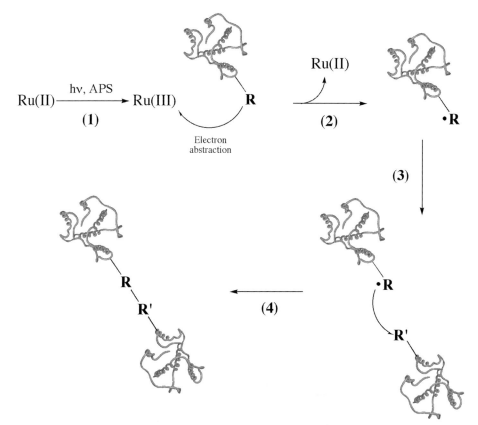

Fig. 2. Simplified reaction mechanism. (1) A Ru(II) complex is photooxidized, producing Ru(III)—a powerful one-electron abstraction agent. (2) The Ru(III) complex abstracts an electron from a nearby reactive group, R, creating a radical intermediate. (3) Adjacent reactive side-chains, R' (e.g., His, Tyr, Met) may react with the radical. (4) When the reaction is intermolecular, this produces a covalent complex.

3.1. Analytical Crosslinking

The analytical method is employed for small scale (final sample volume: 30–180 μL) studies on samples of concentration approx 25 μ*M* (*see* **Note 7**).

1. Isolate the peptide or protein sample as appropriate. Here, for example (**Fig. 1**), LMW Aβ was isolated according to published procedures (*see* **ref. *10*** and Chapter 6).
2. Transfer an 18 μL aliquot to a PCR tube.
3. Add 1 μL of Ru(bpy) and 1 μL of APS and mix by drawing up and expelling solution from a pipet tip.
4. Place in the illumination chamber (bellows) and irradiate for 1 s (*see* **Note 8**).

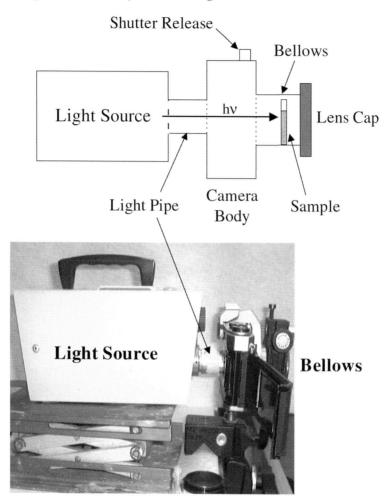

Fig. 3. Reaction apparatus scheme and actual setup. A light source is positioned directly behind the open film chamber of a 35-mm SLR camera. The sample tube is inserted in a bellows attached to the lens aperture and the end of the bellows then is capped. Activation of the shutter release allows light to enter the bellows for a predetermined interval (set using the camera body's exposure time mechanism).

5. Quench immediately by mixing either 10 μL β-ME in sample buffer or 1 μL DTT with the sample (*see* **Note 9**).

3.2. Preparative Crosslinking

The preparative crosslinking method has been used on a large scale for up to 10 mL of a 2 mg/mL solution. It is appropriate when separation and purification of individual oligomers is desired.

1. Isolate the peptide or protein sample as appropriate. Here, LMW Aβ was prepared using 7 mg of peptide at 2 mg/mL in 10 mM sodium phosphate, pH 7.4, by filtration (*see* Chapter 1).
2. Place the protein sample in a 35 × 10 mm Petri dish.
3. Place the Petri dish on a flat surface 10 cm below a miniature 50 W halogen lamp.
4. Add 700 μL each of APS and Ru(bpy). Swirl gently to mix after each addition.
5. Irradiate for 8 s.
6. Quench with 700 μL of DTT (*see* **Note 10**).

4. Notes

1. Both a 150 W Xe lamp and a 200 W incandescent lamp have been used successfully *(15,17)*. Other sources of light can be used. In these cases, exposures must be adjusted empirically to maximize crosslinking efficiency. Other groups have employed the method of filtering the light beam through distilled water to prevent sample overheating *(17)*. We have found this to be unnecessary when short (≤ 8 s) exposure times are used.
2. Studies show that other combinations of water-soluble metal-ion complexes and electron acceptors are possible, such as palladium (II) polyphorins and/or cobalt (III) pentamine chloride *(17)*. The combination of Ru(III) and ammonium persulfate (APS) may not be ideal, depending on the experimental parameters. When using proteins containing cysteine and methionine, which are particularly susceptible to oxidation, it may be preferable to use Co(III) as the electron accepting reagent during longer incubations. For studies involving living cells, it should be noted that APS is not cell-permeable.
3. Ru(bpy) requires vortexing until the solution is transparent to the eye. The Ru(bpy) solution is light-sensitive and should be protected from ambient light. A simple method is to use aluminum foil to wrap the tube containing the Ru(bpy). The APS and Ru(bpy) reagent solutions can be used for up to 48 h following preparation.
4. The studies discussed here have used low molecular weight Aβ *(10)*. However, the method is readily applied to the analysis of other peptides and proteins. The most important factors that must be considered are the reagent stoichiometry, irradiation time, and sample preparation procedure. The former two issues may require empirical optimization experimentation. The latter issue largely determines how the experimental data are to be interpreted. For amyloidogenic proteins in particular, native or time-dependent oligomerization states can only be determined if aggregate-free starting preparations can be produced.
5. The choice of quenching reagent depends upon the purpose of the crosslinking experiment. Samples analyzed using PAGE are quenched with the appropriate sample buffer containing 5% β-mercaptoethanol. Samples analyzed by chromatography or other methods may be quenched with 1 M DTT.
6. Cell media or extracts can be cross-linked using either the analytical or preparative method, provided the reagent concentrations are adjusted. Crosslinking of these types of samples requires higher concentrations of reagents, up to 1 M APS and

100 mM Ru(bpy). Upon addition of reagents to the sample, some precipitate forms. This precipitate does not appear to interfere with crosslinking and can be removed by centrifugation or dissolved upon addition of sample buffer after the procedure is complete.

7. The concentrations of reagents in this procedure are intended to be used with peptide samples of concentration approx 25 µM. For less concentrated samples, the crosslinking reagents should be diluted proportionately to maintain the specified molar stoichiometry (1:2:40; protein:Ru(bpy):APS). Under these conditions, longer irradiation may be necessary for the same crosslinking yield to be obtained.

8. The PCR tube containing the sample is placed in a glass vial in order for it to be free-standing when positioned in the bellows.

9. For the analytical crosslinking procedure, the sample, reagent, and quencher volumes can be increased up to six times without affecting reaction efficiency.

10. Crosslinked samples may be stored in a −20°C freezer for several days prior to analysis. Longer storage of samples will result in decreased resolution on a gel.

Acknowledgments

This work was supported by grants AG18921, AG44147, and NS38328 from the National Institutes of Health (DBT), by the Foundation for Neurologic Diseases (DBT), and by Grant 1042312909A1 from The Massachusetts Alzheimer's Disease Research Center (GB).

References

1. Kirkitadze, M. D., Bitan, G., and Teplow, D. B. (2002) Paradigm shifts in Alzheimer's disease and other neurodegenerative disorders: the emerging role of oligomeric assemblies. *J. Neurosci. Res.* **69,** 567–577.
2. Koo, E. H., Lansbury, P. T., and Kelly, J. W. (1999) Amyloid diseases: abnormal protein aggregation in neurodegeneration. *Proc. Natl. Acad. Sci. USA* **96,** 9989–9990.
3. Selkoe, D. J. (1999) Translating cell biology into therapeutic advances in Alzheimer's disease. *Nature* **399,** A23–A31.
4. Klein, W. L., Krafft, G. A., and Finch, C. E. (2001) Targeting small Aβ oligomers: the solution to an Alzheimer's disease conundrum? *Trends Neurosci.* **24,** 219–224.
5. Walsh, D. M., Klyubin, I., Fadeeva, J. V., et al. (2002) Naturally secreted oligomers of amyloid β protein potently inhibit hippocampal long-term potentiation *in vivo. Nature* **416,** 535–539.
6. Baskakov, I. V., Legname, G., Baldwin, M. A., Prusiner, S. B., and Cohen, F. E. (2002) Pathway complexity of prion protein assembly into amyloid. *J. Biol. Chem.* **277,** 21140–21148.
7. Lashuel, H. A., Petre, B. M., Wall, J., et al. (2002) α-synuclein, especially the Parkinson's disease-associated mutants, forms pore-like annular and tubular protofibrils. *J. Mol. Biol.* **322,** 1089–1102.
8. Sanchez, I., Mahlke, C., and Yuan, J. Y. (2003) Pivotal role of oligomerization in expanded polyglutamine neurodegenerative disorders. *Nature* **421,** 373–379.

9. Roher, A. E., Chaney, M. O., Kuo, Y. M., et al. (1996) Morphology and toxicity of Aβ(1-42) dimer derived from neuritic and vascular amyloid deposits of Alzheimer's disease. *J. Biol. Chem.* **271,** 20631–20635.

10. Walsh, D. M., Lomakin, A., Benedek, G. B., Condron, M. M., and Teplow, D. B. (1997) Amyloid β-protein fibrillogenesis—detection of a protofibrillar intermediate. *J. Biol. Chem.* **272,** 22364–22372.

11. Lomakin, A., Chung, D. S., Benedek, G. B., Kirschner, D. A., and Teplow, D. B. (1996) On the nucleation and growth of amyloid β-protein fibrils: detection of nuclei and quantitation of rate constants. *Proc. Natl. Acad. Sci. USA* **93,** 1125–1129.

12. Garzon-Rodriguez, W., Sepulveda-Becerra, M., Milton, S., and Glabe, C. G. (1997) Soluble amyloid Aβ-(1-40) exists as a stable dimer at low concentrations. *J. Biol. Chem.* **272,** 21037–21044.

13. Huang, T. H., Yang, D. S., Plaskos, N. P., et al. (2000) Structural studies of soluble oligomers of the Alzheimer β-amyloid peptide. *J. Mol. Biol.* **297,** 73–87.

14. Zagorski, M. G., Shao, H., Ma, K., et al. (2000) Aβ(1-40) and Aβ(1-42) adopt remarkably stable monomeric and extended structures in water solution at neutral pH. *Neurobiol. Aging* **21,** S10–S11 (Abstract 48).

15. Bitan, G., Lomakin, A., and Teplow, D. B. (2001) Amyloid β-protein oligomerization: prenucleation interactions revealed by photo-induced cross-linking of unmodified proteins. *J. Biol. Chem.* **276,** 35176–35184.

16. Fancy, D. A. and Kodadek, T. (1999) Chemistry for the analysis of protein-protein interactions: rapid and efficient cross-linking triggered by long wavelength light. *Proc. Natl. Acad. Sci. USA* **96,** 6020–6024.

17. Fancy, D. A., Denison, C., Kim, K., et al. (2000) Scope, limitations and mechanistic aspects of the photo-induced cross-linking of proteins by water-soluble metal complexes. *Chem. Biol.* **7,** 697–708.

3

In Vitro Preparation of Prefibrillar Intermediates of Amyloid-β and α-Synuclein

Hilal A. Lashuel and Dolors Grillo-Bosch

Summary

Elucidating the structural properties of early intermediates (protofibrils) on the fibril formation pathway of Aβ and α-synuclein, the structural relationship among the different intermediates and their relationship to the structure of the amyloid fibrils is critical for understanding the roles of amyloid fibril formation in the pathogenesis of Alzheimer's and Parkinson's diseases. In this chapter we discuss several methods, developed by different laboratories, that enable the preparation and stabilization of amyloid-β and α-synuclein protofibrillar species of defined morphologies for biochemical, biophysical and toxicity studies.

Key Words: Alzheimer's disease (AD); Parkinson's disease (PD); amyloid; fibrils; protofibrils; oligomers; amyloid-β (Aβ)-derived diffusible ligand (ADDLS); amylospheriods; annular structures; pores; size exclusion chromatography (SEC); electron microscopy (EM); atomic force microscopy (AFM); analytical ultracentrifugation (AU); scanning transmission electron microscopy (STEM); arctic variant (E22G); wild-type (WT).

1. Introduction

Amyloid fibril formation is a process by which one of 20 distinct proteins undergoes a conformational change either before or coincident with its self-assembly into highly ordered β-sheet rich aggregates known as amyloid fibrils *(1,2)*. Initially, the amyloid hypothesis supported by strong genetic, pathologic, and biochemical evidence implicated amyloid fibrils as the main cause for neurodegeneration and/or organ dysfunction in several human diseases including Alzheimer's disease (AD), Parkinson's disease (PD), Huntington's disease, and other amyloidoses. The identification and characterization of potential neurotoxic quaternary structure intermediates, referred to as protofibrils, that precede fibril formation and the finding that several pathogenic mutations promote

From: *Methods in Molecular Biology, vol. 299: Amyloid Proteins: Methods and Protocols*
Edited by: E. M. Sigurdsson © Humana Press Inc., Totowa, NJ

protofibril formation suggest that the protofibrils rather than the fibrils are the pathogenic species *(3–5)*. However, the mechanism by which protofibrils contribute to the pathogenesis of these diseases and the nature of the pathogenic species remain the subject of intense investigation and debate.

Biophysical studies have shown that several amyloid forming proteins self-assemble into protofibrils of heterogeneous size and morphology, including spheres, chain-like structures, annular pore-like structures, and large granular structures (**Scheme 1**) *(6–11)*. To date, most of the studies on the biochemical and biological properties of protofibrils employ heterogeneous mixtures of protofibrillar species, making it difficult to decipher which of these species is the toxic species. Therefore, developing solution conditions for stabilizing protofibrils in general and/or protofibrils of defined size and morphology is critical for investigating the relationship between monomer, protofibrillar intermediates, and fibrils in the pathogenesis of neurodegenerative diseases (e.g., AD and PD) and the identification of targets for therapeutic intervention.

Over the past years, several biophysical approaches were developed to stabilize, isolate, and characterize different protofibrillar intermediates on the pathway of amyloid formation. These approaches take advantage of complementary biophysical techniques to guide the preparation, purification, and characterization of protofibrils of defined molecular mass distribution and morphologies *(6,7,12)*. Generally, the first step involves the identification of optimal conditions for protofibril formation. Depending on the protein of interest, these conditions could involve manipulation of solution conditions (pH, temperature, ionic strength, incubation media, or buffer) or use of mutations that are known to promote protofibril formation and/or the formation of particular protofibril morphology. However, one needs to keep in mind that while one can manipulate solution conditions to alter the molecular mass and morphology distribution of protofibrils in favor of a particular protofibrillar species, preparation of protofibrils that are homogeneous in size and morphology remains a challenging task. In this chapter, we will discuss several methods that take advantage of the propensity of disease associated mutation, known to promote oligomerization of the amyloidogenic proteins amyloid-β and α-synuclein, to prepare and stabilize protofibrillar species of different morphologies for biochemical, biophysical, and toxicity studies.

2. Materials

1. α-Synuclein: Recombinant α-synuclein was expressed and purified as described previously (9).
2. Aβ Peptides: $A\beta 40_{WT}$ and the Arctic variant, $A\beta 40_{ARC}$ (E22G), were purchased as TFA salts from the Biopolymer Facility at Brigham and Women's Hospital. All peptides were dissolved using their true peptide weight.

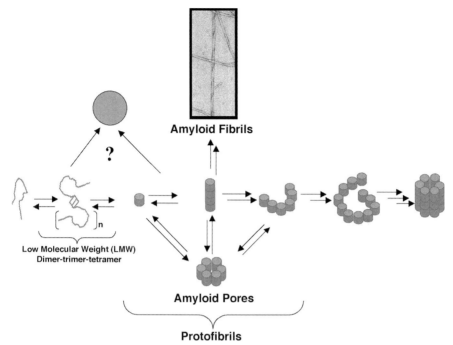

Scheme 1. A schematic representation summarizing our current understanding of the mechanism of oligomerization and amyloid fibril formation of Aβ and α-synuclein in vitro *(5)*.

3. Size exclusion columns.
   a. Superose 6 HR 10/30 (Amersham Pharmacia no. 17-0537-01).
   b. Superdex 75 HR 10/30 (Amersham Pharmacia no. 17-1047-01).
   c. Superdex 200 HR 10/30 (Amersham Pharmacia no. 17-1088-01).
4. Buffer A: 10 mM Tris-HCl, 150 mM NaCl, pH 7.4.
5. Buffer B: 5 mM Tris-HCl, 70 mM NaCl, pH 7.4.
6. Buffer C: 1X PBS (without Ca^{2+} and Mg^{2+}) pH 7.4 (GIBCO no. 70011-044).
7. Hexafluoroisopropanol (HFIP) (Sigma no. H8508).
8. Dry DMSO (Aldrich no. 276855).
9. Ham's F12 media without Phenol Red (Biosource).
10. Phenol Red (Sigma no. P4633).
11. 0.22 μm nylon spin centrifuge tube filter (Costar no. 8160).
12. Ultrafree-15 centrifugal filter device (5 kDa) (Millipore no. UFV2BCC40).

3. Methods

3.1. Preparing α-Synuclein Protofibrils

Genetic, neuropathologic, and biochemical evidence strongly suggest that the process of α-synuclein fibrillization plays a central role in the etiology of (PD).

PD is a neurodegenerative movement disorder that is characterized by the loss of dopaminergic neurons from the *substantia nigra*, and the formation of fibrillar α-synuclein intraneuronal inclusions (called Lewy bodies) *(13,14)*. Two mutations in the α-synuclein gene (A30P and A53T) have been linked to autosomal dominant early-onset PD. Both mutations promote the formation of transient protofibrils (prefibrillar oligomers), suggesting that protofibrils are linked to cytotoxicity. The PD-linked mutations (A30P and A53T) were observed to affect both the morphology and the size distribution of α-synuclein protofibrils (measured by analytical ultracentrifugation and scanning transmission electron microscopy). The A30P variant was observed to promote the formation of annular, pore-like protofibrils, whereas A53T promotes formation of annular and tubular protofibrillar structures. Wild-type (WT) α-synuclein also formed annular protofibrils, but only after extended incubation.

3.1.1. Preparing Crude α-Synuclein Protofibrils

1. To induce protofibril formation, incubate WT α-synuclein or any of the PD-linked variants (A30P and A53T) in 1X PBS buffer (pH 7.4) at concentrations of ≥300 μM for 30 min at 4°C. The amount of protofibrils formed under these conditions varies among the different α-synuclein variants, but the amount of protofibrils formed (based on absorbance values) is usually ≤10% of the total protein in solution. Attempts to populate more protofibrils by further incubation have not been successful.
2. Filter the protein solution through 0.22-μm filters to remove any particles.
3. To separate the protofibrils (5–10%) from the monomer (majority of the sample), inject the α-synuclein sample onto a superdex 200 HR SEC column equilibrated with buffer A. **Figure 1A** shows a typical SEC chromatogram of A53T α-synuclein. Elute at a flow rate of 0.5 mL/min and collect fractions of 0.5 mL. α-Synuclein elutes as two peaks corresponding to protofibrillar oligomeric species and monomer (*see* **Note 1**).

Fig. 1. (*Opposite page*) (**A**) Crude preparation of α-synuclein protofibrils. Size exclusion chromatogram showing separation of protofibrils (heterogeneous mixture) (V) from monomer (M) using a superdex 200 HR 10/30 SEC column. Circular dichroism (CD) spectra of the purified fractions demonstrate that the protofibrils are rich in β-sheet structures, whereas the monomeric fraction exhibits a CD signal characteristic of a random coil. The presence of protofibrils in the void fraction (V) is verified by AFM imaging, which reveals heterogeneous distribution of spherical aggregates. **Figure 1A** was adapted from *(5)*. (**B**) α-Synuclein annular protofibril preparation. Relative separation of α-synuclein protofibrils by fractionation of the protofibril's peak using a superose 6 HR 10/30 SEC column. Size exclusion chromatogram of α-synuclein (A53T) stock solution (350 μM) in buffer A, inset, shows how the oligomeric peaks were fractionated. (**C**) Negative stain electron microscopy (EM) images of the protofibril fractions in the inset of **Fig. 1B**. **Figure 1B** was adopted from **ref. 6**.

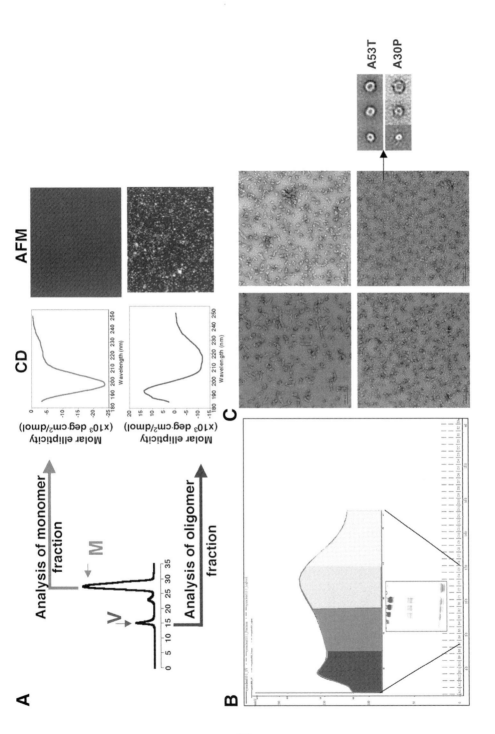

4. Collect the fractions corresponding to α-synuclein monomers and protofibrils separately.

5. Concentrate each species to the desired concentration using an ultrafree-15 centrifugal filter device (5 kDa). For storage purposes, we recommend that the monomer and oligomer concentrations should not exceed 100 μM and 30 μM, respectively, unless future experiments demand so.

6. Filter the samples through 0.2-μm nylon membrane spin filter.

7. Protein concentration should be determined by amino acid analysis (AAA) or by optical absorbance at 280 nm (OD_{280} for 1 mg/mL = 0.354).

3.1.2. Preparing Annular Protofibrils of α-Synuclein

1. To obtain protofibrillar preparations of more defined molecular size distribution and morphology, inject the α-synuclein samples from step 2 onto a Superose 6 HR 10/30 SEC column. Unlike the Superdex 200, where the protofibrils elute as a single peak in the excluded volume, in Superose 6, the majority of α-synuclein protofibrils is eluted as a broad peak in the included volume (**Fig. 1B**).

2. To achieve relative separation of α-synuclein protofibrils, fractionate the protofibril peaks into 3 to 4 fractions. While the first and second fractions remain heterogeneous, the third and fourth fractions tend to be more homogeneous and dominated by annular protofibrils (>85%) of a narrow size distribution as determined by electron microscopy (EM), analytical ultracentrifugation (AU), and scanning transmission electron microscopy (STEM) (**Fig. 1B**) *(6)*.

3. Concentrate each fraction using an ultrafree-15 centrifugal filter device.

4. Fractions 3 and 4 are rich in annular (pore-like) protofibrils having an average diameter of 11 ± 1 nm, and an inner core of 2 to 3 nm, whereas fractions 1 and 2 are heterogeneous and contains significant amounts of ring and tube-like protofibrils (width = 10–12 nm, length = 13–24 nm).

3.2. Amyloid-β (Aβ) and Alzheimer's Disease

Alzheimer's disease (AD) is a progressive neurodegenerative disease that is characterized by the presence of extracellular amyloid plaques and intraneuronal neurofibrillary tangles in the brain *(15–17)*. Biochemical analysis of amyloid plaques revealed that the main constituent is fibrillar aggregates of a 39–42 residue peptide referred to as the amyloid-β protein (Aβ) *(18)*. Several lines of evidence point toward a central role for the process of Aβ fibril formation in the etiology of AD. Transgenic animals overexpressing mutant forms of its precursor, the amyloid precursor protein (APP), develop amyloid plaques comprising fibrillar Aβ *(19)*. Several pathogenic AD mutations have been shown to affect the processing of APP resulting in increased Aβ levels, in particular, the more amyloidogenic variant Aβ42 *(20)*. Finally, active or passive vaccination against Aβ reduces the amyloid burden and reverses behavioral and cognitive deficits in AD transgenic animals *(21,22)*.

3.2.1. Solubilization of Aβ and NaOH Method Aβ Preparation

Several methods have been described for preparing protofibrils of Aβ, Herein, we will focus on the NaOH method used in our laboratory to generate protofibrils of Aβ40 (WT and Arctic). Based on our experience, the NaOH method results in accelerated protofibril formation by Aβ peptides. Synthetic Aβ peptides, prepared as trifluoroacetic acid (TFA) salts, are highly acidic, and care must be excersized to properly buffer the peptide in order to stabilize the effects of pH on morphology and neurotoxicity.

1. Dissolve Aβ40 (TFA salt) at a concentration of 1 mM (4.3 mg/mL, based on total weight of peptide) in 1 mM NaOH plus phenol red (0.1 mg/mL to monitor pH), pH approx 3.0.
2. To minimize isoelectric precipitation of Aβ (pH 5.5), add 10 mM NaOH (140–155 μL NaOH/mg of peptide, this can vary between manufactures and specific lots) to achieve a rapid transition to a pH of approx 7.0 to 7.5 (orange to red). This is the value of the phenol red, you can monitor the pH (yellow = too acidic and purple = too basic).
3. Additional acid or base can be added to achieve the proper pH.
4. Dilute the peptide solution to 500 μM in water and PBS (final concentration: 70 mM NaCl, 1.35 mM KCl, 5 mM NaH$_2$PO$_4$/Na$_2$HPO$_4$).

3.2.2. Preparation of Crude Aβ Protofibrils

1. To prepare Aβ protofibrils, dissolve the lyophilized synthetic Aβ (WT, ARC, or equimolar mixture of WT and ARC) as described previously to obtain a total concentration of 100 μM. To induce protofibril formation, incubate the peptide solution prepared by the NaOH method at room temperature (RT) for 16 to 24 h. The incubation time required for protofibril formation varies depending on the batch of peptide. Generally, we perform analytical SEC experiments to probe the kinetics of protofibril formation every time we receive a new batch of material. The protofibrils formed under these conditions are heterogeneous in terms of their size (80–1000 kDa) and morphology (spheres, chain-like protofibrils, granular aggregates are usually observed) (*see* **Note 2**).
2. Centrifuge the samples at 13,000g for 5 min to remove any insoluble particle(s).
3. Filter the samples through 0.2-μm filter.
4. Load the supernatants onto either a superdex 75 HR (Amersham Pharmacia) SEC column (used to separate protofibrils from LMW Aβ species [monomers-dimers]) equilibrated with protein buffer B or a superdex 6 HR column (used to fractionate Aβ protofibrils). Elute proteins at a flow rate of 0.5 mL/min and collect fractions of 0.5 mL volume size.
5. Collect the fractions corresponding to the oligomeric and LMW peaks. The amount of protofibrils formed ranges from 30 to 60% depending on the incubation time and concentration of the peptide. Using 1 mg of peptide, we generally collect 1.5 mL of protofibrillar Aβ at concentrations ranging form 15 to 20 μM. In the case of

Aβ, we strongly recommend that samples in this concentration range should not be concentrated further.

6. Store purified protofibril fractions at 4°C. Samples should be used within 1 to 10 h.

3.2.3. Preparing Annular Protofibrils of Aβ40 (Amyloid Pores)

The increased propensity of the Arctic variant (E22G) of Aβ (Aβ40$_{ARC}$), which was shown to lead to early onset AD *(12,23)*, to form protofibrils presents researchers with an opportunity to generate significant quantities of protofibrils very rapidly. We have reported that the arctic variant (ARC) seems to more rapidly form annular and pore-like protofibrillar structures ("amyloid pores"), reminiscent of those observed for α-synuclein and bacterial pore-forming toxins *(6,7)*. We have discovered that mixing wild-type Aβ with the arctic variant alters the distribution of Aβ protofibrils and enhances greatly the stability of the formed protofibrils. Purified Aβ$_{ARC}$ protofibrils formed in the presence of WT are stable for 1 to 4 d at 4°C (as judged by SEC and EM) *(12)*, providing sufficient time to study and characterize these intermediates. Therefore, a 1:1 mixture of Aβ$_{ARC}$ and Aβ$_{WT}$ (100 μ*M*) is used to prepare annular protofibrils of Aβ$_{ARC}$ (**Fig. 2**). Attempts to separate Aβ protofibrils prepared from Aβ40 or Aβ40$_{ARC}$ alone on superose 6 have failed, as the protofibrils have been observed to interact with the column matrix and/or undergo dissociation to monomers.

1. Prepare a mixture of Aβ40$_{WT}$ and Aβ40$_{ARC}$ at 100 μ*M* using the NaOH method described in **Subheading 3.2.1.** To generate sufficient amount of protofibrils for biophysical and toxicity studies, 2.5 mg of each variant is used.
2. Incubate at room temperature for 16 to 20 h.
3. Centrifuge at 16000*g* and filter the supernatant through a 0.22-μm membrane. Generally, 15 to 25% of the peptides is converted into amyloid fibrils under these conditions with remaining peptides existing as mixture of protofibrils and low molecular weight species.
4. To separate the protofibrils, inject the peptide solution from step 3 onto a superose 6 column and elute with buffer A. Use a flow rate of 0.5 mL/min and collect fractions of 0.5 mL volume size (**Fig. 2**). To achieve optimal separation, load no more than 800 μL of the Aβ mixtures per run.

3.2.4. Preparation of Aβ Amylospheriod

Recently, Hoshi et al. described procedures for reproducible preparation of spherical oligomers of Aβ40 or Aβ42, termed amylospheriod *(24)*. These spherical aggregates of Aβ (with average diameters ranging from 3 to 20 nm) can be prepared reproducibly by slow rotation of Aβ40 or Aβ42 peptide solution. These aggregates form in the absence of chain-like or pore-like Aβ aggregates. Purification of spherical aggregate of various diameters can be accomplished by using glycerol-gradient (15–30%) centrifugation (86,000*g*) (**Fig. 3**). Hosi et

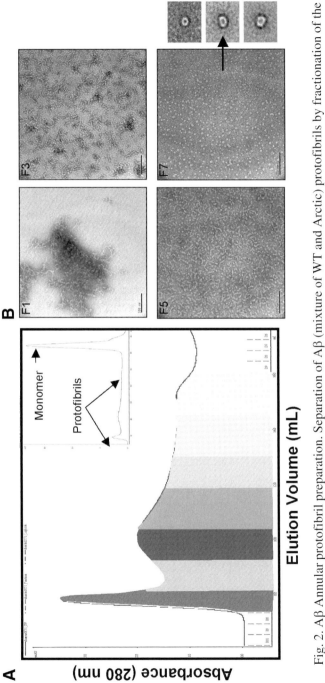

Elution Volume (mL)

Fig. 2. Aβ Annular protofibril preparation. Separation of Aβ (mixture of WT and Arctic) protofibrils by fractionation of the protofibril peak on a superose 6 HR 10/30 SEC column. (**A**) SEC chromatogram illustrating how the Aβ protofibril peak, which was obtained by incubating Aβ40$_{ARC}$ and Aβ40$_{WT}$ at equimolar concentration (100 μ*M*) for 16 h at room temperature, was divided into seven fractions (F1–F7). The inset shows illustrates the amount of protofibrils and monomers present after the 16 h incubation (RT). (**B**) Negative stain EM images of the protofibril fractions F1, F3, F5, and F7. **Figure 2B** was adapted from *ref. 12.*

al. demonstrated that toxicity was highly dependent on the average diameter of
the spherical aggregate. Spherical oligomers with an average diameter >10 nm
were shown to exhibit significantly higher toxicity than those with diameter <10
nm. Aβ42 was shown to form spherical aggregates of similar size (d >10 nm)
more rapidly (8–10 h rotation vs 5–7 d for Aβ40) and at lower concentrations
(0.01–1 u*M*) than Aβ40. Spherical aggregates (d >10 nm) formed by Aβ42
exhibited a 100-fold greater toxicity to neuronal cultures than those formed by
Aβ40. The structural and molecular basis for this apparent difference in toxicity of identical morphologies of Aβ40 and Aβ42 remains unknown. The procedure for preparing amyloispheriod is outlined in **Fig. 3**.

3.2.5. Preparation of 3–5 nm Globular Oligomers of Aβ42

Using atomic force microscopy (AFM), several groups have demonstrated
that small globular oligomers of Aβ42 with an average height of 3–5 nm could
be prepared reproducibly using the protocols developed by Lambert et al. and
Stine et al. *(25–27)*. The procedure for preparing 3 to 5 nm globular oligomers
is summarized in **Fig. 4** as described by Dahlgren et al. Similar structures with
a slightly higher average height of 5.0 + 0.3 nm, referred to as Aβ-derived
diffusible ligand (ADDLS) were first described by Lambert et al. *(25,28)*. The
exact molecular mass distribution of these 3 to 5 nm structures in solution has
not been determined, but on denaturing SDS-PAGE, ADDLS exhibits a distribution of oligomers ranging in size from trimer to tetramer and, in some cases,
higher oligomers (6–12 mer) *(25,27,28)* (*see* **Note 3**). However, it remains to be
determined whether such globular oligomers of similar height, but formed under
different conditions, are structurally and biologically (toxicity) equivalent. The
protocol in **Fig. 4** has been shown to yield preparation of Aβ42 oligomers (3–5
nm) with reproducible structural and neurotoxic properties *(27,29–31)*.

3.3. Preparation of Aβ and α-Synuclein Fibrils

Fibrillar Aβ can be generated by incubation of monomeric Aβ fractions from
SEC at 37°C for 1 to 4 d, depending on the concentration of the protein and the
Aβ variant used. α-Synuclein fibrils can also be prepared by incubation of
monomeric α-synuclein (100–200 μ*M*) from SEC at 37°C for 1 to 2 wk or for
1 to 4 d by agitating the sample or placing the tube containing the sample in a
rotator at 37°C. The fibrils formed under these conditions are identical to those
formed under stagnant conditions (based on EM and AFM). The fibrillar solutions are spun at 15,000*g* during which the fibrils pellet to the bottom of the
tube. The supernatant is removed and the fibrils are washed several times with
buffer (10 m*M* Tris-HCl, pH 7.4) to remove soluble protein. Prior to use, resuspend the fibrils in the desired buffer. EM should be used to verify the structure
of the fibrils and the absence of soluble oligomers in these preparations.

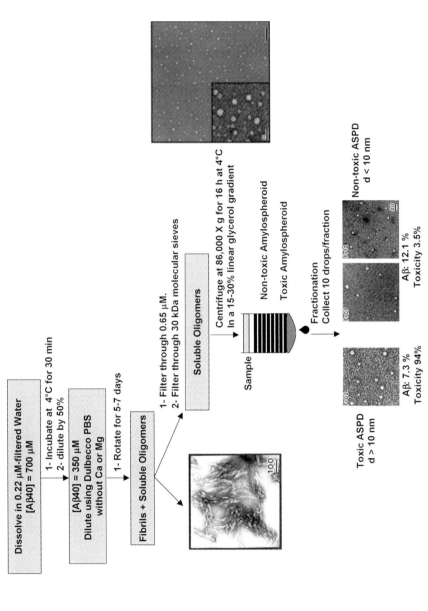

Fig. 3. Amylospheriods preparation. A schematic diagram summarizing the procedures used to prepare Aβ-Amylospheriods of various diameters. Adapted from Hoshi et al. (*24*) with permission from proceedings of the national academy of science.

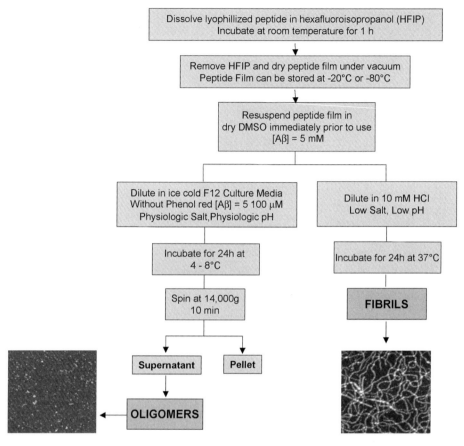

Fig. 4. Preparation of 3 to 5 nm globular oligomers of Aβ42. A schematic diagram summarizing the procedures used to prepare globular oligomers (3–5 nm) of Aβ42 as described by Lambert et al. *(25,28)*, and Dahlgren et al. *(27)*. AFM images were adopted from Stein et al. *(26)*.

4. Notes

1. Various buffers (pH 7.4) can be used in the SEC experiments, including PBS, 10 mM Tris-HCl (pH 7.4), and 10 mM HEPES. All buffers used need to contain 100 to 150 mM NaCl.

2. We did observe some variation in protofibril stability between different preparations. In some cases, a shift in the size distribution of the protofibrils toward higher MW protofibrillar species was observed during the 16 to 24 h procedure. Shortening both the incubation time for inducing protofibril formation and the fractionation time could minimize these changes. Partial separation of the protofibrillar species was possible, in particular, separation of the chain-like protofibrils from the spheres and annular species.

3. These globular oligomers resemble, based on size and morphology, pseudo-spherical oligomers (3.9 ± 0.5 nm diameter) observed by EM and AFM as the first forming oligomers during the in vitro fibrillization of Aβ40 *(10,12,32–34)*.

4. During our studies of Aβ and α-synuclein (both form spherical and annular protofibrils *[7]*), we have observed that spherical oligomers of both Aβ and α-synuclein seem to have the highest affinity for the mica surface, whereas annular protofibrils have the lowest affinity; very few could be detected using enriched annular fractions (unpublished data). In contrast, most of Aβ protofibrillar species appear to adsorb equally to carbon-coated grids used for EM and STEM studies. These findings underscore the importance of supplementing imaging data (AFM or EM) with data from complementary techniques (SVAU and STEM).

Acknowledgments

We thank Dr.Yichin Liu for critical review of the manuscript. We would like to thank Drs. Peter T. Lansbury, Tomas Ding, and Mary Jo LaDu for providing the images used in **Figs. 1** and **4**. Also, we would like to thank Dr. Dean Hartley for assistance with the Aβ solubilization protocols. Support was also derived from the NIH (AG08470).

References

1. Kelly, J. W. (1998) The alternative conformations of amyloidogenic proteins and their multi-step assembly pathways. *Curr. Opin. Struct. Biol.* **8(1)**, 101–106.

2. Rochet, J. C. and Lansbury, P. T. Jr. (2000) Amyloid fibrillogenesis: themes and variations. *Curr. Opin. Struct. Biol.* **10(1)**, 60–68.

3. Goldberg, M. S. and Lansbury, P.T. Jr. (2000) Is there a cause-and-effect relationship between alpha-synuclein fibrillization and Parkinson's disease? *Nat. Cell Biol.* **2(7)**, E115–E119.

4. Caughey, B. and Lansbury, P. T. (2003) Protofibrils, pores, fibrils, and neurodegeneration: separating the responsible protein aggregates from the innocent bystanders. *Annu. Rev. Neurosci.* **26**, 267–298.

5. Volles, M. J. and Lansbury, P.T. Jr. (2003) Zeroing in on the pathogenic form of alpha-synuclein and its mechanism of neurotoxicity in Parkinson's disease. *Biochemistry* **42(26)**, 7871–7878.

6. Lashuel, H., et al. (2002) alpha-Synuclein, especially the Parkinson's disease-associated mutants, forms pore-like annular and tubular protofibrils. *J. Mol. Biol.* **322(5)**, 1089.

7. Lashuel, H. A., et al. (2002) Neurodegenerative disease: amyloid pores from pathogenic mutations. *Nature* **418(6895)**, 291.

8. Ding, T. T., et al. (2002) Annular alpha-synuclein protofibrils are produced when spherical protofibrils are incubated in solution or bound to brain-derived membranes. *Biochemistry* **41(32)**, 10209–10217.

9. Conway, K. A., et al. (2000) Acceleration of oligomerization, not fibrillization, is a shared property of both alpha-synuclein mutations linked to early-onset Parkin-

son's disease: implications for pathogenesis and therapy. *Proc. Natl. Acad. Sci. USA* **97(2)**, 571–576.

10. Harper, J. D., et al. (1999) Assembly of A beta amyloid protofibrils: an in vitro model for a possible early event in Alzheimer's disease. *Biochemistry* **38(28)**, 8972–8980.

11. Harper, J. D., Lieber, C. M., and Lansbury, P. T. Jr. (1997) Atomic force microscopic imaging of seeded fibril formation and fibril branching by the Alzheimer's disease amyloid-beta protein. *Chem. Biol.* **4(12)**, 951–959.

12. Lashuel, H., et al. (2003) Mixtures of wild-type and "Arctic" Abeta40 in vitro accumulate protofibrils, including amyloid pores. *J. Mol. Biol.* **332(4)**, 795–808.

13. Pollanen, M. S., Dickson, D. W., and Bergeron, C. (1993) Pathology and biology of the Lewy body. *J. Neuropathol. Exp. Neurol.* **52(3)**, 183–191.

14. Forno, L. S. (1996) Neuropathology of Parkinson's disease. *J. Neuropathol. Exp. Neurol.* **55(3)**, 259–272.

15. Selkoe, D. J. (1997) Alzheimer's disease: genotypes, phenotypes, and treatments. *Science* **275(5300)**, 630–631.

16. Castano, E. M. and Frangione, B. (1988) Biology of disease human amyloidosis, Alzheimer disease and related disorders. *Lab. Invest.* **58(2)**, 122–132.

17. Selkoe, D. J. (2000) Toward a comprehensive theory for Alzheimer's disease. Hypothesis: Alzheimer's disease is caused by the cerebral accumulation and cytotoxicity of amyloid beta-protein. *Ann. NY Acad. Sci.* **924**, 17–25.

18. Wang, R., et al. (1996) The profile of soluble amyloid beta protein in cultured cell media. Detection and quantification of amyloid beta protein and variants by immunoprecipitation-mass spectrometry. *J. Biol. Chem.* **271(50)**, 31894–31902.

19. Janus, C., et al. (2001) New developments in animal models of Alzheimer's disease. *Curr. Neurol. Neurosci. Rep.* **1(5)**, 451–457.

20. Lichtenthaler, S. F., et al. (1997) Mutations in the transmembrane domain of APP altering gamma-secretase specificity. *Biochemistry* **36(49)**, 15396–15403.

21. Janus, C., et al. (2000) A beta peptide immunization reduces behavioural impairment and plaques in a model of Alzheimer's disease. *Nature* **408(6815)**, 979–982.

22. Morgan, D., et al. (2000) A beta peptide vaccination prevents memory loss in an animal model of Alzheimer's disease. *Nature* **408(6815)**, 982–985.

23. Nilsberth, C., et al. (2001) The 'Arctic' APP mutation (E693G) causes Alzheimer's disease by enhanced Abeta protofibril formation. *Nat. Neurosci.* **4(9)**, 887–893.

24. Hoshi, M., et al. (2003) Spherical aggregates of beta-amyloid (amylospheroid) show high neurotoxicity and activate tau protein kinase I/glycogen synthase kinase-3beta. *Proc. Natl. Acad. Sci. USA* **100(11)**, 6370–6375.

25. Lambert, M. P., et al. (2001) Vaccination with soluble Abeta oligomers generates toxicity-neutralizing antibodies. *J. Neurochem.* **79(3)**, 595–605.

26. Stine, W. B. Jr., et al. (2003) In vitro characterization of conditions for amyloid-beta peptide oligomerization and fibrillogenesis. *J. Biol. Chem.* **278(13)**, 11612–11622.

27. Dahlgren, K. N., et al. (2002) Oligomeric and fibrillar species of amyloid-beta peptides differentially affect neuronal viability. *J. Biol. Chem.* **277**, 32046–32053.

28. Lambert, M. P., et al. (1998) Diffusible, nonfibrillar ligands derived from Abeta1-42 are potent central nervous system neurotoxins. *Proc. Natl. Acad. Sci. USA* **95(11),** 6448–6453.
29. Klein, W. L. (2002) ADDLs & protofibrils—the missing links? *Neurobiol. Aging* **23(2),** 231–235.
30. Klein, W. L., Krafft, G. A., and Finch, C. E. (2001) Targeting small Abeta oligomers: the solution to an Alzheimer's disease conundrum? *Trends Neurosci.* **24(4),** 219–224.
31. Wang, H. W., et al. (2002) Soluble oligomers of beta amyloid (1-42) inhibit long-term potentiation but not long-term depression in rat dentate gyrus. *Brain Res.* **924(2),** 133–140.
32. Harper, J. D., et al. (1997) Observation of metastable Abeta amyloid protofibrils by atomic force microscopy. *Chem. Biol.* **4(2),** 119–125.
33. Walsh, D. M., et al. (1997) Amyloid beta-protein fibrillogenesis. Detection of a protofibrillar intermediate. *J. Biol. Chem.* **272(35),** 22364–22372.
34. Walsh, D. M., et al. (1999) Amyloid beta-protein fibrillogenesis. Structure and biological activity of protofibrillar intermediates. *J. Biol. Chem.* **274(36),** 25945–25952.

4

Purification of Recombinant Tau Protein and Preparation of Alzheimer-Paired Helical Filaments In Vitro

Stefan Barghorn, Jacek Biernat, and Eckhard Mandelkow

Summary

The tau protein is a neuronal microtubule-associated protein. Apart of its physiological function—the binding to and stabilization of microtubules—tau is found in Alzheimer's disease brain as insoluble fibers, the so-called "paired helical filaments" (PHFs). Investigating the fundamentals of tau polymerization is indispensable for identifying inhibitory conditions or compounds preventing PHF formation, which may slow down or even stop the degeneration of neurons in Alzheimer's disease. In this chapter, we describe the methods necessary for studying the characteristics of tau polymerization to PHFs. These include: a purification protocol for recombinantly expressed tau; a general method for the polyanion induced polymerization of tau to PHFs; the quantitation of PHFs by a fluorescence-based assay; the imaging and verification of PHFs by negative stain transmission electron microscopy.

Key Words: Alzheimer's disease; Tau protein; paired helical filament; purification; polyanion-induced polymerization; thioflavine-S assay; electron microscopy; amyloid.

1. Introduction

Alzheimer's disease is characterized by two types of pathological protein deposits. On the one hand, the Alzheimer amyloid fibers consisting of the amyloid-β (Aβ) peptide, which lead to the amyloid plaques, and on the other hand, fibers of tau protein (the paired helical filaments [PHFs]) which lead to the neurofibrillary tangles (NFTs). The Aβ peptide has a partially hydrophobic character *(1–4)*, whereas tau is highly soluble owing to its hydrophilic character. In both cases, the building principles of the fibers are still a matter of debate. Tau occurs mainly in the axonal compartment of neurons. In the human central nervous system it exists as 6 isoforms derived from alternative splicing of a single gene on chromosome 17. Tau's function is to bind and to stabilize microtubules, the tracks for motor proteins during axonal transport. The N-terminal

From: *Methods in Molecular Biology, vol. 299: Amyloid Proteins: Methods and Protocols*
Edited by: E. M. Sigurdsson © Humana Press Inc., Totowa, NJ

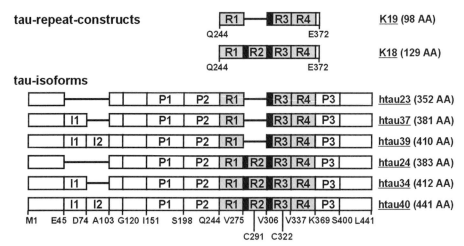

Fig. 1. Diagram of tau isoforms and constructs. In the human CNS htau40 is the largest isoform (441 residues). The C-terminal half contains 3 or 4 pseudo-repeats (~31 residues, R1-R4) which together with their proline-rich flanking regions constitute the microtubule-binding domain. Repeat R2 and the two near N-terminal inserts (I1 and I2) may be absent owing to alternative splicing. The hexapeptide motifs PHF6* ([275]VQIINK[280]) and PHF6 ([306]VQIVYK[311]) at the beginning of R2 and R3 (black) are important for PHF formation by inducing β-structure. Construct K18 comprises only the four repeats (R1–R4, residues Q244-E372; 129 residues); construct K19 (98 residues) comprises only three repeats (R1, R3, R4). The two cysteines of tau are located in the third repeat (C322 in R3) and the second repeat (C291 in R2), which is absent in 3-repeat isoforms.

half of tau projects away from the microtubule surface, whereas the C-terminal half binds to microtubules. The C-terminal domain contains 3 or 4 pseudo-repeats of approx 31 residues (*see* **Fig. 1**) that are important for microtubule binding, as well as for the pathological aggregation into PHFs (for overviews *see* **refs. 5–8**).

The secondary structure of monomeric tau in solution and especially of tau in PHFs has been for a long time a matter of debate. In summary, monomeric tau exhibits a circular dichroism (CD) spectrum which is dominated by a random coil pattern, consistent with the hydrophilic amino acid composition *(9–11)*. Some peptides from the tau sequence can adopt α-helical conformation, but only in non-physiological buffers including helix-inducing agents *(12–15)*. Earlier studies on tau in PHFs using X-ray fiber diagrams of NFTs or reassembled tau fibers show only weak or no contributions from β-structure *(11,16)*. Additionally, neurofibrillary tangles and tau fibers stain poorly with Congo red, a dye considered indicative of cross-β-structure, compared with thioflavin-S

(17). Only recently, it was shown that upon polymerization of tau to PHFs a shift from a random coil structure to an increased β-structure content occurs *(18–22)*.

The preparation of PHFs in vitro resembling those of Alzheimer's disease was until recently hampered by a slow aggregation rate and a low yield. This problem can be overcome by different means: (a) the addition of certain poly-anionic cofactors which increase reaction rates, such as polyanions (e.g., heparin, poly-Glu and RNA *[23–25]*) or fatty acids (e.g., arachidonic acid *[26,27]*), (b) a restriction of tau to certain domains, e.g. the repeat domain alone aggregates more readily into bona fide PHFs than full-length tau *(28–30)*, (c) by introducing mutations that occur in frontotemporal dementias (FTDP-17 *(31–34)*, such as ΔK280 and P301L, which lead to an accelerated or increased PHF aggregation *(19,35–37)*. Overall, pathological aggregation of tau can be described as a nucleation-elongation reaction *(38)* that involves the formation of β-structure around some hexapeptide motifs in the repeat domain *(18–21)*. The repeat domain of tau is sufficient for forming PHFs and forms the core of tau fibers reconstituted from recombinant tau, similar to the fibers from Alzheimer brain *(39–41)*. Current aggregation assays make use of either polyanions or fatty acids, and they are performed in oxidative or reductive conditions. In this chapter, we will describe the general method of purifying recombinantly expressed tau and a general method of preparing PHFs from tau isoforms. We will further highlight on the PHF formation from tau repeat constructs and the influence of reducing and oxidizing conditions.

Methods are now available that yield good estimates of PHF aggregation in solution and in real-time, such as the fluorescence of added reporter dyes (thioflavine-S [ThS]) or light scattering *(30,42)*. We will focus in this chapter on the ThS assay for monitoring PHF polymerization.

The judgment of realistic assembly conditions depends on electron microscopy, which shows whether fibers are "paired helical" or not, because this feature, with its typical 80 nm crossover repeat and a width of 10 to 20 nm *(43)*, is characteristic of Alzheimer PHFs. Straight filaments can be formed as well *(41)*, but in this case, the discrimination from other types and pathways of aggregation is less clear-cut. The verification and imaging of the overall appearance of PHFs by electron microscopy will be the last method described in this chapter. In summary, the main principles of the purification of recombinant tau protein and the generation of Alzheimer-like paired helical filaments in vitro and its verification are described.

2. Materials

2.1. Protein Expression and Purification

1. General HPLC equipment: e.g., HPLC-apparatus, sample loop (~50–150 mL), sample collector, etc.

2. Gel permeation chromatography column (e.g., HiLoad 16/60 Superdex™200 for tau isoforms and Superdex™75 for short tau constructs; Amersham Biosciences, Freiburg, Germany).
3. Cation-exchange chromatography column (e.g., self-packed XK16 column with ~20 mL SP-Sepharose fast flow chromatography material; Amersham Biosciences).
4. Dialysis tubing (3.5 kDa molecular weight cut off).
5. French Press.
6. SDS-PAGE equipment.
7. Bacterial cell pellet resuspension buffer: 20 mM MES, 1 mM EGTA, 0.2 mM MgCl$_2$, 5 mM dithiothreitol (DTT), 1 mM PMSF (phenylmethylsulfonylfluoride), 10 µg/mL leupeptin, 2 mM benzamidin, 10 µg/mL pepstatin A, pH 6.8. Add DTT, PMSF, leupeptin, benzamindin and pepstatin A just prior to use.
8. Cation exchange chromatography buffer A: 20 mM MES, 50 mM NaCl, 1 mM EGTA, 1 mM MgCl$_2$, 2 mM DTT (freshly added), 0.1 mM PMSF (freshly added), pH 6.8.
9. Cation exchange chromatography buffer B: 20 mM MES, 1 M NaCl, 1 mM EGTA, 1 mM MgCl$_2$, 2 mM DTT (freshly added), 0.1 mM PMSF (freshly added), pH 6.8.
10. Gel-filtration buffer: PBS (137 mM NaCl, 3 mM KCl, 10 mM Na$_2$HPO$_4$, 2 mM KH$_2$PO$_4$, pH 7.4) with 1 mM DTT (freshly added), pH 7.4.

2.2. Preparation and Verification of Paired Helical Filaments

1. Heparin (MW ≈ 6000 Da, Sigma Aldrich).
2. Protease inhibitors: PMSF, EDTA, EGTA, leupeptin, aprotinin, and pepstatin A
3. Thioflavine-S (Sigma Aldrich).
4. If parallel processing of ThS-assay is desired: Microplate capable spectrofluorimeter (e.g., Fluoroskan Ascent spectrofluorimeter, Labsystems, Helsinki, Finland) and black microtiter-plates (e.g., 386-well plates, Cliniplate, Labsystems).
5. Equipment for negative stain transmission electron microscopy (transmission electron microscope, electron microscope sample grids [e.g., 600-mesh carbon-coated copper grids], fine tip tweezer [DuMont no. 5], 2% uranyl acetate).

3. Methods

This subheading will cover the following topics: 1) The purification of tau protein from recombinant expression in *E. coli*. 2) A general method of the preparation of bona fide paired helical filaments from tau protein. 3) A brief outline of the methods used to verify and monitor the kinetics of the formation of paired helical filaments from tau protein.

3.1. Purification of Tau Protein

The tau protein can be recombinantly expressed in high quantities in the *E. coli* strain BL21 (DE3) *(44)*. As expression vector, the pNG2 vector *(45)* is used, a derivative of the pET-3 vector *(44)*. From a culture volume of 10 L, a yield of approx 10 to 100 mg protein, depending on the isoform or the tau con-

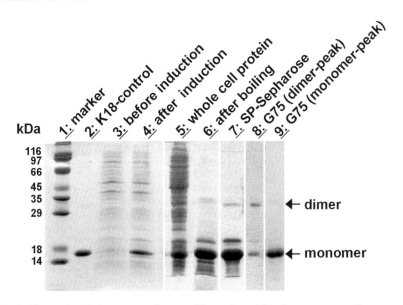

Fig. 2. Example of the expression profile and purification process of a tau repeat construct (K18). The samples were mixed with SDS-sample buffer and separated on a 15% SDS-PAGE gel and stained with Coomassie blue. In addition to the protein size marker (lane 1) a control tau K18 protein from a previous preparation was used (lane 2). After the induction of expression (lane 4) the tau protein is the major protein expressed (compare with nearly no expression before IPTG-induction [lane 3]). After disruption of the cells (lane 5) the cell suspension was adjusted to 500 mM NaCl, boiled for 20 min and pelleted. By this treatment most of the non-tau proteins are removed (lane 6). A further purification and concentration of K18-tau was achieved by a cation-exchange chromatography step (here a SP-Sepharose column was used, lane 7). In a last purification step, tau breakdown products and higher aggregation forms (mainly dimers) were removed. This resulted in a mixed dimer/monomer peak (lane 8) and a monomer peak with tau protein of approx 99% purity.

struct, with a purity of > 95% after purification can be expected. For the expression in a 10 L culture scale (or higher) a biofermenter can be used, although conventional expression in Erlenmeyer flasks gives also good results.

1. After harvesting the bacterial cells by centrifugation, the cell pellet is resuspended in ice cold bacterial cell pellet resuspension buffer. Take an aliquot and analyze for tau protein by SDS-PAGE (*see* **Fig. 2, lane 4**).
2. The resuspended cells can either be directly extracted to at least the stage of dialysis (step 6) before HPLC or frozen at −20°C and stored for further processing.

3. Disrupt the cells with a French pressure cell twice (assure ice cold conditions at any time—proteins in lysed cells are rapidly degraded!). Take an aliquot and analyze for tau protein by SDS-PAGE (*see* **Fig. 2, lane 5**).

4. Add NaCl to a final concentration of 500 m*M* and boil for 20 min. By this treatment, nearly all proteins are denatured apart from tau protein, which stays in solution and maintains its physiological function *(46–48)*.

5. Pellet the denatured proteins and cell debris by centrifugation at 127,000*g* for 40 min at 4°C. Take an aliquot of the supernatant and the pellet and analyze for tau protein by SDS-PAGE (*see* **Fig. 2, lane 6**).

6. Fill the supernatant into dialysis tubings (3.5 kDa molecular weight cut-off) and dialyze overnight against two changes of cation exchange chromatography buffer A (*see* **Subheading 2.**) at 4°C under constant stirring (the first buffer change should be not before 2 h of dialyzing time, *see* **Note 1**).

7. Clear the dialysate by centrifugation at 127,000*g* for 40 min at 4°C. Take an aliquot and analyze for tau protein by SDS-PAGE.

8. Apply the clear supernatant onto a cation-exchange chromatography column (*see* **Note 2**).

9. Wash out unspecific proteins with 3 to 5 column volumes cation exchange chromatography buffer A until UV-absorption reaches a stable value.

10. Elute the tau protein with a linear gradient of 60% final concentration of cation exchange chromatography buffer B over 6 column volumes and fractionate (*see* **Note 3**). Take an aliquot of the eluted fractions containing protein as judged by UV-absorption and analyze for tau protein by SDS-PAGE (*see* **Fig. 2, lane 7**).

11. Pool samples containing tau protein and concentrate by ultrafiltration devices (e.g., Ultrafree, Millipore; 10 kDa MWCO for tau isoforms and 5 kDa MWCO for short tau repeat constructs) to a final volume of 0.5 to 1 mL containing 5 to 10 mg tau (as determined by Bradford assay). Take an aliquot and analyze for tau protein by SDS-PAGE.

12. Apply the tau protein concentrate onto a gel filtration column with a low flow rate (e.g., 0.5 mL/min) of gel-filtration buffer. After elution of the protein, the flow rate can be increased for a faster completion of the run. Take an aliquot of the eluted fractions containing protein as judged by UV-absorption and analyze for tau protein by SDS-PAGE (*see* **Fig. 2, lanes 8** and **9**).

13. Pool suitable tau protein fractions with desired purity regarding amount of degradation (*see* **Note 4**) and tau-dimers (*see* **Note 5**). Determine protein concentration (*see* **Note 6**) and store aliquoted at −80°C. Check an aliquot of the pool by SDS-PAGE for its final purity and as a reference.

3.2. Preparation of Paired Helical Filaments From Tau Protein

In **Subheading 3.2.1.**, we will describe the preparation of PHFs examplified by the longest tau isoforms htau40 (a 4-repeat isoform) and the shortest tau isoform htau23 (a 3-repeat isoform) (*see* **Fig. 1**) under reducing conditions. In the second part (*see* **Subheading 3.2.2.**) we will focus on the preparation of

PHFs from a 3-repeat and a 4-repeat tau repeat-domain construct under reducing or oxidizing conditions. Depending on whether tau isoforms or tau repeat domain constructs are used, the time frame for the in vitro preparation of bona fide paired helical filaments ranges from up to two weeks to hours, respectively. In either case, the PHF polymerization is induced by the cofactor heparin which greatly accelerates the rate and velocity of tau polymerization (*see* **Note 7**).

3.2.1. Preparation of Paired Helical Filaments From Tau Isoforms

1. Prior to any kinetic PHF aggregation analysis, check the used tau proteins for equal concentrations, degree of purity, degradation and possible dimerization (oligomerization) by SDS-PAGE. This greatly influences the efficiency of tau polymerization to PHFs (*see* **Notes 5** and **6**).
2. The conditions for tau polymerization are: 50 µM tau isoform protein, 12.5 µM heparin and a protease inhibitor cocktail of 1 mM PMSF, 1 mM EDTA, 1 mM EGTA, 1 µg/mL leupeptin, 1 µg/mL aprotinin and 1 µg/mL pepstatin. As a buffer PBS pH 7.4 containing 2 mM DTT is used.
3. Incubate at 37°C.
4. Check for PHF assembly at appropriate times (*see* **Fig. 3**) by ThS fluorescence (*see* **Subheading 3.3.1.**) and transmission electron microscopy (*see* **Subheading 3.3.2.**). It might take up to 2 wk to reach the final value of tau PHF polymerization (*see* **Fig. 3**).
5. If a 4-repeat tau construct is to be polymerized, maintain reducing conditions by adding 1 mM DTT each day (*see* **Note 6**).

3.2.2. Preparation of Paired Helical Filaments From Tau Repeat Domain Constructs

1. Prior to any kinetic PHF aggregation analysis, check the used tau proteins for equal concentrations, degree of purity, degradation, and possible dimerization or oligomerization by SDS-PAGE. This greatly influences the efficiency of tau polymerization to PHFs (*see* **Notes 5** and **6**).
2. The conditions for tau repeat domain polymerization are: 20 µM tau, 5 µM heparin in PBS pH 7.4 with or without 1 mM DTT (*see* **Note 6**).
3. Incubate at 37°C.
4. Check for PHF assembly in regular times by ThS fluorescence (*see* **Subheading 3.3.1.**) and transmission electron microscopy (*see* **Subheading 3.3.2.**). It might take up to 3 d to reach the final value of tau PHF polymerization (*see* **Fig. 4**).

3.3. Validation of Paired Helical Filament Aggregation

At first, we will describe a means of quantifying the amount of tau protein polymerized into PHFs, allowing for the kinetic investigation of tau PHF polymerization (*see* **Subheading 3.3.1.**). In the next part the proof of PHF forma-

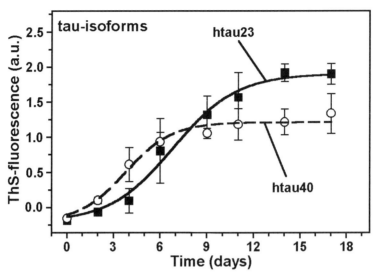

Fig. 3. Aggregation of tau isoforms to PHFs. The kinetics of aggregation of the shortest and the longest human tau isoform (htau23 and htau40) was measured by the ThS assay. Tau protein concentrations were 45 μM, with 11.25 μM heparin as an inducing cofactor in PBS pH 7.4, 2 mM DTT. To maintain reducing conditions over long periods, 1 mM DTT was added to the samples every day (*see* **Note 6**). The 4-repeat human tau isoform htau40 (open symbols, dashed line) and the 3-repeat tau isoform htau23 (closed symbols, solid line) aggregate with a comparable velocity over a time period of about 2 wk. The efficiency of formed PHFs is nearly twice as high for htau23, probably owing to a partly folding of the 4-repeat htau40 into a PHF incompetent structure of a "compact monomer" (*see* **Note 6**).

tion by imaging them with negative stain transmission electron microscopy is described (*see* **Subheading 3.3.2.**).

3.3.1. Quantitation of Paired Helical Filaments Using the Thioflavine-S Fluorescence Assay

The thioflavine-S fluorescence assay can be used in order to prove the polymerization of tau to PHFs *(30)*. The dye binds to PHFs, which results in a shift in the fluorescence emission spectra. With a fluorescence spectrometer, the intensity of emitted fluorescence can be determined, which gives a quantitative measure of the amount of PHFs in the assay solution. Tau in a monomeric-, dimeric-, or unstructered higher aggregate-form does not lead to this fluorescence shift. Therefore, the ThS assay can be used to verify tau PHF polymerization even in a quantitative manner suitable for the study of aggregation kinetics (*see* **Figs. 3** and **4**). For the difference between ThT, which is often used for quantification of Aβ fibrils in solution, and ThS *see* **Note 8**.

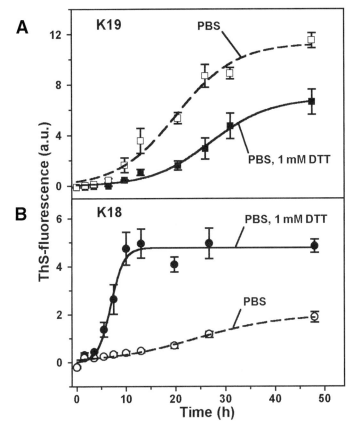

Fig. 4. Aggregation of repeat constructs of tau to PHFs in reducing and oxidative conditions. The kinetics of aggregation was measured by the ThS fluorescence assay (30). The concentrations of tau protein were 20 μM with 5 μM heparin as an inducing cofactor in PBS pH 7.4 containing either 1 mM DTT (closed symbols, solid lines) or no DTT (open symbols, dashed lines). Note that for the short repeat constructs PHF-polymerization occurs within hours compared to days for tau isoforms (*see* **Fig. 3**). **(A)** In oxidizing conditions (no DTT) the aggregation of the 3-repeat construct K19 (containing only C322 in R3) is increased, compared to reducing conditions (with DTT), because covalent dimers (by inter-molecular disulfide bridges) promote aggregation. **(B)** By contrast, the 4-repeat construct K18 (containing C322 in R3 and C291 in R2) shows reduced filament formation under oxidizing conditions because compact monomers (by intra-molecular disulfide bridges) are inhibitory for aggregation.

1. Prepare a total volume of 50 µL assay solution in a black 386-microtiter-plate well containing 10 µM of ThS, 5–15 µM total tau from the PHF-aggregation sample in PBS, pH 7.4.
2. As a control for background fluorescence, light scattering and fluorescence interaction of monomeric tau or heparin with the ThS-assay, prepare the following con-

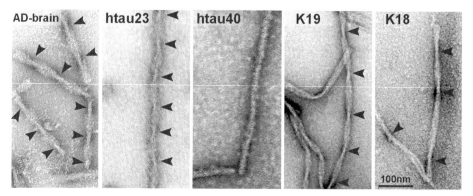

Fig. 5. Negative stain electron microscopy of tau PHFs extracted from Alzheimer brain and polymerized in vitro. The PHFs were negatively stained with uranyl acetate and viewed with a transmission electron microscope at 100 kV. For comparison of the ultrastructure of PHFs, a preparation of Alzheimer brain extracted PHFs is shown (left side). The Alzheimer brain extracted PHFs exhibit the typical characteristics of PHFs. The arrowheads point to the characteristic twist of the filaments with a crossover repeat of approx 80 nm. The filaments have a typical width of 10 to 20 nm. Three-repeat tau, like the isoform htau23 and the repeat domain construct K19 form predominantly paired helical filaments. In contrast, 4-repeat tau as the isoform htau40 exhibits predominantly straight filaments and for the repeat domain construct K18 straight and paired helical filaments are found.

    trols by adding adequate amounts to a ThS-assay sample: (a) buffer only, (b) tau protein only, (c) heparin only (it is best to prepare and incubate these controls simultaneously with the tau-polymerization sample).
3. Mix the ThS assay samples and incubate for 45 min in the dark at room temperature to allow for quantitative binding of ThS to the PHFs.
4. Measure the fluorescence with an excitation at 440 nm and an emission wavelength of 510 nm.
5. Calculate the PHF-specific ThS fluorescence by substracting the measured end values of the background (buffer) from the PHF-aggregation sample.

3.3.2. Imaging of Paired Helical Filaments by Negative Stain Transmission Electron Microscopy

The measurement of the ThS fluorescence change upon PHF formation is a good indicator of PHF formation from tau protein. Nevertheless, ThS fluorescence is not an absolute proof of PHF formation owing to cases where tau aggregates into amorphous structures which also leads to an increase in ThS fluorescence. Therefore, and to investigate the overall appearance of the PHFs, the analysis of the PHF aggregation sample by negative stain electron microscopy (EM) is necessary (*see* **Fig. 5**). It is beyond the scope of this chapter to describe the preparation of EM-grids, use of the EM, possible staining artifacts etc. We can only briefly describe the negative staining procedure of PHFs using uranyl acetate itself.

1. Withdraw an aliquot from the PHF aggregation sample and adjust the volume with H_2O to 10 μL containing 1 to 10 μM total tau protein.
2. Pipet onto a clean parafilm surface 10 μL of the following solutions: 1 drop of tau protein PHF-solution, 2 drops of H_2O, and 1 drop of 2% uranyl acetate (*see* **Note 9**).
3. Using a fine-tip tweezer (e.g., DuMont no. 5) take a freshly glow-discharged 600-mesh carbon coated copper grid and place it onto the PHF solution, the carbon film facing down and let the PHFs adsorb for 45 s. Wick the protein solution with filter paper from the side of the grid.
4. Wash the grid twice for 15 s by transferring it onto the H_2O drop, each time wicking the solution with filter paper.
5. Transfer onto the 2% uranyl acetate drop and stain for 45 s. Wick the solution with filter paper and let the grid dry for 15 min before viewing it in the electron microscope (*see* **Note 10**).

4. Notes

1. As general advice, or definitely if dialysis time is reduced, the ionic strength of the dialysate should be checked by conductivity measurement before application to the cation exchange column. It should have reached approximately the same value as the cation exchange chromatography buffer A (~5 mS/cm), otherwise the protein will not bind to the column material. For example, tau proteins elute from a SP-Sepharose cation exchange column at an ionic strength as measured by conductivity of approx 10 mS/cm (for tau isoforms and tau repeat constructs).
2. In this preparation, no DNase is used to degrade the *E. coli* DNA. Therefore, after the centrifugation, the DNA is sitting as a jelly material on top of the cell-debris pellet. Carefully take the supernatant using a pipette without sucking in the DNA. Do not try to filter DNA containing solutions through a 0.45 μm or smaller pore silze filter, as it will be clogged quickly. If desired, the DNA can also be degraded by DNase.
3. The isoelectric points for tau isoforms and tau constructs differ to some extent. We found that the cation exchange chromatography buffer A and B work for all tau isoforms and also for tau repeat constructs (*see* **Fig. 1**) without significant performance loss.
4. The purity of tau is most critical for its polymerization to PHF. Compared to full length isoforms, the N- or C-terminal truncated forms of tau show a much accelerated aggregation kinetics, as long as they still contain the essential repeat domain (*see* **Fig. 1**, compare **Figs. 3** and **4** and *[28–30,37,52]*). Therefore, for kinetic experiments, the contamination by degradation products of tau should be minimal.
5. Tau protein contains two cysteines, C291 and C322. C322 is present in all isoforms, but C291 is present only in 4-repeat isoforms because it lies in R2 (*see* **Fig. 1**).

The cysteines are the basis for tau's property to form dimers via a disulfide bond. These dimers are important in facilitating PHF formation *(28,29,52)*. Dimerization can be minimized in tau preparations by carefully removing them in the gel-filtration step during preparation (*see* **Subheading 3.1., step 6**). A subsequent formation of dimers can be effectively prevented by maintaining reducing conditions in all buffers (e.g., 1 m*M* DTT) and avoiding any concentrating steps. Although it must be kept in mind that DTT is rather rapidly degraded and consumed. In a typical PHF aggregation kinetics (conditions as in **Fig. 4**), we determined a half time of DTT of 9 h. Therefore, for long-term incubation under reducing conditions, DTT must be added at regular intervals (as in **Fig. 3**). For tau 3-repeat and 4-repeat forms, the redox conditions have opposite effects. For 3-repeat tau reducing conditions reduce both the rate and extent of aggregation. By contrast, the reverse behavior is found with the 4-repeat construct K18 (*see* **Fig. 4**). This apparent contradiction is explained by the fact that 4R-tau preferably forms intra-molecular disulfide crosslinks in oxidizing conditions (C291-C322) which do not participate and even inhibit aggregation because they are locked in the wrong conformation ("compact monomer" *[29,52]*). However, when the two cysteines are kept in a reduced state, aggregation readily takes place in the presence of poly-anions. In conclusion, tau dimers, formed by disulfide crosslinking of 3R-tau, promote aggregation, whereas intramolecular crosslinking of 4R-tau into compact monomers inhibits aggregation.

6. The Bradford assay for determination of the protein concentration is widely used in laboratories owing to its quickness and easiness. But, it contains some pitfalls in the case of tau protein. The Bradford-dye (coomassie brilliant blue) binds relatively unspecific to cationic and unpolar, hydrophobic side chains of proteins. Most important are interactions with arginine. Tau repeat constructs especially contain very few arginines and exhibit a very low relative Bradford absorbance compared to standard proteins like BSA. Therefore, concentrations of tau repeat constructs should be determined by the absorption of the peptide bond at 214 nm. For the A214 method, it is an absolute necessity to use a clean quartz cuvet, filtered solutions, and absolutely comparable buffer conditions for the standard calibration curve (e.g., BSA) and the tau sample. However, we calculated from the absorption values of single amino acids *(49–51)* that the tau concentration is underestimated by about 20% when BSA is used as a standard protein. This is owing to the low content of aromatic residues which also contribute significantly to the A214 nm. Measuring the absorption at 205 nm (the maximum of absorption for the peptide bond) does not circumvent this problem but leads to increased technical demands and problems, owing to the very short wavelength. In order to measure the absolute tau protein concentration, one solution is to perform a standard protein concentration curve using known amounts of lyophilized tau protein and to calculate the extinction coefficient of tau.

7. The preparation of PHFs described in this chapter uses the polyanion heparin as an inducing agent. Other polyanions like poly-glutamic acid and RNA are also known to induce PHF formation *(23–25)* but are not commonly used. Another well-described PHF aggregation method described by Binder and coworkers uses fatty

acids, like arachidonic acid, to induce PHF formation *(26,27,53,54)*. Their buffer conditions are similar to ours (~neutral pH, ~0.15 *M* ionic strength, 37°C), but emphasis is placed on reducing conditions. PHF aggregation of tau isoforms under these conditions is dependent on the presence of at least one of the N-terminal repeats. The only known case where tau aggregates readily into PHFs even in the absence of any inducing agent is a 4-repeat tau construct (K18, *see* **Fig. 1**), lacking K280 *(19,37)* which is one of the mutations found in FTDP-17 *(31)*. Earlier attempts to polymerize tau in the absence of inducing agents to PHFs used a repeat domain tau construct (3-repeat form), around 20-fold higher concentrations and polymerization took several weeks of incubation time *(28)*.

8. In contrast to the quantitation of Aβ-fibrils, where thioflavine-T is used, it was shown for tau PHFs that thioflavine-S is more reproducible and independent of solvent characteristics *(30)*.

9. Uranyl acetate is a radioactive compound (a low intensity emitter of α-particles). Follow general safety precautions for handling radioactive compounds.

10. PHFs are notoriously poorly stained by negative staining procedures. PHFs on one and the same grid may show a positive stain or no staining at all. The best advice is to search for nicely negatively stained areas on the grid, to retry staining, or to use a different grid charge.

Acknowledgments

We thank Dr. Martin von Bergen and Dr. Eva-Maria Mandelkow for helpful discussions throughout this work. We also thank Dr. Peter Davies for providing purified Alzheimer brain-derived PHFs. This project was supported by a grant from the Deutsche Forschungsgemeinschaft.

References

1. Bond, J. P., Deverin, S. P., Inouye, H., El-Agnaf, O. M., Teeter, M. M., and Kirschner, D. A. (2003) Assemblies of Alzheimer's peptides Abeta25-35 and Abeta 31-35: reverse-turn conformation and side-chain interactions revealed by X-ray diffraction. *J. Struct. Biol.* **141(2)**, 156–170.

2. Petkova, A. T., Ishii, Y., Balbach, J. J., et al. (2002) A structural model for Alzheimer's beta-amyloid fibrils based on experimental constraints from solid state NMR. *Proc. Natl. Acad. Sci. USA* **99(26)**, 16742–16747.

3. Torok, M., Milton, S., Kayed, R., et al. (2002) Structural and dynamic features of Alzheimer's Abeta peptide in amyloid fibrils studied by site-directed spin labeling. *J. Biol. Chem.* **277(43)**, 40810–40815.

4. Rochet, J. C. and Lansbury, P. T. Jr. (2000) Amyloid fibrillogenesis: themes and variations. *Curr. Opin. Struct. Biol.* **10(1)**, 60–68.

5. Garcia, M. L. and Cleveland, D. W. (2001) Going new places using an old MAP: tau, microtubules and human neurodegenerative disease. *Curr. Opin. Cell Biol.* **13(1)**, 41–48.

6. Goedert, M., Spillantini, M. G., and Davies, S. W. (1998) Filamentous nerve cell inclusions in neurodegenerative diseases. *Curr. Opin. Neurobiol.* **8(5)**, 619–632.

7. Buee, L., Bussiere, T., Buee-Scherrer, V., Delacourte, A., and Hof, P. R. (2000) Tau protein isoforms, phosphorylation and role in neurodegenerative disorders. *Brain Res. Rev.* **33(1)**, 95–130.

8. Reed, L. A., Wszolek, Z. K., and Hutton, M. (2001) Phenotypic correlations in FTDP-17. *Neurobiol. Aging* **22(1)**, 89–107.

9. Cleveland, D. W., Hwo, S. Y., and Kirschner, M. W. (1977) Physical and chemical properties of purified tau factor and the role of tau in microtubule assembly. *J. Mol. Biol.* **116(2)**, 227–247.

10. Lee, G., Cowan, N., and Kirschner, M. (1988) The primary structure and heterogeneity of tau protein from mouse brain. *Science* **239**, 285–288.

11. Schweers, O., Schonbrunn-Hanebeck, E., Marx, A., and Mandelkow, E. (1994) Structural studies of tau protein and Alzheimer paired helical filaments show no evidence for beta-structure. *J. Biol. Chem.* **269(39)**, 24290–24297.

12. Berry, R. W., Abraha, A., Lagalwar, S., et al. (2003) Inhibition of tau polymerization by its carboxy-terminal caspase cleavage fragment. *Biochemistry* **42(27)**, 8325–8331.

13. Minoura, K., Tomoo, K., Ishida, T., Hasegawa, H., Sasaki, M., and Taniguchi, T. (2002) Amphipathic helical behavior of the third repeat fragment in the tau microtubule-binding domain, studied by (1)H NMR spectroscopy. *Biochem. Biophys. Res. Commun.* **294(2)**, 210–214.

14. Esposito, G., Viglino, P., Novak, M., and Cattaneo, A. (2000) The solution structure of the C-terminal segment of tau protein. *J. Pept. Sci.* **6(11)**, 550–559.

15. Yanagawa, H., Chung, S. H., Ogawa, Y., et al. (1998) Protein anatomy: C-tail region of human tau protein as a crucial structural element in Alzheimer's paired helical filament formation in vitro. *Biochemistry* **37(7)**, 1979–1988.

16. Kirschner, D. A., Abraham, C., and Selkoe, D. J. (1986) X-ray diffraction from intraneural paired helical filaments and extraneural amyloid fibers in Alzheimer disease indicates cross-β conformation. *Proc. Natl. Acad. Sci. USA* **83**, 503–507.

17. Iqbal, K., Braak, H., Braak, E., and Grundke-Iqbal, I. (1993) Silver labeling of Alzheimer neurofibrillary changes and brain beta-amyloid. *J. Histotechnology* **16(N4)**, 335–342.

18. von Bergen, M., Friedhoff, P., Biernat, J., Heberle, J., Mandelkow, E. M., and Mandelkow, E. (2000) Assembly of tau protein into Alzheimer paired helical filaments depends on a local sequence motif ((306)VQIVYK(311)) forming beta structure. *Proc. Natl. Acad. Sci. USA* **97(10)**, 5129–5134.

19. von Bergen, M., Barghorn, S., Li, L., et al. (2001) Mutations of tau protein in frontotemporal dementia promote aggregation of paired helical filaments by enhancing local beta-structure. *J. Biol. Chem.* **276(51)**, 48165–48174.

20. Giannetti, A. M., Lindwall, G., Chau, M. F., Radeke, M. J., Feinstein, S. C., and Kohlstaedt, L. A. (2000) Fibers of tau fragments, but not full length tau, exhibit a cross beta- structure: implications for the formation of paired helical filaments. *Protein Sci.* **9(12)**, 2427–2435.

21. Barghorn, S., Davies, P., and Mandelkow, E. (2004) Tau paired helical filaments from Alzheimer's disease brain and assembled in vitro are based on beta-structure in the core domain. *Biochemistry* **43(6)**, 1694–1703.
22. Berriman, J., Serpell, L. C., Oberg, K. A., Fink, A. L., Goedert, M., and Crowther, R. A. (2003) Tau filaments from human brain and from in vitro assembly of recombinant protein show cross-beta structure. *Proc. Natl. Acad. Sci. USA* **100(15)**, 9034–9038.
23. Perez, M., Valpuesta, J. M., Medina, M., Montejo de Garcini, E., and Avila, J. (1996) Polymerization of tau into filaments in the presence of heparin: the minimal sequence required for tau-tau interaction. *J. Neurochem.* **67(3)**, 1183–1190.
24. Goedert, M., Jakes, R., Spillantini, M. G., Hasegawa, M., Smith, M. J., and Crowther, R. A. (1996) Assembly of microtubule-associated protein tau into Alzheimer-like filaments induced by sulphated glycosaminoglycans. *Nature* **383(6600)**, 550–553.
25. Kampers, T., Friedhoff, P., Biernat, J., and Mandelkow, E. M. (1996) RNA stimulates aggregation of microtubule-associated protein-tau into Alzheimer-like paired helical filaments. *FEBS Letters* **399(3)**, 344–349.
26. Wilson, D. M. and Binder, L. I. (1997) Free fatty acids stimulate the polymerization of tau and amyloid beta peptides. In vitro evidence for a common effector of pathogenesis in Alzheimer's disease. *Am. J. Pathol.* **150(6)**, 2181–2195.
27. King, M. E., Ahuja, V., Binder, L. I., and Kuret, J. (1999) Ligand-dependent tau filament formation: implications for Alzheimer's disease progression. *Biochemistry* **38(45)**, 14851–14859.
28. Wille, H., Drewes, G., Biernat, J., Mandelkow, E. M., and Mandelkow, E. (1992) Alzheimer-like paired helical filaments and antiparallel dimers formed from microtubule-associated protein tau in vitro. *J. Cell Biol.* **118**, 573–584.
29. Schweers, O., Mandelkow, E. M., Biernat, J., and Mandelkow, E. (1995) Oxidation of cysteine-322 in the repeat domain of microtubule-associated protein tau controls the in vitro assembly of paired helical filaments. *Proc. Natl. Acad. Sci. USA* **92(18)**, 8463–8467.
30. Friedhoff, P., Schneider, A., Mandelkow, E. M., and Mandelkow, E. (1998) Rapid assembly of Alzheimer-like paired helical filaments from microtubule-associated protein tau monitored by fluorescence in solution. *Biochemistry* **37(28)**, 10223–10230.
31. Foster, N. L., Wilhelmsen, K., Sima, A. A., Jones, M. Z., D'Amato, C. J., and Gilman, S. (1997) Frontotemporal dementia and parkinsonism linked to chromosome 17: a consensus conference. *Ann. Neurol.* **41(6)**, 706–715.
32. Hutton, M., Lendon, C. L., Rizzu, P., et al. (1998) Association of missense and 5'-splice-site mutations in tau with the inherited dementia FTDP-17. *Nature* **393(6686)**, 702–705.
33. Clark, L. N., Poorkaj, P., Wszolek, Z., et al. (1998) Pathogenic implications of mutations in the tau gene in pallido-ponto-nigral degeneration and related neurodegenerative disorders linked to chromosome 17. *Proc. Natl. Acad. Sci. USA* **95(22)**, 13103–13107.

34. Poorkaj, P., Bird, T. D., Wijsman, E., et al. (1998) Tau is a candidate gene for chromosome 17 frontotemporal dementia. *Ann. Neurol.* **43(6)**, 815–825.

35. Goedert, M., Jakes, R., and Crowther, R. A. (1999) Effects of frontotemporal dementia FTDP-17 mutations on heparin-induced assembly of tau filaments. *FEBS Lett.* **450(3)**, 306–311.

36. Nacharaju, P., Lewis, J., Easson, C., et al. (1999) Accelerated filament formation from tau protein with specific FTDP-17 missense mutations. *FEBS Lett.* **447(2–3)**, 195–199.

37. Barghorn, S., Zheng-Fischhofer, Q., Ackmann, M., Biernat, J., von Bergen, M., and Mandelkow, E. (2000) Structure, microtubule interactions, and paired helical filament aggregation by tau mutants of frontotemporal dementias. *Biochemistry* **39(38)**, 11714–11721.

38. Friedhoff, P., von Bergen, M., Mandelkow, E. M., Davies, P., and Mandelkow, E. (1998) A nucleated assembly mechanism of Alzheimer paired helical filaments. *Proc. Natl. Acad. Sci. USA* **95(26)**, 15712–15717.

39. Wischik, C. M., Novak, M., Thogersen, H. C., et al. (1988) Isolation of a fragment of tau derived from the core of the paired helical filament of Alzheimer disease. *Proc. Natl. Acad. Sci. USA* **85(12)**, 4506–4510.

40. Novak, M., Kabat, J., and Wischik, C. M. (1993) Molecular characterization of the minimal protease resistant tau-unit of the Alzheimer's-disease paired helical filament. *EMBO J.* **12**, 365–370.

41. Crowther, R. A. (1991) Straight and paired helical filaments in Alzheimer disease have a common structural unit. *Proc. Natl. Acad. Sci. USA* **88(6)**, 2288–2292.

42. Gamblin, T. C., King, M. E., Dawson, H., et al. (2000) In vitro polymerization of tau protein monitored by laser light scattering: method and application to the study of FTDP-17 mutants. *Biochemistry* **39(20)**, 6136–6144.

43. Kidd, M. (1963) Paired helical filaments in electron microscopy of Alzheimer's disease. *Nature (Lond.)* **197**, 192–193.

44. Studier, F. W., Rosenberg, A. H., Dunn, J. J., and Dubendorff, J. W. (1990) Use of T7 RNA polymerase to direct expression of cloned genes. *Methods Enzymol.* **185**, 60–89.

45. Biernat, J., Mandelkow, E. M., Schröter, C., et al. (1992) The switch of tau protein to an Alzheimer-like state includes the phosphorylation of two serine-proline motifs upstream of the microtubule binding region. *EMBO J.* **11**, 1593–1597.

46. Weingarten, M. D., Lockwood, A. H., Hwo, S. Y., and Kirschner, M. W. (1975) A protein factor essential for microtubule assembly. *Proc. Natl. Acad. Sci. USA* **72(5)**, 1858–1862.

47. Herzog, W. and Weber, K. (1978) Fractionation of brain microtubule-associated proteins. Isolation of two different proteins which stimulate tubulin polymerization in vitro. *Eur. J. Biochem.* **92(1)**, 1–8.

48. Gustke, N., Trinczek, B., Biernat, J., Mandelkow, E. M., and Mandelkow, E. (1994) Domains of Tau protein and interactions with microtubules. *Biochemistry* **33**, 9511–9522.

49. Wetlaufer, D. B. (1962) Ultraviolet spectra of proteins and amino acids. *Adv. Protein Chem.* **17,** 303–391.
50. Saidel, L. J. and Lieberman, H. (1958) Ultraviolet absorption spectra of peptides. IV. Alanine residue. *Arch. Biochem. Biophys.* **76,** 401–409.
51. Goldfarb, R. (1953) Absorption spectrum of the peptide bond. II. Influence of chain length. *J. Biol. Chem.* **201,** 317–320.
52. Barghorn, S. and Mandelkow, E. (2002) Toward a unified scheme for the aggregation of tau into Alzheimer paired helical filaments. *Biochemistry* **41(50),** 14885–14896.
53. Gamblin, T. C., King, M. E., Kuret, J., Berry, R. W., and Binder, L. I. (2000) Oxidative regulation of fatty acid-induced tau polymerization. *Biochemistry* **39(46),** 14203–14210.
54. King, M. E., Gamblin, T. C., Kuret, J., and Binder, L. I. (2000) Differential assembly of human tau isoforms in the presence of arachidonic acid. *J. Neurochem.* **74(4),** 1749–1757.

5

Cyclic Amplification of Protein Misfolding and Aggregation

Paula Saá, Joaquín Castilla, and Claudio Soto

Summary

Diverse human disorders, including most neurodegenerative diseases, are thought to arise from the misfolding and aggregation of an underlying protein. We have recently described a novel technology to amplify cyclically the misfolding and aggregation process in vitro. This procedure, named protein misfolding cyclic amplification (PMCA), conceptually analogous to DNA amplification by PCR, has tremendous implications for research and diagnosis. The PMCA concept has been proved on the amplification of prions implicated in the pathogenesis of transmissible spongiform encephalopathies (TSE). In these diseases, there is a tremendous need for early and sensitive biochemical diagnosis to minimize the further spreading of the prion infectious agent through the food chain. In this chapter, we describe the principles behind the PMCA technology, its application, and methodology to detect minute quantities of misfolded prion protein and its potential to be used for amplification of misfolding of other proteins implicated in diverse diseases.

Key Words: Protein conformational disorders; prion; Creutzfeldt-Jakob disease; bovine spongiform encephalopathy; Scrapie; protein misfolding cyclic amplification (PMCA); Alzheimer's disease; amyloid.

1. Introduction

Compelling evidence has accumulated in the last few years to indicate that a hallmark event in several diverse diseases is the misfolding of an otherwise normal protein. These diseases are now grouped together under the name of protein conformational disorders *(1–4)*. This group includes most of the neurodegenerative diseases, such as Alzheimer's disease, transmissible spongiform encephalopathies, Parkinson's disease, Amyotrophic Lateral Sclerosis, and Huntington's disease *(1)*. There are also several systemic disorders in this group, including diabetes type II, serpin-deficiency disorders, hemolytic anemia, cystic

From: *Methods in Molecular Biology, vol. 299: Amyloid Proteins: Methods and Protocols*
Edited by: E. M. Sigurdsson © Humana Press Inc., Totowa, NJ

fibrosis, dialysis-related amyloidosis, and more than 15 other less well-known diseases *(2,4,5)*.

Transmissible spongiform encephalopathies (TSEs), also known as prion diseases, are the prototype of protein conformational disorders, in which the role of protein misfolding is perhaps mostly clear *(6)*. TSEs are fatal and infectious neurodegenerative disorders that affect both humans and animals *(7)*. In these diseases a long incubation period, in which the infectious agent starts its replication in the target organs, is followed by a brief and fatal clinical phase *(8)*. Although these disorders are rare, the description on May 1996 of a new human disease, variant Creutzfeldt-Jakob disease (vCJD) *(9)*, related to the consumption of meat contaminated with the causative agent of bovine spongiform encephalopathy (BSE), has taken the attention of the scientific community *(10,11)*.

Although the etiologic agent of TSEs is still matter of controversy, the most accepted hypothesis "the protein-only" proposes that a misfolded form of a host-encoded protein, named cellular prion protein (PrP^C), is the sole component of the infectious agent and its propagation does not require nucleic acid replication *(12,13)*. The disease-associated isoform, termed PrP^{res} or PrP^{Sc}, differs from the PrP^C in its three-dimensional structure, having much higher β-sheet content, whereas the primary amino acid sequence is the same *(14)*. This structural change confers new biochemical properties to the misfolded protein: insolubility in non-denaturing detergents and partial resistance against digestion with proteinase K *(15,16)*. In this chapter, we will use PrP^{Sc} to refer to the misfolded protease-resistant prion protein that has also been demonstrated to be associated with infectivity and PrP^{res} to refer to the misfolded protein that exhibit protease resistance, but has not been demonstrated yet to be associated with infectivity.

According to the protein-only hypothesis, PrP^{Sc} propagates itself by an autocatalytic reaction *(15)*. The exact mechanism is not well-understood, but some evidence indicates that it involves a close interaction between the endogenous PrP^C and the exogenous PrP^{Sc} *(13,15)*. The notion that the host PrP^C is involved in the generation of infectivity and in the development of the disease is supported by experiments with transgenic mice in which the endogenous *PrP* gene was knocked out. These animals were resistant to infection as well as unable to generate new infectious particles *(17)*. A physical association between the two isoforms during the infectious process is suggested by the primary sequence specificity in prion transmission *(18)* and by the reported in vitro generation of PrP^{res} molecules by mixing purified PrP^{Sc} with PrP^C *(19)*. Nevertheless, from the analysis of the requirements for the process in vitro, it seems that a factor that catalyzes the reaction is necessary for the conversion. Many efforts have been made to identify this factor and, recently, Deleault et al. have shown, using a modified version *(20)* of the protein misfolding cyclic amplification method *(21)*, that the in vitro conversion is enhanced by the presence of stimulatory RNA molecules in the test tube *(22)*. These data suggest that host-encoded RNA molecules might act as cellular cofactors for PrP[res] formation.

Two models have been proposed to explain the conversion mechanism: in the "template-assisted model," the PrP[Sc] isoform acts as a template, and its interaction with the cellular form lowers the activation energy between both states *(23)*. The alternative "nucleation/polymerization model," proposes that both isoforms coexist in a thermodynamic equilibrium in solution *(24–26)*; as the PrP[Sc] monomer is unstable and becomes stabilized by aggregation, the equilibrium is displaced toward the formation of the pathological conformer *(27)*. This model is characterized by a lag phase in which the formation of a PrP[Sc]-nucleus leads to rapid growth of the polymers by the addition, at the ends, of newly converted molecules. The latter is supported by mathematical modeling studies *(26,28)*, quantitative data obtained from in vitro conversion experiments *(27,28)* and by the morphological characterization of prion aggregates as un-branched polymers with a relatively constant diameter *(29,30)*.

1.1. Protein Misfolding Cyclic Amplification

Based on the nucleation/polymerization model, we have recently described a strategy that mimics in vitro the PrP[Sc] conversion process that takes place in vivo, and that amplifies in an exponential fashion minute quantities of PrP[Sc] present in a sample *(21)*. This system is called protein misfolding cyclic amplification (PMCA), and consists of cycles of accelerated prion replication. PMCA is conceptually analogous to DNA amplification by PCR, given that in both systems, a template grows at the expense of a substrate in a cyclic reaction, combining phases of growing and multiplication of the template units *(31)*. In PMCA, each cycle is composed of two phases. During the first, the samples containing minute amounts of PrP[Sc] and a large excess of PrP[C] are incubated to induce growing of PrP[res] polymers. In the second phase the samples are subjected to ultra-sound in order to break down the polymers, multiplying the number of nuclei. **Figure 1** (right panel) shows a schematic representation of the PMCA principles, illustrating what we envision is happening in each of the phases.

Proof-of-concept experiments were done by using a healthy hamster brain homogenate as a source of PrP[C] and other factors that might be important during prion replication, and a scrapie infected hamster brain homogenate as a source of PrP[Sc]. The latter was diluted serially into the healthy brain homogenate to mimic samples from various tissues or distinct states of the disease where different quantities of PrP[Sc] are expected. Half of these samples were immediately frozen and the other half was subjected to PMCA. After several cycles of

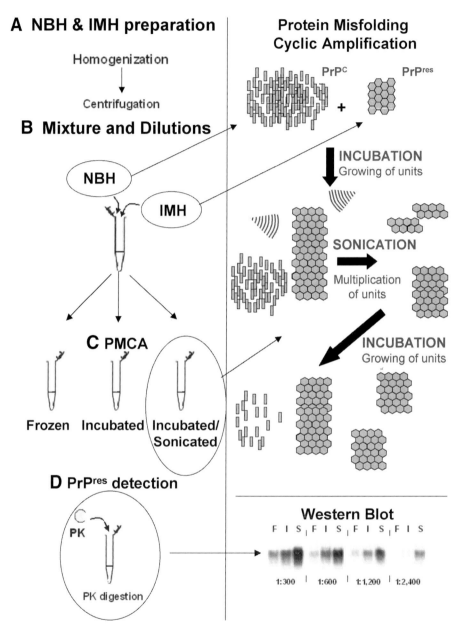

Fig. 1. Schematic representation of the principles and methodology for PMCA. NBH, normal brain homogenate; IMH, infectious material homogenate; PMCA, protein misfolding cyclic amplification; F, frozen sample; S, sonicated sample; PrP^C, cellular prion protein; PrP^{res}, protease-resistant prion protein.

PMCA, all the samples were treated with proteinase K, followed by Western blotting in order to evaluate the $PrP^C \rightarrow PrP^{res}$ conversion. The amount of PrP^{res} was dramatically increased in comparison to the equivalent frozen samples where the signal corresponding to PrP^{res} disappeared after a 3000-fold dilution of the infected brain; whereas in the amplified samples, the signal was still detected after 500,000-fold dilution. Densitometric analysis of the immunoblots indicated that of the total amount of PrP^{res} present in the sample after amplification, >99%, corresponded to newly generated PrP^{res} *(21)*. The conversion depends on the presence of both PrP^{Sc} and PrP^C in the sample, since no amplification was observed when the healthy brain homogenate and diluted PrP^{Sc} were incubated alone under the same conditions, with and without sonication.

2. Materials

PMCA have been applied successfully to a variety of brain samples from different species. In the following subheadings, we describe the equipment required, the technical considerations, and the standard parameters that afford an optimal amplification of PrP^{Sc} from different samples. **Figure 1** illustrates the different aspects of the PMCA procedure.

2.1. Equipment

1. Bandelin Sonoplus HD 2070 for a tip (manual) sonication (Bandelin Electronic, Berlin, Germany).
2. Titanium Microtip MS73 (Bandelin Electronic, Berlin, Germany).
3. Misonix 2020 sonicator (Misonix, Farmingdale, NY) for a water bath (automatic) sonication.
4. ULT 900 −86°C Upright Freezer (Thermo Electron Co., Marietta, OH).
5. pHmeter Model 6171 (Jenco Electronics, LTD., San Diego, CA).
6. Homogenizer: High-viscosity mixer Eurostar PWR BSC S1 (IKA, Wilmington, NC).
7. Surgical equipment.
8. Thermomixer R (Eppendorf, Westbury, NY).
9. Pipets (Gilson, Middleton, WI).
10. Refrigerated centrifuge (Thermo Electron Co., Marietta, OH).

2.2. Biological Samples

1. Normal brain homogenate.
2. Infectious material homogenate.

2.3. Solutions, Reagents, and Buffers

1. PBS 1X (Solution 1): Prepare PBS (ICN, cat no. 2810305) according to manufacturer's instructions in ultra pure water.
2. 5 *M* NaCl (Solution 2): NaCl (58.44 g), deionized water (J. T. Baker, Phillipsburg, NJ, cat. no. 4201) up to 200 mL.

3. Triton X-100 10% (Solution 3): 10% Triton X-100 (0.5 mL), deionized water 4.5 mL.
4. Conversion Buffer (Solution 4): 5 M NaCl (1.5 mL), 5 mM EDTA, 10% Triton X-100 (5 mL), 1X Protease Inhibitor Cocktail (Roche, Basel, Switzerland, cat. no. 1836145), PBS 1X up to 50 mL.
5. SDS 4% (Solution 5): SDS (4 g), deionized water up to 100 mL.
6. Proteinase K 1 µg/µL (Solution 6): Proteinase K (Roche, Basel, Switzerland, cat. no. 0745723) (0.1 mg), deionized water (100 µL).
7. Phenylmethylsulfonyl Fluoride (PMSF) 50 mM (Solution 7): PMSF (VWR, West Chester, PA, cat. no. 80055-380) (0.044 g), methanol (5 mL).

3. Methods

3.1. Preparation of Sample Homogenates (Fig. 1A)

3.1.1. 10% (w/v) Normal Brain Homogenate (NBH)

1. The tissue has to be fresh or frozen without fixation (*see* **Note 1**).
2. 0.5 g of normal brain is homogenized at 4°C in nine volumes of solution 4 and centrifuged at 5000g for 30 s to remove cell debris (*see* **Notes 2–4**). The pellet was discarded and the supernatant was aliquoted and stored at –80°C.

3.1.2. 10% (w/v) Infectious Material Homogenate (IMH)

1. The infectious brain was processed in the same way as NBH.
2. Infected tissues other than brain can be used as IMH, however, the homogenization and the centrifugation speed should be adapted to their specific properties.

3.2. Cyclic Amplification (Fig. 1B,C)

1. Aliquots of 10% IMH were serially diluted in 240 µL of 10% NBH from the same species. Eighty microliters of these samples (named frozen samples) were immediately frozen at –80°C and 80 µL (incubated samples) were incubated in parallel with the 80 µL (sonicated samples) exposed to incubation-sonication cycles (**Fig. 1B**).
2. The frozen and the incubated samples were used as different controls of the amplification.

3.2.1. Incubation

The standard cycles begin with 1 hour incubation at 37°C and 450 rpm shaking in the Thermomixer. Nevertheless, owing to the differences in the conformation/aggregation state among the PrP^{res} strains, the incubation time and temperature, as well as the number of cycles, have to be adjusted for optimal amplification.

3.2.2. Sonication

Because of the aforementioned differences, the ultrasound strength of the sonication should be adapted to the corresponding PrP^{res} strain. The aim of the sonication is to fragment the PrP^{res} aggregates, in order to increase the amount of seeds present in the sample without affecting their ability to act as conversion nucleus. Therefore, the sonication strength must be carefully chosen given that a weak sonication might be inefficient in cutting the aggregates, and an excessive sonication might damage the protein polymer.

PMCA can be performed efficiently either manually or automatically using any two of the following procedures (*see* **Note 5**):

3.2.2.1. MANUAL SONICATION

This step was carried out with a microtip sonicator, and different conditions were tested to determine the best amplification parameters. After one hour pre-incubation at 37°C and 450 rpm, the samples were subjected to cycles of sonication-incubation.

We saw that cycles of 1 h incubation at 37°C and 450 rpm shaking in the Thermomixer, followed by 10 sonication pulses of 0.1 s and 30 watts power, are appropriate to amplify PrP^{Sc} obtained from hamster brain infected with the 263K scrapie strain; but the amplification of the ME7 mouse strain requires a lower potency of 10 watts/pulse and 3 h incubation in each cycle to obtain optimal results.

3.2.2.2. AUTOMATIC SONICATION

As this sonication was performed in a bath sonicator, in which the ultrasound is not directly applied to the samples, a higher power had to be set. Because of the automation of the system, the microplate horn was kept in an incubator set at 37°C during the whole process. In this way, the incubation was performed without shaking. The best results for hamsters 263K were obtained when the sonicator was programmed to supply 40-s pulses with an output level of 10.0 (in a range from 0.0 to 10.0) after 1 h incubation at 37°C.

3.3. PrP^{res} Detection (Fig. 1D)

After PMCA, the remaining PrP^{C} was removed from all the samples (frozen, incubated, and sonicated) by proteinase K treatment (*see* **Note 6**).

Eighteen microliters of each hamster sample were incubated with 1 µL of 4% SDS and 1 µL of 1 µg/µL Proteinase K for 1 h at 45°C and 450 rpm. The reaction was stopped by adding 2 µL of 50 m*M* PMSF and 8 µL of 4X SDS sample buffer (Invitrogen, Carlsbad, CA).

The samples were denatured for 5 min at 100°C and protein electrophoresis was performed on a NuPAGE 12% Bis-Tris Gel (Invitrogen, Carlsbad, CA).

The proteins were transferred to a Hybond-ECL nitrocellulose membrane (Amersham, Piscataway, NJ), and the immunoblot was developed with the ECL plus Western Blot Detection System (Amersham Biosciences, Piscataway, NJ) after incubation with the mAb 3F4 (Signet, Dedham, MA).

3.4. Applications of PMCA

3.4.1. Basic Science

Because of the exceptional nature of the infectious agent, the exact mechanism of these diseases is still not completely understood. Many attempts have been made to find a strategy that helps to understand the underlying biology of prions. In vitro conversion of PrP[res] has been reported by Caughey and colleagues under different conditions *(32)*. This system has been useful to study the molecular mechanism of PrP conversion *(33)*, the species-barrier phenomenon *(34)*, and to identify and evaluate inhibitors of PrP transition *(35,36)*. But until the PMCA was reported in 2001, the low-efficiency conversion of the pre-existing in vitro models ruled out the characterization of the structural and infectious properties of newly generated PrP[res] *(37)*. After several cycles of PMCA, the newly generated PrP[res] in the test tube is >99% of the total amount of PrP[res]; in this way, infectious and structural studies can be performed and the properties of the PrP[res] can be monitored.

As PMCA mimics in the test tube the pathogenic process in an accelerated manner, it is a useful tool for the identification of other factors that may be implicated in prion conversion, for the discovery of novel drug targets and for testing PrP conversion inhibitors. Species-barrier studies with PMCA are less time-consuming, and can be a complement to those studies performed in animal models.

3.4.2. Diagnosis

At present there is not an accurate diagnosis for TSE. The clinical diagnosis of sporadic CJD (sCJD) is currently based upon the combination of subacute progressive dementia (less than 2 yr), myoclonus, and multifocal neurological dysfunction, associated with a characteristic periodic electroencephalogram *(38)*. However, vCJD, most of the iatrogenic forms of CJD, and up to 40% of the sporadic cases, do not show electroencephalogram abnormalities *(39)*. The definitive diagnosis is established only by neuropathological examination and detection of PrP[Sc] either by immunohistochemistry, histoblot or Western blot *(38,40)*. Presymptomatic detection of sCJD or vCJD in living people is not possible. This problem, in a scenario of substantial number of people incubating vCJD, raises an enormous concern of iatrogenic propagation of vCJD *(41, 42)*. A pre-symptomatic diagnosis of CJD is also very important for treatment, because it is likely that potential therapies would require intervention before symptoms appear. In animal TSEs, the situation is no better. Although several tests have been developed to diagnose BSE in postmortem brain tissue *(43)*, there is still no reliable way to identify cattle early after infection *(44)*. Therefore, it is possible that infected animals without symptoms and negative by the biochemical tests enter into the food chain, imposing a risk for human health. All these issues indicate that the development of tests that can effectively detect animals and people incubating the disease is a top priority *(44–46)*.

PrP^{Sc} is not only the major component of the infectious agent and the most likely cause of TSE, but also the only validated surrogate marker for the disease *(13)*. Detection of PrP^{Sc} in tissues and cells correlates widely with the disease and with the presence of TSE infectivity, and treatments that inactivate or eliminate TSE infectivity also eliminate PrP^{Sc} *(7)*. The identification of PrP^{Sc} in human or animal tissues is considered key for TSE diagnosis. The problem for a diagnosis based on detecting PrP^{Sc} is that the pathological form of PrP is abundant only in its primary target organ: the brain. However, infectivity studies have shown that prions are also present in minute amounts in peripheral tissues, such as lymphoid organs and blood *(41,47–49)*.

Most of the efforts to develop a diagnostic system for prion diseases have been focused on the increase of sensitivity of the current detection methods. PMCA offers the opportunity to enhance existing methods by amplifying the amount of PrP^{Sc} in the sample. Combining the strategy of reproducing prions in vitro with any of the high sensitive detection methods, the early diagnosis of TSE may be achieved. The aim would be not only to detect prions in the brain in early pre-symptomatic cases, but also to generate a test to diagnose living animals and people. For this purpose, a tissue other than brain is required, and in order to have an easier non-invasive method, detection of prions in body fluids such as urine or blood is the best option. A blood test for CJD, might have many applications, including screening of blood banks, identification of populations at risk, reduction of iatrogenic transmission of CJD, and early diagnosis of the disease *(44,45)*.

3.4.3. Applications Beyond Prion Diseases

At present, it has not been shown that any of the other protein conformational disorders has an infectious origin nor that the misfolded protein has the ability to infect an individual and induce the conversion of the natively folded protein. However, in a similar way to PrP^{Sc} in TSE, the protein conformational changes associated with the pathogenesis of these diseases result in the formation of abnormal proteins rich in β-sheet structure, partially resistant to proteolysis and with a high tendency to aggregate *(2–4,50)*. Indeed, a common feature of several protein conformational disorders (including TSE) is the aggregation and deposition of the misfolded protein in different organs in the form of amyloid plaques. The available data indicate that amyloid deposition follows a seeding-nucleation mechanism *(24,25)*. Analogous to a crystallization process, amyloid formation depends on the slow interaction between misfolded protein monomers to form oligomeric nuclei around which a faster phase of elongation takes place. The limiting step in this process is the nuclei formation and the extent of amyloidosis will depend on the number of seeds produced *(24,51)*.

The conditions to produce amyloid fibrils in vitro have been optimized for most of the proteins implicated in the formation of amyloid plaques *(2,52)*. However, the experimental procedures usually involve highly concentrated and nonphysiological protein solutions to speed up the nuclei formation and induce aggregation in a short time. It is likely that a small amount of misfolded oligomers may be present in tissues and biological fluids of individuals much before the appearance of clinical symptoms. These endogenous nuclei could be used to induce the aggregation of seed-free low concentration of the amyloid protein. Combining phases of incubation to allow protein–protein interaction resulting in the elongation of seeds with phases of sonication to multiply the number of nuclei should produce a cyclic amplification of protein aggregation. Thus, the principles of PMCA may be used to detect low concentrations of early intermediates of amyloid formation in tissues or biological fluids of people incubating some of the amyloid-related disorders. Therefore, PMCA may have broader applications for research and diagnosis of diseases where protein misfolding and aggregation is implicated.

4. Notes

1. Brains have to be taken in the shortest possible time after death and washed in cold PBS before homogenization to reduce, as much as possible, the amount of blood that can interfere in further steps. Indeed, it is better to perfuse the animals prior to the sacrifice with PBS containing 5 mM EDTA.
2. If the tissue is not immediately homogenized, it must be stored at −80°C. The homogenization has to be performed at 4°C.
3. A high centrifugal force is not recommended because it might remove important membrane components implicated in $PrP^C \rightarrow PrP^{res}$ conversion.
4. The samples should to be thawed at 4°C; frequent freezing–thawing of the NBH reduces significantly the amplification.
5. The samples are prepared in the same way for either the manual or the automatic sonication; nevertheless, the employ of thin-wall tubes is recommended for the latter to allow a better ultrasound transmission.
6. The conditions for proteinase K digestion (concentration and temperature) are different for each PrP^{res} strain, and have to be carefully chosen. If PrP^C is not completely degraded, it can interfere in the quantification of PrP^{res}. The activity of proteinase K is affected by temperature, decreasing when the aliquots are frozen and thawed several times, and can be inhibited by certain blood components.

References

1. Soto, C. (2003) Unfolding the role of protein misfolding in neurodegenerative diseases. *Nature Rev. Neurosci.* **4,** 49–60.

2. Soto, C. (2001) Protein misfolding and disease; protein refolding and therapy. *FEBS Lett.* **498,** 204–207.
3. Dobson, C. M. (1999) Protein misfolding, evolution and disease. *Trends Biochem. Sci.* **24,** 329–332.
4. Carrell, R. W. and Lomas, D. A. (1997) Conformational disease. *Lancet* **350,** 134–138.
5. Kelly, J. W. (1996) Alternative conformations of amyloidogenic proteins govern their behavior. *Curr. Opin. Struct. Biol.* **6,** 11–17.
6. Soto, C. and Saborio, G. P. (2001) Prions: disease propagation and disease therapy by conformational transmission. *Trends Mol. Med.* **7,** 109–114.
7. Prusiner, S. B. (1991) Molecular biology of prion diseases. *Science* **252,** 1515–1522.
8. Roos, R., Gajdusek, D. C., and Gibbs, C. J. Jr. (1973) The clinical characteristics of transmissible Creutzfeldt-Jakob disease. *Brain* **96,** 1–20.
9. Will, R. G., Ironside, J. W., Zeidler, M., et al. (1996) A new variant of Creutzfeldt-Jakob disease in the UK. *Lancet* **347,** 921–925.
10. Cousens, S. N., Vynnycky, E., Zeidler, M., Will, R. G., and Smith, P. G. (1997) Predicting the CJD epidemic in humans. *Nature* **385,** 197–198.
11. Bruce, M. E., Will, R. G., Ironside, J. W., et al. (1997) Transmissions to mice indicate that 'new variant' CJD is caused by the BSE agent. *Nature* **389,** 498–501.
12. Soto, C. and Castilla, J. (2004) The controversial protein-only hypothesis of prion propagation. *Nat. Med.* **10 Suppl,** S63–S67.
13. Prusiner, S. B. (1998) Prions. *Proc. Natl. Acad. Sci. USA* **95,** 13363–13383.
14. Pan, K. M., Baldwin, M., Nguyen, J., et al. (1993) Conversion of alpha-helices into beta-sheets features in the formation of the scrapie prion proteins. *Proc. Natl. Acad. Sci. USA* **90,** 10962–10966.
15. Cohen, F. E. and Prusiner, S. B. (1998) Pathologic conformations of prion proteins. *Annu. Rev. Biochem.* **67,** 793–819.
16. Baldwin, M. A., Cohen, F. E., and Prusiner, S. B. (1995) Prion protein isoforms, a convergence of biological and structural investigations. *J. Biol. Chem.* **270,** 19197–19200.
17. Bueler, H., Aguzzi, A., Sailer, A., Greiner, R. A., Autenried, P., Aguet, M., and Weissmann, C. (1993) Mice devoid of PrP are resistant to scrapie. *Cell* **73,** 1339–1347.
18. DeArmond, S. J. and Prusiner, S. B. (1995) Prion protein transgenes and the neuropathology in prion diseases. *Brain Pathol.* **5,** 77–89.
19. Kocisko, D. A., Come, J. H., Priola, S. A., et al. (1994) Cell-free formation of protease-resistant prion protein. *Nature* **370,** 471–474.
20. Lucassen, R., Nishina, K., and Supattapone, S. (2003) In vitro amplification of protease-resistant prion protein requires free sulfhydryl groups. *Biochemistry* **42,** 4127–4135.
21. Saborio, G. P., Permanne, B., and Soto, C. (2001) Sensitive detection of pathological prion protein by cyclic amplification of protein misfolding. *Nature* **411,** 810–813.
22. Deleault, N. R., Lucassen, R. W., and Supattapone, S. (2003) RNA molecules stimulate prion protein conversion. *Nature* **425,** 717–720.

23. Cohen, F. E. (1999) Protein misfolding and prion diseases. *J. Mol. Biol.* **293**, 313–320.

24. Jarrett, J. T. and Lansbury, P. T. Jr. (1993) Seeding "one-dimensional crystallization" of amyloid: a pathogenic mechanism in Alzheimer's disease and scrapie? *Cell* **73**, 1055–1058.

25. Harper, J. D. and Lansbury, P. T. Jr. (1997) Models of amyloid seeding in Alzheimer's disease and scrapie: mechanistic truths and physiological consequences of the time-dependent solubility of amyloid proteins. *Annu. Rev. Biochem.* **66**, 385–407.

26. Masel, J., Jansen, V. A., and Nowak, M. A. (1999) Quantifying the kinetic parameters of prion replication. *Biophys. Chem.* **77**, 139–152.

27. Caughey, B., Kocisko, D. A., Raymond, G. J., and Lansbury, P. T. Jr. (1995) Aggregates of scrapie-associated prion protein induce the cell-free conversion of protease-sensitive prion protein to the protease-resistant state. *Chem. Biol.* **2**, 807–817.

28. Masel, J. and Jansen, V. A. (2001) The measured level of prion infectivity varies in a predictable way according to the aggregation state of the infectious agent. *Biochim. Biophys. Acta* **1535**, 164–173.

29. Prusiner, S. B., McKinley, M. P., Bowman, K. A., et al. (1983) Scrapie prions aggregate to form amyloid-like birefringent rods. *Cell* **35**, 349–358.

30. Jeffrey, M., Goodbrand, I. A., and Goodsir, C. M. (1995) Pathology of the transmissible spongiform encephalopathies with special emphasis on ultrastructure. *Micron.* **26**, 277–298.

31. Soto, C., Saborio, G. P., and Anderes, L. (2002) Cyclic amplification of protein misfolding: application to prion-related disorders and beyond. *Trends Neurosci.* **25**, 390–394.

32. Bessen, R. A., Raymond, G. J., and Caughey, B. (1997) In situ formation of protease-resistant prion protein in transmissible spongiform encephalopathy-infected brain slices. *J. Biol. Chem.* **272**, 15227–15231.

33. Horiuchi, M. and Caughey, B. (1999) Prion protein interconversions and the transmissible spongiform encephalopathies. *Structure Fold. Des.* **7**, R231–R240.

34. Kocisko, D. A., Priola, S. A., Raymond, G. J., Chesebro, B., Lansbury, P. T. Jr., and Caughey, B. (1995) Species specificity in the cell-free conversion of prion protein to protease-resistant forms: a model for the scrapie species barrier. *Proc. Natl. Acad. Sci. USA* **92**, 3923–3927.

35. Chabry, J., Caughey, B., and Chesebro, B. (1998) Specific inhibition of in vitro formation of protease-resistant prion protein by synthetic peptides. *J. Biol. Chem.* **273**, 13203–13207.

36. Caughey, B., Raymond, G. J., and Bessen, R. A. (1998) Strain-dependent differences in beta-sheet conformations of abnormal prion protein. *J. Biol. Chem.* **273**, 32230–32235.

37. Aguzzi, A. and Weissmann, C. (1997) Prion research: the next frontiers. *Nature* **389**, 795–798.

38. Weber, T., Otto, M., Bodemer, M., and Zerr, I. (1997) Diagnosis of Creutzfeldt-Jakob disease and related human spongiform encephalopathies. *Biomed. Pharmacother.* **51,** 381–387.

39. Steinhoff, B. J., Racker, S., Herrendorf, G., et al. (1996) Accuracy and reliability of periodic sharp wave complexes in Creutzfeldt-Jakob disease. *Arch. Neurol.* **53,** 162–166.

40. Budka, H., Aguzzi, A., Brown, P., et al. (1995) Neuropathological diagnostic criteria for Creutzfeldt-Jakob disease (CJD) and other human spongiform encephalopathies (prion diseases). *Brain Pathol.* **5,** 459–466.

41. Collinge, J. (2001) Prion diseases of humans and animals: their causes and molecular basis. *Annu. Rev. Neurosci.* **24,** 519–550.

42. Frosh, A., Joyce, R., and Johnson, A. (2001) Iatrogenic vCJD from surgical instruments. *BMJ* **322,** 1558–1559.

43. Soto, C. (2004) Diagnosing prion diseases: needs, challenges, and hopes. *Nature Rev. Microbiol.,* in press.

44. Schiermeier, Q. (2001) Testing times for BSE. *Nature* **409,** 658–659.

45. Anonymous (2001) Scientists race to develop a blood test for vCJD. *Nat. Med.* **7,** 261.

46. Ingrosso, L., Vetrugno, V., Cardone, F., and Pocchiari, M. (2002) Molecular diagnostics of transmissible spongiform encephalopathies. *Trends Mol. Med.* **8,** 273–280.

47. Aguzzi, A. (2000) Prion diseases, blood and the immune system: concerns and reality. *Haematologica* **85,** 3–10.

48. Brown, P., Cervenakova, L., and Diringer, H. (2001) Blood infectivity and the prospects for a diagnostic screening test in Creutzfeldt-Jakob disease. *J. Lab. Clin. Med.* **137,** 5–13.

49. Wadsworth, J. D., Joiner, S., Hill, A. F., et al. (2001) Tissue distribution of protease resistant prion protein in variant Creutzfeldt-Jakob disease using a highly sensitive immunoblotting assay. *Lancet* **358,** 171–180.

50. Carrell, R. W. and Lomas, D. A. (1997) Conformational disease. *Lancet* **350,** 134–138.

51. Lomakin, A., Chung, D. S., Benedek, G. B., Kirschner, D. A., and Teplow, D. B. (1996) On the nucleation and growth of amyloid beta-protein fibrils: detection of nuclei and quantitation of rate constants. *Proc. Natl. Acad. Sci. USA* **93,** 1125–1129.

52. Teplow, D. B. (1998) Structural and kinetic features of amyloid beta-protein fibrillogenesis. *Amyloid* **5,** 121–142.

6

X-Ray Diffraction Studies of Amyloid Structure

O. Sumner Makin and Louise C. Serpell

Summary

Elucidation of the underlying core structure of amyloid fibrils is essential for understanding the mechanism by which amyloid fibrils are formed and deposited. Conventional methods of X-ray crystallography and NMR cannot be used, since the fibers are insoluble and heterogeneous. X-ray fiber diffraction is one method that has been successfully used to examine the structure of these insoluble fibers. The procedure involves the formation of suitable, ordered amyloid fibrils and characterization (by electron microscopy), partial alignment of fibers, X-ray data collection, data analysis, and finally, model building.

Key Words: Amyloid; fiber; X-ray diffraction; texture; peptide; β-sheet; structure; conformation.

1. Introduction

Amyloid fibrils are involved in a number of fatal diseases known collectively as the "Amyloidoses." These diseases include Alzheimer's disease, the spongiform encephalopathies, and familial amyloidotic polyneuropathy. Each disease is characterized by an individual protein that aggregates to form the deposited amyloid fibrils (1). Recently, several other diseases have been shown to involve the deposition of amyloid-like fibrils intracellularly. These diseases include Parkinson's disease and Huntington's disease, in which α-synuclein and huntington are deposited respectively. Each type of fibril is composed of a normally soluble protein or polypeptide that undergoes conformational change associated with fibrillogenesis.

Amyloid fibrils are highly ordered, very stable aggregates. X-ray diffraction has been used to examine the repeating structure within these fibrous structures and has revealed that the fibrils are composed of a predominantly β-sheet structure, which is hydrogen bonded along the length of the fibers. Additional

From: *Methods in Molecular Biology, vol. 299: Amyloid Proteins: Methods and Protocols*
Edited by: E. M. Sigurdsson © Humana Press Inc., Totowa, NJ

information about each type of amyloid fibril may be obtained from high-quality diffraction data, such as the parallel or antiparallel arrangement of the β-structure *(2)* and the stacking of the β-sheets, as well as some side-chain packing information.

2. Materials

1. Protein or peptide for fibril formation.
2. Amyloid fibrils: high concentration in suitable low salt buffer or water.
3. Carbon coated, electron microscope grids (AGAR Scientific, Stansted, UK).
4. Uranyl acetate (2%), 0.2 μm filtered.
5. Stretch frame plus glass capillaries (Harvard apparatus Ltd, Fircroft way, Edenbridge, Kent, UK), (1.5 mm OD x 1.17 ID borosilicate) to be filled with a wax plug.
6. Magnet (2 Tesla) plus 0.7-mm diameter X-ray glass capillaries (GLAS, W. Muller, D-13503, Berlin, Germany).
7. Goniometer (Hampton), X-ray set (in house).
8. Computer workstation running Mosflm or other suitable software.

3. Methods

3.1. Fibril Formation

The formation of amyloid fibrils has to be carried out according to established protocols for each protein or peptide. A high concentration of protein will be necessary in order to provide many amyloid fibrils for the alignment procedure and to produce a large enough sample for diffraction. The exact concentration required will vary for different amyloid proteins (concentrations approx 5 mg/mL are a good starting point). It is preferable to use a low salt buffer or water. This is important for data collection, since high salt buffers will give salt rings in the diffraction pattern, which may obscure reflections. Some fibril formation procedures may require high salt concentration. In these cases, it may be useful to consider washing or dialysing the mature fibrils against a low salt buffer or water.

3.2. Examining Fibrils Suitability for Fiber Diffraction

Fibril formation is often monitored using spectroscopic techniques such as Thioflavin T fluorescence, circular dichroism, turbidity assays, light-scattering, etc. Many of these methods are discussed elsewhere in this volume. Although these methods indicate the presence of ordered amyloid fibrils to varying degrees, it is essential to examine the sample using a visual technique such as electron microscopy or atomic force microscopy. It is important to know that your aggregates resemble amyloid fibrils before carrying out X-ray diffraction, that they are well-ordered and preferably straight and that the concentration is high

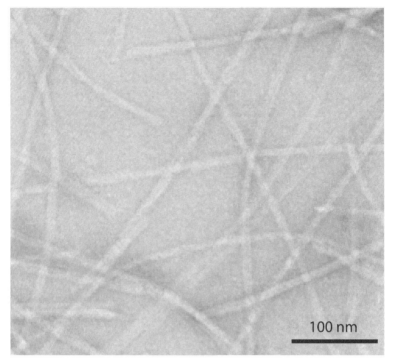

Fig. 1. Negative stain, transmission electron micrograph showing long, straight amyloid fibrils formed in vitro from islet amyloid polypeptide. The sample is diluted 1/10 compared to the concentration used for the preparation of a fiber diffraction sample. The micrograph was taken on a Philips 208 microscope with a magnification of 63,000 times. A scale bar is shown.

enough to see a clear distribution of amyloid fibrils on the electron microscope grid (example shown in **Fig. 1**).

Ensure that the fibrils are long, straight, and unbranching (**Fig. 1**). Establish the covering and distribution of fibrils on the grid. Examine the fibril morphology. If you have short fibrils, you may find that alignment is difficult. If lateral alignment of the fibrils is evident on the grid, this may indicate that alignment may be successful.

3.2.1. Electron Microscopy

Grids may be prepared in the following way:

1. Clean 400 mesh copper grids using acetone.
2. Discard acetone and allow the grids to dry.
3. Make up a 0.5% w/v Formvar solution in chloroform for the plastic coating.

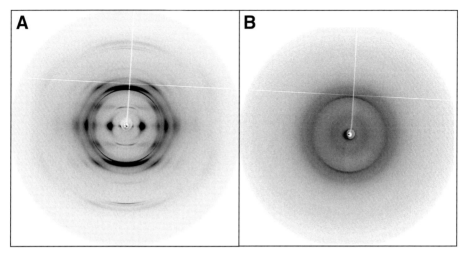

Fig. 2. X-ray fiber diffraction from well-oriented (**A**) and partially oriented amyloid fibrils (**B**). It is clear that the well-aligned sample (**A**) gives more information, revealing reflection peaks on layer lines.

4. Dip a clean, glass slide into the Formvar, remove, and allow to dry.
5. Cut an outline around the plastic on the glass slide using a razor blade and float off the plastic layer on the surface of a small water bath.
6. Place grids at regular intervals on the floating plastic and finally pick up the flat plastic using a piece of filter paper on the surface of the water.
7. Allow to dry.
8. The grids should then be coated in carbon using an Edwards coater.
9. When the grids are required, the plastic can be removed with a small amount of chloroform dropped onto the filter paper around (not on) the grids, leaving carbon coated grids.
10. Holding a grid in a pair of fine forceps, place 4 µL of your protein/fibril solution on a glow-discharged grid and blot away the excess.
11. Wash twice with water (4 µL) (filtered 0.2 µm) blotting between each addition and then add freshly filtered Uranyl acetate (1–2%) (4 µL) twice and finally blot excess and allow to air dry.
12. Examine grid using a transmission electron microscope.

3.3. Fiber Alignment

The higher the degree of alignment of the fibers, the more information can be obtained from the resulting diffraction image (*see* **Fig. 2**).

The diffraction pattern from a sample with no alignment will show reflection rings, while partial alignment will yield arced reflections *(3)* (**Fig. 2A**). Higher degree of alignment may yield very well-oriented diffraction patterns

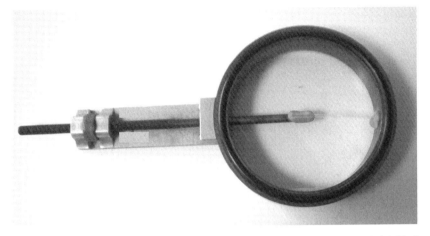

Fig. 3. The stretch frame apparatus used for partial alignment of amyloid fibrils.

containing considerably more information (**Fig. 2B**) *(2)*. The degree of alignment may depend on the nature of the sample. For example, a sample containing many long fibrils will be viscous and perhaps very amenable to stretch frame alignment *(3)*. Other samples may form small crystallites that can be oriented as they grow using magnetic alignment *(2,4,5)*. Samples containing fibrils that are laterally aggregated may be partially oriented by placing a suspension of fibrils in a capillary tube and then allowing them to dry down to form a disk. Several of the following methods should be attempted depending on facilities available.

3.3.1. Stretch Frame Alignment

The stretch frame apparatus (**Fig. 3**) allows the mounting of capillaries and their movement apart during drying. The use of a stretch frame is not essential and a petri dish or similar can be used instead. A stretch frame will allow movement of the specimen during alignment. The method for stretch frame alignment is detailed in the following.

1. Plug the borosilicate glass capillaries with a short piece of wax. This is achieved by dipping the glass capillaries into melted wax (such as candle wax). If the correct temperature is used, the capillary will draw up a small length (1–2 mm) of wax. This leaves an almost flat surface for the droplet to rest on.
2. Cut the tubes to a length of about 2 to 3 cm using a diamond pen. Two of these wax-plugged capillaries should be placed about 0.5 to 1 mm apart (with the wax ends facing). They may be secured to a stretch frame or to a simple petri dish using Plasticine.

3. A droplet of 10 to 20 μL of fibril solution/suspension should be carefully placed between the wax ends. A successful technique will yield a droplet held between the two capillary ends (but not running along the glass tube).
4. Cover to prevent dust contamination and allow to air dry (this may take several hours).

Experimentation with air drying at different temperatures may improve alignment. If a stretch frame is being used, it may be possible to separate the ends of the capillaries slightly to increase alignment. The stretch-frame method has been successfully used to examine a number of ex vivo amyloid fibrils *(3,6,7)* and also amyloid fibrils formed from peptides *(2)*.

3.3.2. Alignment of Amyloid Fibrils in a Glass Capillary

The object of alignment in a glass capillary is to obtain a disk of protein fibers. A glass capillary used for X-ray data collection with 0.7-mm diameter should be used. The use of siliconized capillary tubes (so that the meniscus is flat), can improve disk formation.

1. Using a 1-mL syringe fitted with a small piece of rubber tubing, fix to the wide end of the glass capillary and suck up about 2 to 3 cm of fibril solution/suspension.
2. Seal the end of the capillary using melted wax, leaving the top of the tube open. Fix the capillary to a surface so that it stands vertical and allow to dry.
3. When drying is complete, seal the top end using wax.

3.3.3. Magnetic Alignment of Amyloid Fibrils

A permanent 2.4 Tesla magnet (Hummingbird instruments, Arlington, MA) (**Fig. 4**) is used for magnetic alignment of fibrils within a glass capillary tube.

1. The preparation of the sample follows the instructions given above for preparation of a glass capillary sample.
2. Place the capillary vertically in the magnet, with the bottom of the sample level with the magnetic poles.
3. Allow to dry over several weeks, then seal the top end using wax.

The Kirschner group was the first to use this technique as a means of amyloid fiber alignment *(4)*. They examined amyloid fibrils formed from a number of different Alzheimer's Aβ peptides. Since then, the method has been used for aligning other amyloid fibrils *(2,5,8)*.

3.3.4. Alignment of Fibrils as a "Mat" or Thin Film

The preparation of a mat or thin films may require a larger volume of protein than the previously described methods. The resulting sample should be a thin, flat film that can be mounted such that the X-ray beam passes parallel to the plane of the film. In the resulting sample, the fiber alignment is in the plane of

Fig. 4. A 2 Tesla magnet used for mounting a capillary tube containing a solution of amyloid proteins for magnetic fibril alignment.

the film and the diffraction through the face of the film will yield diffraction rings.

The sample can be dried onto a glass slide and then carefully lifted off. This method was used to prepare mats of polyamino-acids by Fandrich and Dobson *(9)*. However, it can be difficult to lift the material off the glass possibly owing to the presence of short fibers and/or the sample being non-viscous. Other materials that can be used as a suitable surface to make a film are Parafilm or Teflon. This method is commonly used for polymer alignment and was also used to obtain diffraction patterns from PolyQ peptide by Perutz and co-workers *(10)*.

We have recently developed a method to make thin films. This involves the use of cryo-loops normally used for freezing single crystals (Serpell and Makin, personal communication). A cryo loop (Hampton) is dipped into a solution containing the fibrils, lifted out and allowed to dry. The result should be a thin, flat film (**Fig. 5C**).

The arrangement of fibrils resulting from each alignment method is shown schematically in **Fig. 5**.

3.3.5. Examining the Sample

An indication of how well a sample has aligned can be obtained by examining the sample under cross-polarized light in a light microscope. The presence of birefringence can indicate that the fibers are aligned to some degree. This method was used by Malinchik et al. *(5)* for examining the alignment of Aβ amyloid fibrils, and allowed the authors to find that some peptides formed fibrils more amenable to alignment than others.

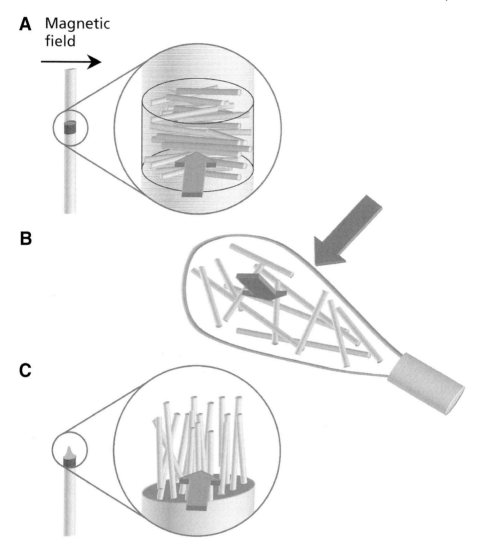

Fig. 5. A schematic showing fiber orientation for **(A)** a magnetically aligned, capillary disk, **(B)** a mat in a cryo-loop, and **(C)** a stretch frame aligned fiber. Block arrows indicate the possible X-ray beam directions.

3.4. Data Collection

The use of a calibrant such as calcite (3.035 Å) is recommended. The partially aligned samples produced by the previously mentioned methods should be mounted on a goniometer using Plasticine to secure in place and placed in the X-ray beam. The rotation axis will be the around the fiber axis for a stretch frame aligned fiber. In a sample produced using the magnet, the fibers tend to align across the diameter of the tube (*see* **Fig. 5A**). A mat will have two distinct directions (parallel to the plane of the film and perpendicular to it) (*see* **Fig. 5B**).

A diffraction pattern maybe collected using in house X-ray equipment such as Rigaku rotating anode (CuKα) and MARresearch image plate. Exposure times would be expected to be from 5 min to 2 h depending on the quality of the sample (a well-aligned sample will diffract more strongly) and the intensity of the X-ray beam.

To collect a full, useful set of data, it is necessary to examine each result before deciding how to proceed. However, these are very general rules:

First, collect a single frame (a still or small oscillation) with the specimen to detector distance set with a maximum resolution of about 2 to 3 Å. This will ensure that you will collect all the data necessary for observation of a cross-β pattern. If collecting from a film, collect the first pattern with the X-ray beam parallel to the plane of the film. Then rotate through 90° around the fiber and collect another pattern for the same exposure time. You may find that the specimen is thicker in one direction and this may yield an improved diffraction pattern. For a film, collect through the face of the thin-film. You expect reflection rings and the pattern will be much weaker. A magnetically aligned sample may have a preferred orientation and yield two distinct patterns *(2)*. Finally, collect a low-angle diffraction pattern to examine the lower resolution data if possible. This may give you information about the packing of protofilaments or crystallites within your sample. A low-angle diffraction pattern is obtained by moving the detector and backstop away from the sample. The backstop must be moved, otherwise the shadow of the backstop remains the same, relative to the diffraction pattern. To collect low angle data, an increase in the exposure time may be necessary to compensate for the extra air-scatter and lower flux density. A helium chamber can be fitted between the sample and detector to prevent the air scatter affecting the X-rays.

Data may also be collected using synchrotron X-rays. The advantages of using synchrotron radiation are that the beam will be more focused yielding sharper reflections. The wavelength may be tuneable, in which case it would be possible to use a wavelength of 0.99 Å, thus preventing air-scatter, which can be a problem in house using a wavelength 1.5419 Å. The background will then be lower, allowing observation of weaker spots.

3.5. Data Analysis

Details of the physics of diffraction can be found in texts such Holmes and Blow *(11)*. Substantive information can be extracted without such resort and a précis follows.

A diffraction pattern is formed as a result of X-rays interacting with the electrons in the sample. The sample's structure is three-dimensional and therefore so is its electron density. The interaction between the X-rays and the electron density can be described as a three-dimensional diffraction pattern. Mathematically, this is the intensity of a Fourier transform of the electron density. What is seen on the X-ray detector plate is two-dimensional, and hence only part of the three-dimensional pattern is seen. The part we see can be visualized by imagining a hollow sphere intersecting with the three-dimensional diffraction pattern. At certain spots, this sphere touches the three-dimensional pattern, leaving a two-dimensional pattern on the surface of the sphere. This two-dimensional pattern is then projected on to the film or X-ray detector plate in the same way as an illuminated globe might project a map of the earth onto the wall of a room. In order to view another part of the three-dimensional diffraction pattern, it is necessary to move the hollow sphere; this is achieved by rotating the sample.

A crystal is made up of identical repeating units, each occupying a point on a periodic lattice. This lattice is defined mathematically as being all the possible points that can be generated by adding any whole number combination of three vectors **a**, **b**, and **c**, together. These three vectors are known as the lattice vectors. Diffraction from crystals results in a diffraction pattern comprising of sharp spots described by the reciprocal lattice vectors **a***, **b***, and **c***. All crystals with the same lattice will have a diffraction pattern with spots in identical places, even though some spots might be missing. The intensity of spots is determined by the structure of the repeating unit. Indexing is the process by which each diffraction spot is assigned three whole numbers [*hkl*], known as Miller indices so each spot has a position given by **r*** = *h***a*** + *k***b*** + *l***c***.

Under certain conditions, amyloid proteins do not crystallize but form amyloid fibers consisting of units that repeat along the fiber axis. Fibers are cylindrically symmetric, so the sharp spots are spread out into three-dimensional arcs. The appearance of these arcs depends on the arrangement of crystallites (microstructure) within the material, known as the texture. In fibers, the crystallites generally have no particular rotational orientation with respect to the fiber axis, so the three-dimensional arcs take the form of circles centered about the fiber axis. This results in each quadrant of the diffractogram being symmetrical with each other. Simply, it is useful to be aware that the meridional direction of a fiber diffraction pattern will contain the crystalline dimension data, whereas the equatorial direction will combine the other two directions, since a fiber is by its nature, cylindrically symmetric. This means that the cell-dimension (a) corresponding to the fiber axis can be indexed simply by measuring the reflections on the meridian. The equatorial direction (b and c) will be com-

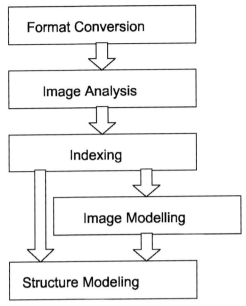

Fig. 6. A flowchart showing the procedure for processing high-quality diffractograms.

plicated by the presence of overlapping reflections. The spacing of reflections may indicate a particular type of crystallite packing (e.g., hexagonal lattice).

There are a number of fiber diffraction processing programs available to facilitate processing of fiber diffraction data. CCP13 *(12)* is a collection of programs designed to analyze high quality diffraction patterns. Fit2d (www.esrf.fr/computing/expg/subgroups/data_analysis/FIT2D) can be used to analyze one- and two-dimensional data sets. The CCP4 suite *(13)* is intended for protein crystallography, but many of its components, such as ipdisp and mosflm can be utilized for examining the diffraction pattern. Fxplor (www.molbio.vanderbilt.edu/fiber/software.html) is an extension of X-Plor, a program useful for atomic model refinement.

A method for processing a high-quality, oriented, X-ray fiber-diffraction pattern is detailed below and summarized in **Fig. 6**.

1. The data from the detector must be converted into a format compatible with the other programs in the chain. Examples of such programs are XCONV (CCP13) *(12)*, marcvt (www.marresearch.com/software.htm), and Denzo *(14)*.
2. The image must be centred to avoid a systematic error whilst measuring spot locations. Mosflm allows this to be done by manually moving the image with respect to a series of overlaid concentric circles. One may also wish to correct for the shape of the image plate and determine the sample's tilt and rotation *(15)*.

3. Background subtraction is beneficial owing to detector fog, white radiation, and X-ray scattering from air, camera components, the sample holder, and amorphous material in the specimen such as solvent and disordered polymer *(16,17)*.

4. An initial survey can be carried out by noting down spot locations and resolutions. This is straightforward in Mosflm, since the resolutions are calculated by clicking on spot maxima using a zoomed image. Resolution calculation can also be done by hand using an approximation to Bragg's law. This is the result of the product of the wavelength of the X-rays and the distance between the sample and the image plate, divided by the distance of the diffraction spot from the center of the diffraction pattern. An accurate determination of spot locations follows. Circularly average an angular segment of the image containing the spot to obtain a one-dimensional image. Then an approximate position can be found by means of differentiation and final determination achieved by fitting an inverted parabola to each peak on the one-dimensional image.

5. Index the diffraction pattern; that is, assign a set of Miller indices to each spot on the pattern. Automated indexing programs such as those used in powder diffraction may be useful, although, in practice, indexing is normally done by hand. Experience is helpful, but tutorials are available (www.matter.org.uk/diffraction/electron/ratio_technique.htm).

6. The diffraction image may then be modelled in terms of a sum of functions describing peaks. Such modeling is not necessary, but may be beneficial, as it cleans up the diffractogram and aids the next stage. LSQINT (CCP13) *(12)* is such a program, but requires the pattern layer lines that do not overlap (that is, the sets of points lying perpendicular to the fiber axis and sharing a common Miller index are distinct from one another). This is rarely observed in diffractograms from amyloid fibril specimens.

7. Finally. a molecular structure may be modelled using software such as LALS *(18)* or Cerius2 (http://www.accelrys.com/cerius2/cerius248/) to obtain a simulated diffractogram that matches the empirical data.

Often, the fiber is not well-oriented, so there is insufficient information in the diffraction pattern to complete all the steps in the aforementioned chain. In this case, the initial survey and indexing steps should be carried out as far as possible and then followed by a far simpler analysis. Amyloid fibrils exhibit a characteristic diffraction pattern, described as cross-β; comprised of two major, mutually orthogonal reflections (**Fig. 7**). Spots in the meridional direction carry information about how the structure repeats in the direction of the fiber. In amyloid, β-strands are repeated along the fiber. Therefore, the 4.7 Å spacing between strands corresponds to a substantial spot along the meridian, at a resolution of 4.7 Å. If adjacent β-strands are rotated 180° with respect to one another (an antiparallel conformation), then a spot at 9.4 Å may be observed. The lack of such a spot does not necessarily imply a parallel arrangement; since a 2_1 symmetry would mean that the spot is systematically absent (for a full expla-

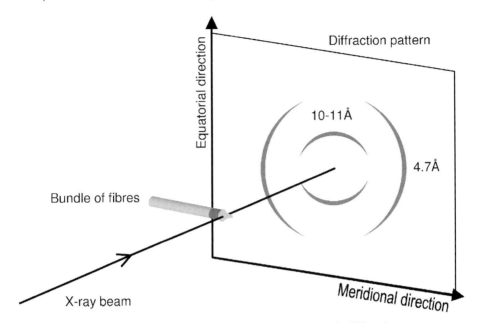

Fig. 7. A schematic describing the features of the cross-β diffraction pattern.

nation see *[2]*). Perpendicular to the fiber axis is the equatorial direction. Here, the directions along the β-strand (6.9 Å for each pleat of the β-strand) and between the sheets (approx 10 Å) are combined. There may also be information close to the center of the diffraction pattern (at low angle) about the size of the crystallites or protofilaments making up the sample.

X-ray fiber diffraction analysis can be particularly effective if used in combination with other methods such as cryo-electron microscopy and electron diffraction.

Acknowledgments

LCS is a Wellcome RCDF. OSM is funded by MRC, UK. The authors would like to thank Dr. T. Stromer for her contribution and critical reading of the manuscript.

References

1. Sunde, M. and Blake, C. (1998) From the globular to the fibrous state: protein structure and structural conversion in amyloid formation. *Quarterly Reviews of Biophysics* **31(1),** 1–39.
2. Sikorski, P., Atkins, E. D. T., and Serpell, L. C. (2003) Structure and textures of fibrous crystals of the Abeta(11-25) fragment from Alzheimer's Aβ amyloid protein. *Structure* **11,** 915–926.

3. Sunde, M., Serpell, L., Bartlam, M., Fraser, P., Pepys, M., and Blake, C. (1997) Common core structure of amyloid fibrils by synchrotron X-ray diffraction. *J. Mol. Biol.* **273**, 729–739.

4. Inouye, H., Fraser, P., and Kirschner, D. (1993) Structure of β-crystallite assemblies by Alzheimer β amyloid protein analogues: analysis by X-ray diffraction. *Biophys. J.* **64**, 502–519.

5. Malinchik, S., Inouye, H., Szumowski, K., and Kirschner, D. (1998) Structural analysis of Alzheimer's β(1-40) amyloid: protofilament assembly of tubular fibrils. *Biophys. J.* **74**, 537–545.

6. Blake, C. and Serpell, L. (1996) Synchrotron X-ray studies suggest that the core of the transthyretin amyloid fibril is a continuous β-helix. *Structure* **4**, 989–998.

7. Inouye, H., Domingues, F. S., Damas, A. M., et al. (1998) Analysis of x-ray diffraction patterns from amyloid of biopsed vitreous humor and kidney of transthyretin (TTR) Met30 familial amyloidotic polyneuropathy (FAP) patients: axially arrayed TTR monomers constitute the protofilament. *Amyloid* **5(3)**, 163–174.

8. Inouye, H. and Kirschner, D. A. (1997) X-ray diffraction analysis of scrapie prion: intermediate and folded structures in a peptide containing two putative α-helices. *J. Mol. Biol.* **268**, 375–389.

9. Fandrich, M. and Dobson, C. (2002) The behaviour of polyamino acids reveals an inverse side-chain effect in amyloid structure formation. *EMBO J.* **21**, 5682–5690.

10. Perutz, M., Johnson, T., Suzuki, M., and Finch, J. (1994) Glutamine repeats as polar zippers: their possible role in inherited neurodegenerative diseases. *PNAS USA* **91**, 5355–5358.

11. Holmes, K. C. and Blow, D. M. (1980) *The use of X-ray diffraction in the study of protein and nucleic acid structure.* Robert Krieger Publishing Co., New York.

12. Squire, J., Al-Khayat, H., Arnott, S., et al. (2003) New CCP13 software and the strategy behind further developments: stripping and modelling of fiber diffraction data. *Fiber Diffraction Review* **11**, 13–19.

13. CCP4. (1994) The CCP4 Suite Programs for Crystallography. *Acta Cryst.* **D50**, 760–763.

14. Otwinowski, Z. and Minor, W. (1997) Processing of X-ray diffraction data collected in oscillation mode, in *Methods in Enzymology,* vol. 276 (Carter, C. W. Jr. and Sweet, R. M., eds.), Academic Press, New York, pp. 307–326.

15. Fraser, R. D. B., Macrae, T. P., Miller, A., and Rowlands, R. J. (1976) Digital processing of fiber diffraction patterns. *J. Appl. Cryst.* **9**, 81–94.

16. Millane, R. P. and Arnott, S. (1985) Background subtraction in X-ray fiber diffraction patterns. *J. Appl. Cryst.* **18**, 419–423.

17. Ivanova, M. I. and Makowski, L. (1998) Iterative low-pass filtering for estimation of the background in fiber diffraction patterns. *Acta Cryst. A* **54**, 626–631.

18. Okada, K., Noguchi, K., Okuyama, K., and Arnott, S. (2003) WinLALS for a linked-atom least squares refinement program for helical polymers on Windows PCs. *Computational Biol. Chem.* **3**, 265–285.

7

Molecular Electron Microscopy Approaches to Elucidating the Mechanisms of Protein Fibrillogenesis

Hilal A. Lashuel and Joseph S. Wall

Summary

Electron microscopy (EM) has played a central role in our current understanding of the mechanisms underlying the pathogenesis of several amyloid diseases, including Alzheimer's disease, Parkinson's disease, and prion diseases. In this chapter, we discuss the application of various EM techniques to monitor and characterize quaternary structural changes during amyloid fibril formation in vitro and the potential of extending some of these techniques to characterizing ex vivo material. In particular, we would like to bring to the attention of the reader two very powerful molecular EM techniques that remain under utilized by researchers in the amyloid community, namely scanning transmission electron microscopy and single particle molecular averaging EM. An overview of the strength and limitations of these techniques as tools for elucidating the structural basis of amyloid fibril formation will be presented.

Key Words: Electron microscopy (EM); scanning transmission electron microscopy (STEM); single particle analysis; negative staining; α-synuclein, amyloid-β (Aβ); amyloid; fibrils; protofibrils; oligomers; sedimentation velocity analytical ultracentrifugation (SVAU).

1. Introduction

The extracellular and/or intracellular aggregation of a subset of proteins in the form of amyloid fibrils is a defining neuropathological hallmark that is shared by several clinically and pathologically distinct neurodegenerative and systemic diseases, including Alzheimer's disease (AD), Parkinson's disease (PD), Prion diseases, and systemic amyloidoses *(1,2)*. The finding of these deposits in the vicinity of dying tissues and neurons suggested that they play a role in neurodegeneration and/or organ dysfunction. However, the first clues concerning the nature of these deposits became apparent only after the introduction of electron microscopy in the late 1950s *(3–5)*. Electron microscopic examination of ex vivo amyloid fibrils composed of various proteins revealed long and

From: *Methods in Molecular Biology, vol. 299: Amyloid Proteins: Methods and Protocols*
Edited by: E. M. Sigurdsson © Humana Press Inc., Totowa, NJ

unbranching fibrils that are typically 10 nm in diameter. The possibility that soluble prefibrillar oligomeric species, rather than the fibrils, could be the pathogenic species in AD and related amyloid diseases arose as a result of the detection and characterization of prefibrillar β-sheet rich oligomeric species, termed protofibrils, during the fibrillization of Aβ and other amyloid forming proteins in vitro by electron microscopy (EM) and atomic force microscopy (AFM) *(6–8)*. Later observations demonstrating that protofibrillar species alter neuronal function and/or cause cell death *(9–12)* have given rise to the hypothesis that protofibrils might be the primary pathogenic species in AD and related amyloid diseases *(13,14)*. Therefore, in addition to playing a central role in defining the pathology of amyloid diseases, electron microscopy continues to be instrumental in providing key mechanistic insights into the role of amyloid fibril formation and the modes by which this process contributes to the pathogenesis of amyloid diseases, which are critical for developing effective therapeutic strategies to treat and/or prevent these devastating diseases.

The highly associated, non-crystalline, and insoluble nature of amyloid fibrils has precluded high-resolution structure determination by nuclear magnetic resonance (NMR) or X-ray crystallography. In addition, the morphological and size heterogeneity of fibril preparations and the absence of protocols for preparing and isolating stable intermediates along the amyloid pathway in monodisperese form remain major hurdles to understanding the structure and function of protofibrils and fibrils. The structures of amyloid fibrils derived from several proteins have therefore been studied by alternative structural approaches, including EM, AFM, and X-ray fiber diffraction *(15,16)*.

The greatest advantage of EM over other structural techniques is that: (i) it does not require samples to be crystalline, (ii) it enables the study of transient quaternary structural intermediates, (iii) it requires a small amount of sample relative to NMR and X-ray crystallography, and (vi) there is no upper limit for the size of the molecule to be studied. Samples are usually preserved in a heavy metal stain (negative stain) or in a hydrated state at liquid nitrogen temperatures (CryoEM). Samples for scanning transmission electron microscopy (STEM) are usually freeze-dried to minimize biochemical alteration and permit quantitative microscopy. Recent advances in cryo electron microscopy and image processing techniques have enabled detailed structural studies of large macromolecular assemblies, including protofibrils and amyloid fibrils *(17–25)*. In this chapter, we will present a brief summary of the application, strength, and limitations of some EM techniques that include negative staining electron microscopy, scanning transmission electron microscopy, and single particle averaging and image processing as tools for elucidating the structural basis of amyloid fibril formation.

1.1. Negative Staining Electron Microscopy

Negative staining EM is a simple technique that allows rapid and routine morphological and structural examination of biological macromolecules and macromolecular assemblies. Negative staining EM offers several distinct advantages over other techniques: 1) it is technically simple to perform; 2) it produces high contrast images; 3) it is relatively insensitive to damage to the specimen by the electron beam; 4) samples are easy to prepare (2–4 min) (*see* **Note 1**); and 5) it requires only a small amount of sample (2–5 µL). The procedure for negative stain EM involves adsorption onto a surface of a carbon coated formvar film that is attached to an EM grid followed by staining with heavy-metal solution and drying. Contrast is introduced by interaction of the electron beam with the sample regions having differing scattering power. Since macromolecules are composed mainly of atoms (e.g., C, H, O, N) that scatter electrons weakly, the EM images produced are of low contrast. Therefore, image contrast is maximized by embedding the sample in a thin layer of heavy metal stain such as uranyl acetate and uranyl formate (*see* **Note 2**). The thin shell of heavy metal atoms surrounding the sample scatters the electron beam more strongly than the sample itself, giving negative contrast. To improve the adsorption of particles to the grid, the surface of the carbon-coated EM grid is made more hydrophilic by glow discharging prior to use. Some of the problems encountered with negatively stained specimens are that: (i) the resolution is limited to about 25 Å, (ii) the structure is usually flattened owing to drying of the sample, (iii) different quaternary structures exhibit differential adsorption and/or orientation to the carbon coated grid, (iv) uneven staining could result in some image artifacts, and (v) the high ionic strength and non-physiological pH of the stain may alter the specimen.

Cryo-electron microscopy overcomes most of these limitations imposed by negative staining and has been used to achieve near atomic resolution in the most favorable cases. In cryo-EM, the samples are preserved in their hydrated state in ice by quick freezing in liquid nitrogen or ethane, and are imaged at low to moderate dose by EM to minimize damage to the sample (*see* review in *[26]*).

1.2. Single Particle Analysis and Image Processing

In addition to negative staining EM, single particle averaging has recently emerged as a powerful technique in the structural studies of heterogeneous specimens such as those characterstic of protofibrils and fibril preparations of amyloidogenic proteins. **Figure 1** outlines the different steps involved in single particle averaging. Data collection: large sets of electron micrographs are collected and the appropriate images are then digitized for computational image processing. Particle selection: the first step involves manual selection of single

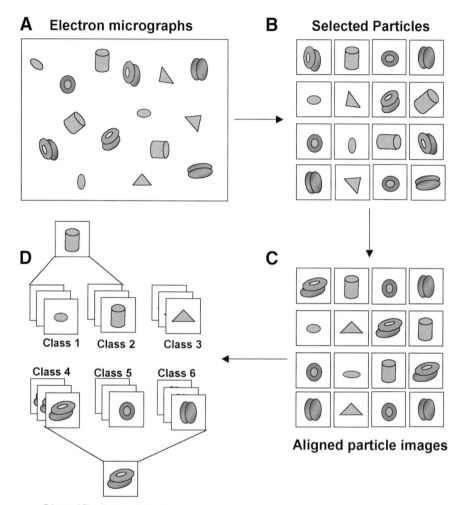

Classified particle images

Fig. 1. Single particle analysis and image processing. (**A**) Cartoon depiction of sample micrograph showing different views of the objects under study. (**B**) Sixteen selected particle images from micrograph shown in **A**. (**C**) Aligned particle images. Each of the sixteen images now positioned to a common view. (**D**) Classified particle images. Sixteen unique views that were present in the original micrograph have resulted in six particle image classes. Classes 4, 5, and 6 represent three different views corresponding to the side, front, and top of the object.

projected electron microscopic particle images from the digitized electron micrographs. Since proteins adsorb on the carbon film in one or a few preferred orientations, the projected images may reflect many views of the protein. Particle alignment: the particles are then centered, aligned with respect to each other, ordered, and subjected to computational classification using one of several commercial image processing programs (e.g., SPIDER image processing package *[27]*). Particle classification: involves an iterative process of several rounds of alignment and classification, in which the aligned images that project the same particle image, but differ in their orientation, i.e., the images are related to each other by a simple rotation in the plane of the image, are sorted into a specific number of output classes. The clustered raw images in each class are then averaged to yield average images for each class. Therefore computational classification is an excellent tool that allows single particle microscopy to work with heterogeneous samples. In cases where a monodisperse specimen can be prepared, the images making up the different classes can be used in 3D reconstruction for each class by random conical tilt approach. Although 3D reconstruction of negatively stained samples is limited to a resolution of 25Å, higher resolution (2–6 Å) structural details can be obtained using cryo EM and single-particle 3D image reconstruction techniques. In Cryo EM, the specimen is visualized in a native-like environment (hydrated), thus eliminating stain and drying artifacts. Furthermore, the structural integrity of the specimens is preserved by imaging at low electron dose. However, Cryo EM of heterogeneous specimens is complicated by low contrast, which makes it difficult to differentiate EM projections of identical particles displaying different orientations from those of different structures.

1.2.1. Characterization of Amyloid Fibrils by Single Particle Averaging of Embedded Sectioned Specimen

Single particle averaging provided the first detailed images of the fibril substructure demonstrating that amyloid fibrils are composed of several thin protofilaments which undergo lateral association and/or intertwine around each other to form the ribbon and cable-like morphologies characteristic of amyloid fibrils. Using single particle averaging of cross-section EM images of different mutant (V30M) transthyretin (TTR) amyloid fibrils from the vitreous humors of patients with familial amyloid polyneuropathy, Serpell et al. demonstrated that TTR fibrils are composed of four protofilaments organized around a electron-lucent central core, indicative of the presence of a hollow center, or central channel (**Fig. 2**) *(19)*. These observations have been confirmed by many subsequent studies. For example, recent cross sectional analysis studies of ex vivo amyloid fibrils composed of amyloidogenic proteins (Amyloid A protein, monoclonal immunoglobulin lambda light chain, Leu60Arg variant of apolipoprotein AI, and Asp67His variant lysozyme, and fibrils prepared from synthetic peptides [TTR 10–19] using single particle averaging *[28]*) revealed that amy-

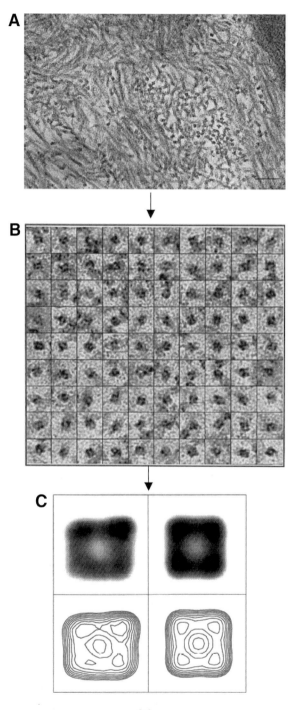

loid fibrils have an electron-lucent center around which the protofilaments making up the fibrils are organized.

1.2.2. Cryo Electron Microscopy and Single Particle 3D Image Reconstruction of Amyloid Fibrils

Recently, combining single particle averaging and cryo-EM methods has enabled 3D reconstruction of amyloid fibrils. This approach has been successful in providing a detailed structural view of the amyloid fibrils composed of SH3 domain of phosphatidylinositol-3'-kinase, revealing a double helix of two protofilament pairs intertwined around a hollow core, with a helical crossover repeat of approx 600 Å and an axial subunit repeat of approx 27 Å (**Fig. 3**) *(20)*. Jimenez et al. (1999) divided the fibril images into individual crossover repeats (total of 890) and treated these as single particles. The cut-out repeats were iteratively aligned and classified by multivariate statistical analysis according to their length (**Fig. 3B**). 3D maps were then calculated from the individual classes using helical reconstruction, yielding a 3D reconstructed structure at 25 Å resolution (**Fig. 3B**). Since amyloid fibrils are long structures formed by repeated structural units, many fewer micrographs would be required for 3D reconstructions, in favorable cases.

1.2.3. Characterization of Protofibrils by Negative Stain EM and Single Particle Averaging

In collaboration with Dr. Thomas Walz and coworkers at Harvard Medical School, we have taken advantage of the capabilities of single particle electron microscopy to study heterogeneous protofibril samples of amyloid-β (Aβ) and α-synuclein. To gain further insight into the structural properties of α-synuclein protofibrils, we analyzed protofibrillar fractions of A30P and A53T by negative stain EM and single particle averaging *(23,24)*. The A30P variant was observed to promote the formation of annular, pore-like protofibrils, whereas A53T promotes formation of annular and tubular protofibrillar structures (**Fig. 4**), suggesting that α-synuclein is capable of forming a pore-like protofibrillar structure. The ring-shaped protofibrils have a diameter of 11 ± 1 nm, and an inner core of 2 to 3 nm, whereas the tube-like protofibrils showed similar diameter

---

Fig. 2. (*Opposite page*)   Single particle analysis provided first glimpses of the substructure of amyloid fibrils. (**A**) Electron micrograph of fibril cross-sections (Scale bar = 100 nm). (**B**) A total of 210 cross-section images were selected and used for averaging, only 100 cross-sections selected are shown. (**C**) Resulting averaged images of fibril cross-section revealing images of structure have a fourfold symmetry, suggesting that V30M-TTR fibril is composed of four protofilaments organized around an electron lucent center. Adapted with permission, from *(19)*.

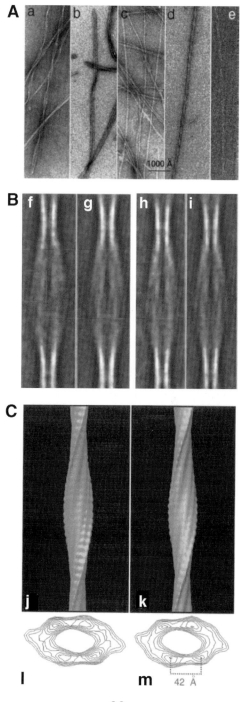

(10–12 nm), but varied in length from 13 to 24 nm. The similar diameter observed for the annular and rectangular particles suggests that the two morphologies represent two different views (top and side view) of the same species. These results provided a possible structural basis for the pathogenic membrane permeabilization and channel activities reported for α-synuclein and Aβ *(29–32)*. Further improvements in the preparation of highly homogeneous samples should enable higher resolution structural studies that would provide us with a powerful basis for understanding the mechanism of protofibril formation and assembly into fibrils, as well as the quaternary and tertiary structural changes required for protofibril interactions and/or insertion into membranes.

1.3. Scanning Transmission Electron Microscopy (STEM)

Several biophysical techniques (e.g., dynamic light scattering [DLS] and analytical ultracentrifugation [AU]) have been employed to monitor quaternary structural changes, which occur during amyloid fibril formation. In the case of heterogeneous samples, as is the case for amyloid forming proteins, AU and DLS provide average molecular masses that reflect the distribution of quaternary structures in solution. However, the morphological and size heterogeneity of the structures populated during protein fibrillogenesis precludes the assignment of MW averages to a particular quaternary structural species. STEM offers unique advantages over other mass determination techniques in its ability to image isolated and unstained biological molecules after freeze drying, allowing for accurate, quantitative dry mass determination of individual molecules as well as macromolecular assemblies (*see* **Note 3**). Furthermore, STEM requires only 2 to 5 μL of samples at concentrations ranging from 10 to 500 μg/ mL (*see* **Note 4**), far less than what is required for AU and DLS experiments. The ability of STEM to provide accurate mass measurements of small quantities of heterogeneous samples provides it with an edge over other techniques when it comes to investigating the molecular mass and structural properties of

Fig. 3. (*Opposite page*) Cryo electron microscopy and single particle 3D image reconstruction of amyloid fibrils. (**A**) Negative-stain electron micrographs demonstrating the polymorphic nature of amyloid fibrils formed by SH3 domain of phosphatidylinositol-3'-kinase formed after several months of incubation at pH 2.0 (a–d). (e) The Cryo EM image of d. (**B**) The class averages of the 580 Å (f, 92 images) and 600 Å (h, 77 images) repeats used for reconstruction, reprojection of 3D reconstructions (g, i). The averages in B represent approx 19% of the entire data set. (**C**) 3D reconstructions and contoured density calculated from the 610 Å (j,l) and 580 Å (k,m) class averages reveal similar fibril structures composed of four protofilaments packed around a hollow core. In addition to the helical twist, the 27 Å subunit repeat is observable on the surface of the fibril structure and is more pronounced on the edge structures. (Adapted with permission, from **ref. 20**.)

A Electron micrographs   **B**   Aligned particle images

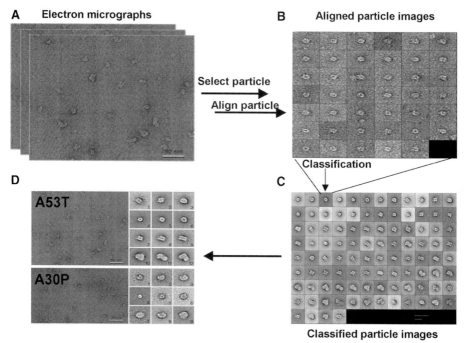

Select particle

Align particle

Classification

D   **C**

A53T

A30P

Classified particle images

Fig. 4. Single particle averaging of A30P and A53T α-synuclein reveals a collection of distinct, yet related, protofibrillar species. (**A**) Representative images of negatively stained samples of A53T and A30P. For image processing, 31 images of A53T and 16 images of A30P were digitized with a Zeiss SCAI scanner using a pixel size of 4.04 Å at the specimen level. From the digitized images, 5815 (A53T), and 5040 (A30P) particles were selected for further computational processing using the SPIDER image processing package 39. The 5000 to 6000 particle images were subjected to ten rounds of alignment (**B**) and classification specifying 100 output classes (**C**). These 100 classes fell into four major groups of structures. A gallery with representative class averages obtained from images of A53T and A30P are shown in (**D**), illustrating the four major groups of protofibril structures. (Adapted from EM images provided by Dr. Thomas Walz and Benjamine Peter and reproduced with permission from **ref. 23**.)

biological samples isolated and purified from human tissues or cellular extracts. Therefore, STEM is an excellent tool for investigating the molecular mass and structural properties of transient quaternary structure intermediates during amyloid formation pathway in vitro or quaternary structures isolated from ex vivo material. Indeed, STEM has proven to be instrumental for investigating the mechanism of self-assembly and amyloid fibril formation of several proteins, including tau *(33,34)*, Aβ *(22,35–37)*, α-synuclein *(23)*, transthyretin *(38)*, and the Ure2p prion protein *(39)*. The ability to monitor changes in morphology and molecular mass during protein fibrillogenesis should bring us one step closer to understanding the structure of the various assembly intermediates on the pathway to amyloid formation and their structural relationship to the final product, the amyloid fibrils.

The STEM produces an image one point at a time by probing the sample with a focused electron beam and counting the emerging electrons on an array of detectors according to how much they are deflected in striking nuclei of atoms in the specimen. The image is formed by scanning the beam over the specimen in a TV-type raster while simultaneously recording the various detector signals, any one of which can be used to give an image. The detector measuring large-angle scattering gives a signal proportional to the mass of atoms in the path of the beam (heavy atoms scatter more strongly in proportion to their mass and interference effects between nearby atoms are minimal). We record the ratio of large angle scattering to total signal for all emerging electrons, so the measured scattering percentage is independent of incident beam current. Therefore the STEM image is essentially a map of the mass in each small element (pixel) of the image.

Ultimate spatial resolution is determined by the probe size, 0.25 nm for BNL STEM1 operating at 40 keV. Practical resolution is determined by radiation damage during imaging and any distortion owing to freeze-drying. STEM1 maintains the specimen at $-150°C$ to minimize damage and detects every electron passing through the specimen so that no information will be wasted.

Mass measurement is performed offline using custom software, PCMass, which runs on a PC running Windows 95 or later. The program is available on the STEM FTP site (ftp.stem.bnl.gov) and provides image viewing, manual mass measurement, automated mass measurement, statistical summaries, and documentation.

The mass measurement procedure involves selecting isolated particles with a circle or rectangle. The program sums pixels within the marked area, computing the total mass of particle and underlying substrate. It then subtracts background on the assumption that the substrate under the particle has the same mass per unit area as that observed in "clean" areas between particles. Absolute accuracy ranges from roughly 10% for 100 kDa particles, to 4% for 1 MDa, and 2% for particles above 5 MDa. The useful range for compact particles is 30 kDa to over 1000 MDa. If the particle is extended in shape, more background is included and must be subtracted, giving slightly larger errors.

The mass calibration of the STEM is checked using tobacco mosaic virus (TMV) included in every freeze dry. This virus is easily distinguishable from most specimens being a rod 300 nm long and 18 nm in diameter with one concave end and one convex end. If the shape of the TMV is not that expected or if the mass per unit length is not 131 kDa/nm, the quality of the included specimen is suspect and additional controls are required. Problems may arise owing to impurities in the specimen or buffers.

STEM measures both mass and length, so it is particularly suitable for studying filaments, giving mass per unit length. Furthermore, the distance of every pixel from the centerline of a filament or center of a particle is known accurately, giving a radial mass profile. If the object has spherical or cylindrical symmetry, the mass profile can be transformed to give a radial density profile.

The STEM mass measurements are most useful for characterizing the uniformity of a preparation, comparing native vs reconstituted structures and locating heavy atom labels. Since the STEM image is a projection with somewhat limited resolution, it does not give direct 3D models. However it is very powerful in distinguishing between several proposed models. An example is the work summarized in **Fig. 5**, which showed the organizing principle of prion filaments, namely that there is one subunit every 0.45 nm, the spacing of a β helix. The filaments have a central core surrounded by head groups whose size can be controlled by genetic manipulation.

STEM has also proven to be excellent tool for characterizing the homogeneity/heterogeneity as well as structural properties of transient quaternary structure intermediates "protofibrils" on the amyloid fibril formation pathway of several amyloid forming proteins, including Aβ *(22,35–37)*, α-synuclein *(23)*, and transthyretin *(38)*. Recently, we employed STEM to characterize the molecular mass distribution of protofibrils and to estimate the stoichiometric composition of the various α-synuclein and Aβ protofibrillar structures (**Fig. 6**) *(22,23)*. Quantitative STEM analysis of α-synuclein particles reveals a mass distribution consistent with that predicted by sedimentation velocity analytical ultracentrifugation (SVAU).

The STEM can image specimens in negative stain, but the added material makes mass measurements difficult to interpret. However, image contrast in the STEM can be enhanced electronically, making it possible to use negative stains with lower contrast as long as their signal to noise ratio is high. One such stain is methylamine vanadate *(40)*, which is less grainy than uranyl salts and stable at pH 7.0. An example of prion filaments prepared with this stain is shown in **Fig. 7**. With its lower contrast, methylamine vanadate does not obscure heavy atom labels, which can be used to mark specific sites.

Fig. 5. (*Opposite page*)   STEM mass analysis of filaments of Ure2p constructs. The upper panel shows dark-field micrographs. Type A (single) and Type B (paired) filaments are indicated. Bar = 100 nm. Left: Mass per unit length measurements for several constructs. Right: Average mass per unit length plotted against subunit mass for type A (full circles) and type B (open circles) filaments. Solid line is mass per unit length expected for one subunit every 4.5 nm. Adapted with permission from *(39)*.

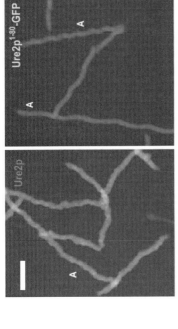

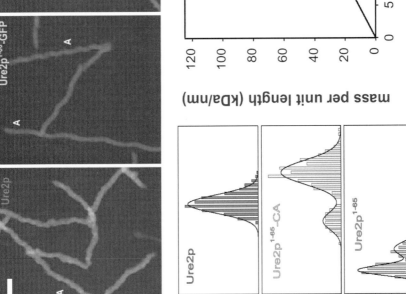

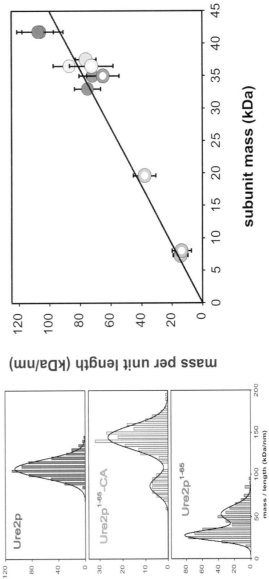

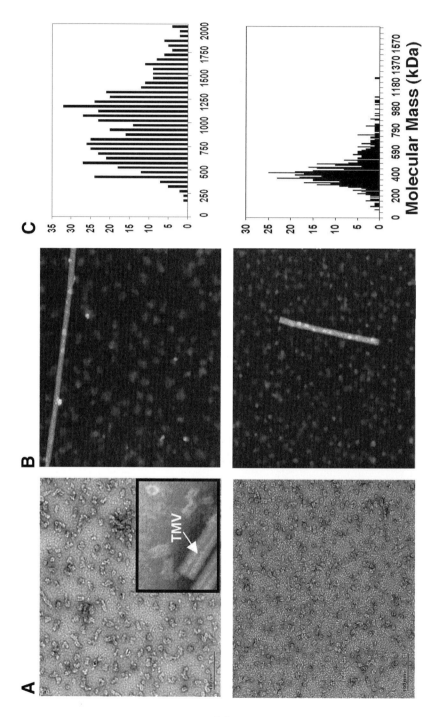

2. Materials

1. Formvar carbon coated cooper grids (Electron Microscopy Sciences no. FCF200-Cu).
2. Carbon coated titanium grids for use in the STEM are prepared by the STEM staff.
3. EMS blocking solution (Electron Microscopy Sciences no. 25595).
4. Uranyl acetate (Electron Microscopy Sciences no. 22400).
5. Uranyl format (Pfaltz Bauer Inc. no. 06708).
6. Methylamine vanadate (NanoVan, NanoProbes, Inc. no. 2011).
7. Numbered grid storage box (Electron Microscopy Sciences no. 71138).
8. Buffer A: 5 mM Tris-HCl, 70 mM NaCl, pH 7.4.

3. Methods

3.1. Negative Staining Electron Microscopy

3.1.1. Preparing Uranyl Formate Solution for Negative Staining (0.6% [w/v])

1. Prepare a solution of 6 mg/mL of uranyl acetate in water.
2. Filter the solution through a 0.2-µm filter. The uranyl formate solution is ready to use.

3.1.2. Preparing Uranyl Formate Solution for Negative Staining (0.7% [w/v])

1. Weigh 37.5 mg of uranyl formate in a beaker and cover it to prevent exposure to light.
2. Bring 5 mL of H$_2$O to boil in a test tube (4.5–4.7 mL).
3. Add the boiling water to the uranyl formate in the beaker. Stir for 5 min.
4. Add 5 mL of NaOH (5 M) (a color change to a darker yellow should occur) (6–7 mL).
5. Stir for additional 5 min.
6. Filter through a 0.2-µm filter. The uranyl formate solution is ready to use.

Fig. 6. (*Opposite page*) Quantitative STEM analysis of two α-synuclein proto-fibril fractions separated by size exclusion chromatography reveals clear separation and mass distributions consistent with that predicted by SVAU. (**A**) Negatively stained (uranyl formate) electron micrographs of the two A53T α-synuclein protofibril fractions. (**B**) Electron micrograph of unstained/freeze-dried A53T-synuclein particles recorded by STEM. Light regions represent areas of high-mass density and dark regions represent areas of low mass density. (**C**) Histograms of molecular mass measurements of A53T protofibril fractions shown in (A). The inset in (A) shows an STEM micrograph of vanadate stained protofibrillar fraction of A53T revealing annular and tubular structures, similar to those seen in uranyl formate stained samples (**Fig. 4**).

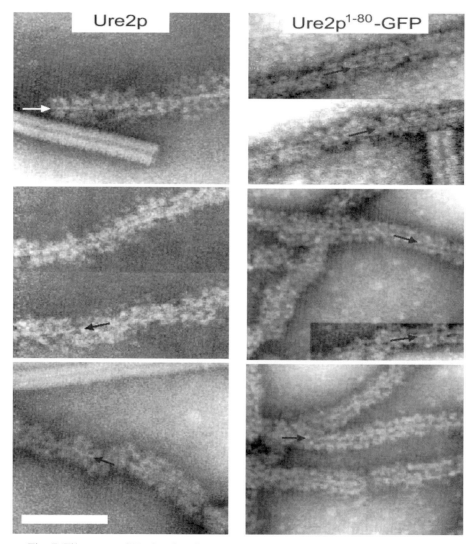

Fig. 7. Filaments of Ure2p visualized by dark-field STEM of vanadate stained specimens. Bar = 50 nm. Arrows indicate regions where the core fiber is clearly visualized.
Adapted with permission from *(39)*.

3.1.3. Staining Samples on Formvar-Carbon Coated Grids

1. Place a drop of the diluted material on a formvar and carbon-coated cooper grid.
2. After 30 to 60 s, blot off the excess liquid with filter paper.
3. Add a drop of uranyl acetate or uranyl formate solution.
4. After 1 min, blot off the excess with filter paper.

5. Allow the grid to air dry or apply a vacuum line to the edge of the grid.
6. Examine in the transmission electron microscope.

3.2. STEM Sample Preparation (Wet Film, Hanging Drop Method)

Specimens for the STEM are prepared by the STEM staff (*see* **Note 5**), but the details will be discussed here to enable the users to identify potential problems before sending samples for analysis to the STEM facility. When possible, it is recommended that samples should be screened in a conventional microscope under conditions used to prepare the sample for examination by the STEM (*see* **Note 6**).

1. The thin (2–3 nm) carbon film, which has been shadowed onto freshly cleaved rock salt, is floated off the crystal onto a dish of STEM water.
2. The titanium grids, covered with holey film (and for the STEM, in rings and caps), are placed face down on the floating carbon for approx 1 h.
3. A grid is picked up from above and retains a hanging droplet of water. It is washed and wicked with STEM water two or three times, never being allowed to dry. Water is applied from one side with an Eppendorf pipette and wicked from the other side with filter paper.
4. Three microliters of TMV at 100 µg/mL is injected into a drop and allowed to adsorb to the thin carbon film for 1 min.
5. The grid is then washed and wicked two or three times with injection buffer for the sample. Three microliters of the specimen is injected into the drop and allowed to adsorb for 1 min.
6. The grid is washed with sample buffer a couple of times, followed by washes (approx 10), ending with washes of 20 to 50 mM ammonium acetate. The intermediate washes depend on the sample buffer (e.g., high-salt buffers require washes with ammonium acetate).
7. For freeze-drying: after the final wash, the grid is briefly blotted between the two pieces of the filter paper, leaving a layer of liquid a few micrometers thick and rapidly plunged into liquid nitrogen slush to freeze it. (At this point, the grids may be stored under liquid nitrogen until STEM time is available for observation.) For negative staining: the last two washes are performed with a solution of 1% methylamine vanadate (NanoVan, NanoProbes, Inc.), then the grid is wicked as above and air dried.
8. For freeze drying: six samples are transferred from liquid nitrogen to slots in the precooled STEM freeze dry cartridge, freeze dried overnight under ultra-high vacuum, and transferred under vacuum to the STEM.

3.2.1. STEM Operation

To assure optimum data quality and efficient use of microscope time, a trained operator usually runs the STEM. Frequent users can obtain training on request, but the alacrity of most STEM experiments makes this unnecessary.

For a detailed description of how the STEM is operated *see* http://www.biology.
bnl.gov/stem/stem.htm.

3.2.2. STEM Data Analysis

The STEM images are stored on a CD or hard drive and can be viewed using
commercial software or PCMass mass-analysis program provided by the
STEM group. The format of the Brookhaven National Laboratory (BNL)
STEM images is a header 4096 bytes long, followed by two 512 × 512, 8-bit
images interleaved. Adobe Photoshop can import these images in the "RAW"
mode for contrast adjustment, annotation, and publication.

4. Notes

1. Most commonly used buffers (Tris-HCl, HEPES, and MOPS) are amenable to
   preparation of stained or frozen samples. Additives such as sucrose, glycerol, or
   dithiothreitol will usually wash off. Phosphate buffers often do not wash off well
   and leave bright spots, which interfere with STEM analyses.
2. The choice of heavy metal stain will depend on the stability of the specimen at the
   pH of the staining solution. Uranyl acetate and uranyl formate (0.5–2.0%) have a
   low pH (pH 4.2–4.5) and would not be recommended for samples that are unstable
   at these pHs. Neutral phosphotungstic acid is a useful stain for such samples, but
   produces less contrast than uranyl acetate. A 1% to 3% solution of phosphotungstic acid is made up in water and the pH is adjusted to 7.0 using sodium hydroxide
   with 1 N NaOH. When working with a new sample, it is recommended to try at
   least two different stains with different staining properties (e.g., uranyl acetate,
   phosphotungstic acid, vandate). Alternatively, macromolecular assemblies can be
   stabilized against pH-induced dissociation by brief exposure (approx 15 min) to
   general fixative and crosslinking reagents such as gluteraldehyde and formaldehyde (approx 0.1%) before applying them to the grid.
3. The expected mass accuracy depends on particle size, shape, and measuring dose.
   A 100 kDa globular particle on a 2-nm carbon substrate measured with a dose of
   10 μl/A^2 should give a standard deviation (SD) of 10%, whereas a 1-MDa particle
   should give 2%. If the particle is extended or the background is dirty, the SD will
   be worse. The TMV reference particles should give mass per unit length of 13.1
   kDa/A with a SD of 1%. When possible, it is critical to compare mass measurements from STEM to those determined by AU and/or DLS to eliminate the possibility that mass estimates by STEM might be biased towards a particular quaternary
   structure owing to differential adsorption to the EM grid.
4. Only small concentrations of samples are needed for examination in the STEM,
   but it is useful if the concentration is relatively high. Ideally 10 to 50 μL of sample
   at 100 to 200 μg/mL is adequate, from which a series of dilutions will be prepared
   and examined.
5. The STEM is a NIH Research Resource. As such, it is available to users with
   appropriate projects free of charge. A project is usually initiated by a discussion

of it on the phone or by e-mail. A trial sample is sent by overnight mail to BNL. The grids are prepared on the day the samples arrive, if it has been prearranged. The grids are stored in a grid fridge under liquid nitrogen (stable for years) until they can be freeze-dried and examined in the microscope. Additional details about the facility can be found on http://www.biology.bnl.gov/stem/stem.html. Additional support is provided by the United States Department of Energy.

6. All samples for STEM must be physically clean. Any contamination from the purification procedures, such as column or gel material, or other physical additives that scatter electrons will interfere with the observation and analysis. Detergent should be avoided if at all possible.

Acknowledgments

We would like to thank Dr. Thomas Walz, Benjamin Petre, Dr. Louise Serpell, Dr. Helen Saibil, and Dr. Alasdair C. Steven for permitting reproduction of their data in this chapter. The molecular EM facility at Harvard Medical School was established by a generous donation from the Giovanni Armenise Harvard Center for Structural Biology and is maintained by funds from NIH grant GM62580. The Brookhaven National Laboratory STEM is an NIH Supported Resource Center, NIH P41-EB2181, with additional support provided by the Department of Energy (DOE) and the Office of biological and environmental research (OBER). Support was also derived from a sabbatical fellowship (H.A.L) from the Harvard Center for Neurodegeneration and Repair and NIH (AG08470).

References

1. Selkoe, D. J. (2003) Folding proteins in fatal ways. *Nature* **426(6968),** 900–904.
2. Rochet, J. C. and Lansbury, P. T. Jr. (2000) Amyloid fibrillogenesis: themes and variations. *Curr. Opin. Struct. Biol.* **10(1),** 60–68.
3. Paul, W. E. and Cohen, A. S. (1963) Electron microscopic studies of amyloid fibrils with ferritinconjugated antibody. *Am. J. Pathol.* **43,** 721–738.
4. Cohen, A. S. (1965) The constitution and genesis of amyloid. *Int. Rev. Exp. Pathol.* **4,** 159–243.
5. Bladen, H. A., Nylen, M. U., and Glenner, G. G. (1966) The ultrastructure of human amyloid as revealed by the negative staining technique. *J. Ultrastruct. Res.* **14(5),** 449–459.
6. Harper, J. D., Lieber, C. M., and Lansbury, P. T. Jr. (1997) Atomic force microscopic imaging of seeded fibril formation and fibril branching by the Alzheimer's disease amyloid-beta protein. *Chem. Biol.* **4(12),** 951–959.
7. Harper, J. D., et al. (1997) Observation of metastable Abeta amyloid protofibrils by atomic force microscopy. *Chem. Biol.* **4(2),** 119–125.
8. Walsh, D. M., et al. (1997) Amyloid beta-protein fibrillogenesis. Detection of a protofibrillar intermediate. *J. Biol. Chem.* **272(35),** 22364–22372.

9. Lambert, M. P., et al. (1998) Diffusible, nonfibrillar ligands derived from Abeta1-42 are potent central nervous system neurotoxins. *Proc. Natl. Acad. Sci. USA* **95 (11)**, 6448–6453.

10. Hartley, D. M., et al. (1999) Protofibrillar intermediates of amyloid beta-protein induce acute electrophysiological changes and progressive neurotoxicity in cortical neurons. *J. Neurosci.* **19(20)**, 8876–8884.

11. Walsh, D. M., et al. (2002) Naturally secreted oligomers of amyloid beta protein potently inhibit hippocampal long-term potentiation in vivo. *Nature* **416(6880)**, 535–539.

12. Bucciantini, M., et al. (2002) Inherent toxicity of aggregates implies a common mechanism for protein misfolding diseases. *Nature* **416(6880)**, 507–511.

13. Goldberg, M. S. and Lansbury, P. T. Jr. (2000) Is there a cause-and-effect relationship between alpha-synuclein fibrillization and Parkinson's disease? *Nat. Cell Biol.* **2(7)**, E115–E119.

14. Caughey, B. and Lansbury, P. (2003) Protofibrils, pores, fibrils, and neurodegeneration: separating the responsible protein aggregates from their innocent bystandards. *Annu. Rev. Neurosci.* **26**, 267–298.

15. Serpell, L. C., Fraser, P. E., and Sunde, M. (1999) X-ray fiber diffraction of amyloid fibrils. *Methods Enzymol.* **309**, 526–536.

16. Serpell, L. C. (2000) Alzheimer's amyloid fibrils: structure and assembly. *Biochim. Biophys. Acta* **1502(1)**, 16–30.

17. Saibil, H. R. (2000) Macromolecular structure determination by cryo-electron microscopy. *Acta Crystallogr. D Biol. Crystallogr.* **56(Pt 10)**, 1215–1222.

18. Radermacher, M., et al. (1987) Three-dimensional reconstruction from a single-exposure, random conical tilt series applied to the 50S ribosomal subunit of Escherichia coli. *J. Microsc.* **146(Pt 2)**, 113–136.

19. Serpell, L. C., et al. (1995) Examination of the structure of the transthyretin amyloid fibril by image reconstruction from electron micrographs. *J. Mol. Biol.* **254(2)**, 113–118.

20. Jimenez, J. L., et al. (1999) Cryo-electron microscopy structure of an SH3 amyloid fibril and model of the molecular packing. *Embo J.* **18(4)**, 815–821.

21. Jimenez, J. L., et al. (2002) The protofilament structure of insulin amyloid fibrils. *Proc. Natl. Acad. Sci. USA* **99(14)**, 9196–9201.

22. Lashuel, H., et al. (2003) Mixtures of wild-type and "Arctic" Abeta40 in vitro accumulate protofibrils, including amyloid pores. *J. Mol. Biol.* **332(4)**, 795–808.

23. Lashuel, H., et al. (2002) alpha-Synuclein, especially the Parkinson's disease-associated mutants, forms pore-like annular and tubular protofibrils. *J. Mol. Biol.* **322(5)**, 1089.

24. Lashuel, H. A., et al. (2002) Neurodegenerative disease: amyloid pores from pathogenic mutations. *Nature* **418(6895)**, 291.

25. Wang, L., et al. (2002) Murine apolipoprotein serum amyloid A in solution forms a hexamer containing a central channel. *Proc. Natl. Acad. Sci. USA* **99(25)**, 15947–15952.

26. Makin, O. S. and Serpell, L. C. (2002) Examining the structure of the mature amyloid fibril. *Biochem. Soc. Trans.* **30(4)**, 521–525.

27. Frank, J., et al. (1996) SPIDER and WEB: processing and visualization of images in 3D electron microscopy and related fields. *J. Struct. Biol.* **116(1)**, 190–199.

28. Serpell, L. C., et al. (2000) The protofilament substructure of amyloid fibrils. *J. Mol. Biol.* **300(5)**, 1033–1039.

29. Volles, M. J., et al. (2001) Vesicle permeabilization by protofibrillar alpha-synuclein: implications for the pathogenesis and treatment of Parkinson's disease. *Biochemistry* **40(26)**, 7812–7819.

30. Volles, M. J. and Lansbury, P. T. Jr. (2002) Vesicle permeabilization by proto-fibrillar alpha-synuclein is sensitive to Parkinson's disease-linked mutations and occurs by a pore-like mechanism. *Biochemistry* **41(14)**, 4595–4602.

31. Kagan, B. L., et al. (2002) The channel hypothesis of Alzheimer's disease: current status. *Peptides* **23(7)**, 1311–1315.

32. Kourie, J. I., et al. (2002) Heterogeneous amyloid-formed ion channels as a common cytotoxic mechanism: implications for therapeutic strategies against amyloidosis. *Cell. Biochem. Biophys.* **36(2-3)**, 191–207.

33. King, M. E., et al. (2001) Structural analysis of Pick's disease-derived and in vitro-assembled tau filaments. *Am. J. Pathol.* **158(4)**, 1481–1490.

34. Ksiezak-Reding, H., et al. (1996) Ultrastructural instability of paired helical filaments from corticobasal degeneration as examined by scanning transmission electron microscopy. *Am. J. Pathol.* **149(2)**, 639–651.

35. Antzutkin, O. N., et al. (2002) Supramolecular structural constraints on Alzheimer's beta-amyloid fibrils from electron microscopy and solid-state nuclear magnetic resonance. *Biochemistry* **41(51)**, 15436–15450.

36. Antzutkin, O. N. (2004) Amyloidosis of Alzheimer's Abeta peptides: solid-state nuclear magnetic resonance, electron paramagnetic resonance, transmission electron microscopy, scanning transmission electron microscopy and atomic force microscopy studies. *Magn. Reson. Chem.* **42(2)**, 231–246.

37. Goldsbury, C. S., et al. (2000) Studies on the in vitro assembly of a beta 1-40: implications for the search for a beta fibril formation inhibitors. *J. Struct. Biol.* **130(2-3)**, 217–231.

38. Cardoso, I., et al. (2002) Transthyretin fibrillogenesis entails the assembly of monomers: a molecular model for in vitro assembled transthyretin amyloid-like fibrils. *J. Mol. Biol.* **317(5)**, 683–695.

39. Baxa, U., et al. (2003) Architecture of Ure2p prion filaments: the N-terminal domains form a central core fiber. *J. Biol. Chem.* **278(44)**, 43717–43727.

40. Tracz, E., et al. (1997) Paired helical filaments in corticobasal degeneration: the fine fibrillary structure with NanoVan. *Brain Res.* **773(1-2)**, 33–44.

8

Time-Lapse Atomic Force Microscopy in the Characterization of Amyloid-Like Fibril Assembly and Oligomeric Intermediates

Claire Goldsbury and Janelle Green

Summary

The atomic force microscope (AFM) images the topography of biological structures adsorbed to surfaces with nanometer to angstrom scale resolution. Amyloid-like fibrils and oligomers can be imaged in buffer solutions, allowing the samples to retain physiological-like properties while temporal changes in structure are monitored, e.g., the elongation of fibrils or the growth of single oligomers. These qualities distinguish AFM from conventional imaging techniques of comparable resolution, i.e., electron microscopy (EM). However, AFM is limited in that the specimen must be firmly attached to a solid support for measurement and that time-lapse imaging of individual assemblies can thus only be achieved for fibrils and oligomers growing on this support. Nevertheless, AFM has provided several insights into the in vitro assembly mechanism and structures of amyloid-like fibrils. The first section of this chapter provides a methodological introduction to AFM, whilst the second details the application of this technique to the investigation of amyloidogenic proteins, specifically amylin and amyloid-β (Aβ) peptides.

Key Words: Amyloid fibril; protofibril; time-lapse atomic force microscopy; Alzheimer's disease; amylin; islet amyloid polypeptide (IAPP); amyloid-β protein (Aβ); fibril assembly; fibril structure; oligomeric intermediates.

1. Introduction

1.1. Principles of AFM

The atomic force microscope (AFM) images the topography of biological structures adsorbed to surfaces with nanometer to angstrom scale resolution. Amyloid-like fibrils and oligomers can be imaged in buffer solutions, allowing the samples to retain physiological-like properties while temporal changes in structure are monitored, e.g., the elongation of fibrils or the growth of single

From: *Methods in Molecular Biology, vol. 299: Amyloid Proteins: Methods and Protocols*
Edited by: E. M. Sigurdsson © Humana Press Inc., Totowa, NJ

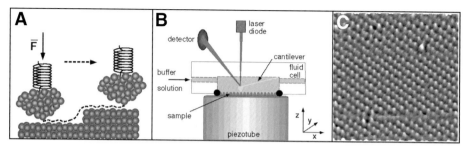

Fig. 1. Scheme showing how AFM images surfaces. (**A**) The sample surface is scanned with a fine tip. The interaction force between the tip and the surface causes the cantilever to bend. (**B**) The sample is mounted in the AFM under a transparent cantilever holder (fluid cell). A laser beam is focused on the end of the cantilever. The deflection of the laser spot, induced by the bending of the cantilever, is measured by a photodiode detector and a feedback loop corrects the vertical distance by adjusting the piezo position. (**C**) The AFM thereby outputs a topographical image of the sample (surface of "constant force"). This figure was obtained with permission from the M.E. Mueller-Institute for Structural Biology, Basel, Switzerland, publication "Imaging, measuring and manipulating native biomolecular systems with the atomic force microscope" prepared by Daniel J. Müller, Ueli Aebi, and Andreas Engel.

oligomers. These qualities distinguish AFM from conventional imaging techniques of comparable resolution, i.e., electron microscopy (EM). However, AFM is limited in that the specimen must be firmly attached to a solid support (such as mica) for measurement. Despite this perceived disadvantage, there have been certain instances in which the use of a mica surface has been of benefit. For example, human amylin (hA) protofibrils were stabilized on a mica surface, whereas they could only seldom be detected on EM grids (*1*). AFM has provided several insights in to the *in vitro* assembly mechanism and structures of amyloid-like fibrils (*1–3*).

AFM imaging is fundamentally different from conventional microscopy: a fine tip attached to a cantilever "spring" is scanned over the sample surface (**Fig. 1**). Cantilever bending caused by the interaction force between sample and tip (typical range: pN to nN) is detected by the deflection of a laser beam focused on the end of the cantilever. The sample is mounted on top of a piezo-electric scanner which scans in the x and y direction and adjusts the vertical distance between sample and tip according to input from the detector feedback control. In "contact mode," the feedback loop between the laser detector and the piezo adjusts the vertical distance from tip to sample such that the cantilever is held at constant deflection. In "dynamic mode" (MAC Mode™, Molecular Imaging, AZ; TappingMode™, Digital Instruments/Veeco, CA), the feedback loop keeps the amplitude of an oscillating cantilever constant by correcting the vertical distance. Owing to the low lateral forces of AFM in TappingMode™, protein structures are not easily damaged by the tip. In summary, surface topographies are obtained and sample heights can be accurately determined. In addition, forces felt by the cantilever as the tip approaches and retracts from the sample can reveal long-range attractive or repulsive forces between the tip and sample surfaces, allowing measurements of electrostatic, chemical, or mechanical properties such as adhesion, elasticity, or binding forces at the nanometer scale. The reviews in **refs. *4–14*** provide comprehensive discussions of the various applications of AFM.

1.2. AFM on Amyloidogenic Proteins

Amyloid peptide complexes can be assembled either in solution (*see* **Subheading 3.1.**) or directly on a substrate surface such as mica (*see* **Subheading 3.2.**). The morphology of mature amyloid fibrils or transient intermediate oligomeric structures formed throughout the assembly process can be characterized. It should be noted that the species within a sample that adsorb to a mica surface are often distinct from those that adsorb to carbon-coated EM grids. Therefore, it can be helpful to complement AFM studies of amyloid fibrils and their assembly by EM. Furthermore, whilst AFM gives accurate height measurements. EM provides information on the width of these species.

Time-lapse AFM is used to monitor individual fibril assembly on a surface in real-time. At a certain protein concentration, fibril growth on the surface may be spontaneous, as is the case for hA fibrils *(1,2)*, or may require initial "seeding" of the substrate with pre-assembled fibrils as for amyloid-β (Aβ) *(15)*. As well as a detailed analysis of the enlargement of oligomers and the fibril elongation process, further applications of time-lapse AFM have enabled the inhibition of fibril growth by certain compounds to be visualized as well as the interaction of amyloid peptides with lipid bilayers deposited on a solid substrate. We discuss these approaches for investigations into amyloid fibril structure and assembly dynamics, with specific reference to the amylin and Aβ peptides.

2. Materials

2.1. Peptides and Chemicals

1. Synthetic hA and Aβ lyophilized peptides are purchased from Bachem (Torrence, CA). Stock solutions are prepared in 100% DMSO, Milli-Q water (Millipore, Schwalbach, Germany), and/or 1,1,1,3,3,3-hexafluoro-2-isopropanol (HFIP) (*see* **Note 1**).
2. Congo red (Sigma Aldrich, Munich, Germany).
3. Basic buffers: PBS (25 mM phosphate, 120 mM NaCl, pH 7.4); 10 mM Tris-HCl, pH 7.3. Up to 300 mM KCl or NaCl and/or divalent cations (For e.g., 20 mM MgCl$_2$)

may be added to basic buffers. Buffers are filtered through 0.2-μm filters (Millipore) to remove any particulate.

2.2. Substrates

1. For Preparation of standard substrates: Thin Teflon discs (diameter 10 mm) are glued to metal stubs using superglue. Discs (diameter 5 mm) of freshly cleaved muscovite mica (Mica New York Corp., NY) are attached to the Teflon using a two-component araldite glue (Ciba-Geigy, AG, Basel, Switzerland, or Devcon Epoxy, ITW Brands, Wood Dale, IL). The mica and Teflon discs are prepared by punching them out of thin sheets of the respective materials. The mica is cleaved by removing a single layer using tape (*see also* **Note 2–4**).
2. For Preparation of lipid-bilayers: 1-Palmitoyl-2-oleoyl-*sn*-glycero-3-phosphocholine (POPC) and 1-palmitoyl-2-oleoyl-*sn*-glycero-3-[phospho-rac-(1-glycerol)] (POPG) are from Avanti Polar Lipids, Inc. (Alabaster, AL). Lipids are dissolved in chloroform to a concentration of 7 mM and stored at $-20°C$. Mica-supported POPC/POPG (3:1) lipid bilayers are produced using the Langmuir-Blodgett technique *(7,16–18)* (*see* **Note 3**).

2.3. Instrumentation and Software

1. Nanoscope IIIa multimode scanning probe workstation with Digital Instruments Nanoscope software version 5.12 (Veeco/Digital Instruments, Santa Barbara, CA). Damping system for microscope.
2. Oxide-sharpened silicon nitride probes with 0.32 N/m spring constant (Veeco, Digital Instruments, Santa Barbara, CA) (*see* **Notes 5–7**).

3. Methods

3.1. Imaging Pre-Assembled Fibrillar and Oligomeric Structures Formed in Solution

An amyloid protein sample is prepared in buffer conditions that favor fibril formation (*see* **Note 1**). Aliquots are removed from the sample at time-points throughout, or at the endpoint, of the fibril assembly process, for immediate adsorption to a freshly cleaved mica surface (*see* **Notes 2** and **4**). The sample is diluted to an appropriate concentration and approx 25 μL is adsorbed on to the substrate for 5 to 30 min. Normally, the lower the concentration, the longer the adsorption time should be. A suggested initial condition could be a peptide concentration of 50 μM with a 15 min adsorption time and the concentration and/or time subsequently adjusted as required. After adsorption, the sample should be washed three times to remove any unadsorbed protein. Initially, it is beneficial to image (*see* **Note 8** and **9**) the sample at a 20 μm^2 scan size to determine the homogeneity of the peptide assemblies on the surface. However, images at 2 to 4 μm^2 provide good overviews of the complexes, with any fibril periodicities normally being visible.

On occasions, fibrils that are present in solution are not adsorbed on to the mica surface. In these instances, either changing the buffer conditions that the fibrils are prepared in, or pelleting the fibrils with resuspension into another buffer should be considered. Another difficulty may be that the protein assemblies adsorbed on the substrate surface are not stable, but continue to grow. When this occurs, imaging under conditions that do not favor fibril formation is suggested. One must take into consideration, however, that heights of protein complexes imaged under low-salt conditions may be inaccurate (*see* **Note 10**).

3.1.1. Characterizing Fibril Polymorphisms: Human Amylin and CCβ-Met Fibrils

Fibril polymorphisms for hA and a *de novo* designed amyloid peptide CCβ-Met were characterised using the method described above *(19,20)*. The hA fibril composed of two protofilaments, having a 25 nm periodicity, had an average height (*see* **Note 10**) of 6.8 nm under 10 m*M* Tris-HCl, pH 7.3 (**Fig. 2A**) *(1,19)*. hA samples are prepared at concentrations of 13 µ*M* to 100 µ*M* in 10 m*M* Tris-HCl, pH 7.3, and incubated for up to 5 d before adsorption onto a mica surface. For CCβ-Met, it was observed that two predominant fibrils existed with either an axial periodicity of 61 nm or 30 nm *(20)* (**Fig. 2B**). Under 10 m*M* Tris-HCl, 100 m*M* NaCl, pH 7.4, the fibrils with the 61-nm periodicity had maximum and minimum heights of 5.9 and 4.3 nm, whereas the 30-nm periodicity fibrils had maximum and minimum heights of 7.4 and 6 nm. Interestingly, these axial periodicities were barely visible when imaged by EM, after unidirectional metal shadowing, illustrating why studies using both microscopy techniques are advantageous. CCβ-Met samples are prepared at concentrations of 500 µ*M* in water and incubated overnight at 37°C. The sample is then diluted fourfold and adsorbed onto a mica surface for 4 min. Although these studies only imaged the mature fibrils, the same technique could be used to image intermediate amyloid structures in solution before the appearance of mature fibrils.

3.2. Direct Assembly of Fibrillar and Oligomeric Structures on a Surface

For investigating fibril assembly occurring directly on a surface, the optimal adsorption time is dependent on the protein concentration. The surface may also enhance fibril formation, as is the case for hA on mica and insulin on lipid surfaces, meaning lower protein concentrations should be used *(1,21)*. An amyloid protein sample is prepared and approx 25 to 60 µL aliquots immediately pipetted on to freshly cleaved mica substrates. These substrates are placed in a moist chamber (i.e., a Petri dish containing moist filter paper) for various incubation times (in order to investigate the effect of adsorption time on fibril

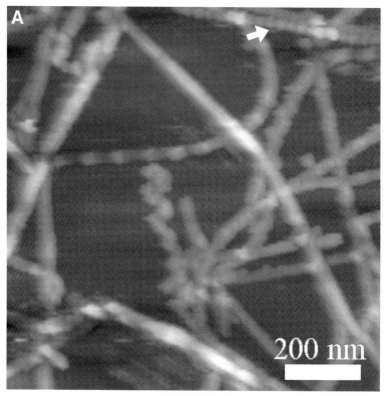

Fig. 2. Amyloid fibril polymorphisms. (**A**) hA fibrils with an axial periodicity of 25 nm, imaged by AFM (white arrow).

formation). After incubation, the substrate is washed three times to remove any excess or unadsorbed protein. *See* **Note 8** for a suggested imaging protocol.

3.2.1. hA Protofibrils and Oligomers

Mature hA fibrils are known to be highly polymorphic and to be composed of a 5-nm wide subunit protofibril with a mass-per-length of 10 kDa/nm *(19)*. As they usually form part of a larger fibril containing multiple subunits, isolated protofibrils are rarely observed when solutions of fibrils are looked at by EM *(19)*. In contrast, the presence of a mica surface changes the assembly process and morphology of the fibrils allowing the protofibril to be visualized (**Fig. 3**) *(1)*. In the case of hA, due to constraints imposed by the mica surface, fibrils growing on it are non-coiled and show a different height distribution compared to fibrils preassembled in solution *(1)*. The protofibril has a smooth

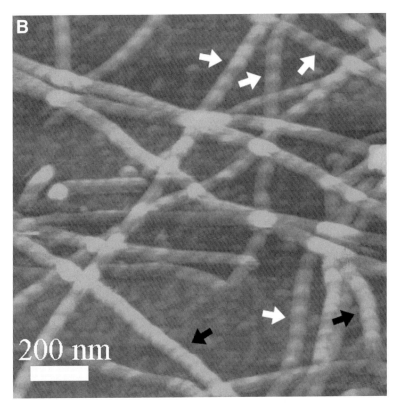

Fig. 2. **(B)** CCβ-Met fibrils imaged by AFM (adapted from *[20]*). The arrows indicate the two prominent fibril populations, which are characterized by 610 ± 70 (white arrows) and 300 ± 40 Å (black arrows) periodicities.

appearance and a height of 2.4 nm on mica in 10 m*M* Tris-HCl, pH 7.3 buffer (note this height increases to approx 3.7 nm upon addition of 50 m*M* KCl to the buffer—*see* **Note 10**). Importantly, protofibrils are not the first structures to be seen on adsorption of a fresh solution of hA to mica. Larger fibrils are initially formed and only when the peptide solution above the surface becomes depleted do the protofibrils appear. This suggests that hA protofibrils are not stable precursor structures, but are subunits of larger fibrils. Their elongation is initially not favored because usually oligomers can be formed which subsequently elongate in to more complex higher order hA fibrils (discussed in **Subheading 3.3.2.**). For protofibril studies, up to 60 µL of an approx 2.5 µ*M* hA solution in 10 m*M* Tris-HCl, pH 7.3 is incubated on a mica surface for up to 5 d in a moist chamber. The peptide is initially dissolved in water (50 µg/mL) and then diluted to the final concentration in buffer. Time-lapse AFM may also be used to measure protofibril growth directly (*see* **Subheading 3.3.1.**).

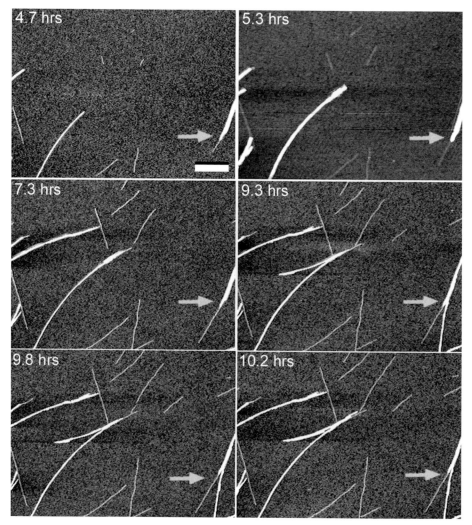

Fig. 3. Morphology of hA protofibrils on mica. Time after injection of peptide indicated on image. Arrow indicates a protofibril growing out of larger fibril. Scale bar = 200 nm.

In order to visualize early oligomeric intermediates in the hA fibril assembly process, the preparation method for making the initial peptide solution is modified. Rather than initially dissolving lyophilized peptide in water (as described previously), the peptide is dissolved in HFIP (*see* **Note 1**). HFIP pre-treatment reduces the amount of aggregated material in the stock solution and enables small oligomeric fibril precursors to be detected. However, it also increases the rate of fibril assembly. HFIP stock solutions of hA are diluted to 10 μM in 10 mM Tris-HCl, pH 7.4 (final HFIP = 1.25%) and adsorbed to mica substrates for 30 to 120 s. The mica is washed and imaged under water, so that further growth of the oligomers is inhibited. Using the single particle analysis software (*see* **Note 10**), the structures formed after 30 s are, on average, 2.3 nm in height and 22.9 nm in length. In contrast, oligomeric structures formed after 60 s have average heights and lengths of 4.6 nm and 46.8 nm respectively. After only 120 s, full-length mature fibrils with an average height of 10.6 nm and length of 202 nm are formed. These experiments demonstrate (further corroborated by time-lapse AFM; *see* **Subheading 3.3.**) that hA oligomers grow into fibrils by a two step process, in which the lateral growth of oligomers is followed by longitudinal growth into mature fibrils (**Fig. 4**) *(3)*.

3.3. Measuring Fibril Growth In Situ by Time-Lapse AFM

A 10 μm^2 scan area of a clean substrate surface is imaged under 100 μL of buffer in TappingMode™ (*see* **Note 8**). A clean area of the scanned surface, from 2 to 5 μm^2 is chosen, or if using a lipid substrate (*see* **Note 3**), only a few lipid bilayer defects detected. Once a stable image is attained, 100 μL of freshly prepared peptide is slowly injected into one of the openings in the front of the fluid cell using a pipet (*see* **Notes 6** and **7**). One hundred microliters are then removed from the other opening in the fluid cell and injected back into the first opening in order to mix the solution in the fluid cell. Mixing should be repeated two or three times, taking care to add and remove the fluid in a slow and even manner. The deflection voltage is adjusted to zero using the photodetector adjustment knobs, should the value have changed during injection of the peptide solution. An initial scan is captured starting at the top of the frame and the time of injection recorded. Should tip contamination have occurred during injection (*see* **Note 7**), scanning is left to continue for one or two frames as the tip may decontaminate itself. Otherwise decontamination may be attempted by increasing the force (in a clean area of the substrate) or increasing the scan frequency. After a stable image has been obtained, scanning is continued either continuously or intermittently (for e.g., every 30 min for 4 or 5 h). If intermittent scanning is preferred, the AFM tip is disengaged and the laser unplugged between scans. Should some drift occur in between scans, the x- and y-offset settings are adjusted.

3.3.1. Determination of Fibril Elongation Rates: hA Protofibrils

Protofibril elongation is measured by injecting approx 5 μM hA (final concentration = ~2.5 μM) in 10 mM Tris-HCl, pH 7.3 and continuously scanning a 4 μm^2 area of the mica surface. hA is diluted from an initial stock solution of approx 50 μg/mL in Milli-Q water. Protofibrils begin to appear after 2 to 3 h,

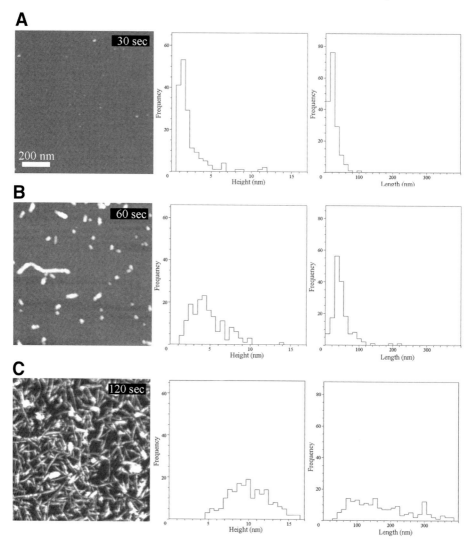

Fig. 4. The effect of adsorption time on hA morphology and assembly on a mica surface. hA species visualised after **(A)** 30 s, **(B)** 60 s, and **(C)** 120 s adsorption. The heights and lengths of oligomers and fibrils were measured for each timepoint. (Adapted with permission from **ref. 3**.)

while larger fibrils appear and elongate in the initial scans. The total growth rate of hA protofibrils is 1.1 ± 0.5 nm/min and occurs from both ends of the fibril at roughly the same rate. This is in contrast to larger hA fibrils which exhibit much faster elongation rates. Protofibrils are also seen growing out from larger fibrils, indicating that they are constituents of the latter *(1)* (**Fig. 3**).

A

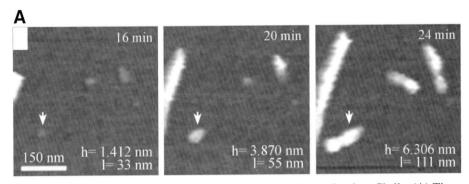

Fig. 5. Oligomers grow in height, prior to extensive elongation into fibrils. **(A)** Time-lapse AFM showing an hA oligomer which grows into a fibril. The height of the oligomer (arrow) increases with each scan. The timepoints and height (h) and length (l) measurements are indicated in each image.

3.3.2. Observing hA Oligomer Growth

Time-lapse AFM was used to follow the assembly of hA fibrils via oligomeric intermediates revealing that prior to elongation of mature fibrils, oligomers increase in height (**Fig. 5A** and **Fig. 6**; *see also* **Subheading 3.2.1.**) *(3)*. To perform these experiments, freshly diluted hA is injected into the fluid cell and the growth of hA oligomers is monitored by repeated scanning (5–10 μ*M* hA, 0–1.25% HFIP, 10 m*M* Tris-HCl, pH 7.4—in this case, hA is initially diluted from a stock solution in HFIP; *see* **Note 1**). When height versus length is plotted for individual species under this assembly regime, it is evident that significant height increases occur early in most cases, often before a substantial increase in length (**Fig. 5B**). When all the heights and lengths from the time-lapse AFM experiments are graphed together, a pattern emerges whereby the oligomers increase in height first and only after reaching approx 6 nm do they start to significantly elongate (**Fig. 5C**).

3.3.3. Comparing Variant Amylin Peptides

To elucidate the role of individual residues in fibril formation, we generated a series of full-length rat amylin (rA) variants and examined their ability to form fibrils in vitro *(2)*. As part of this study, we used time-lapse AFM to compare the nucleation and elongation rates of three peptides, which had one or two amino acid sequence differences (**Fig. 7**) *(2)*. At a concentration of 40 μM, the fastest fibril-forming variant peptide, rA(R18H, L23F) as judged by EM, formed nuclei after 9 min that grew into fibrils after 35 min (**Fig. 7**; top row). Fibril formation by rA(R18H, V26I) and rA(R18H) was slower owing to lower nucleation rates (**Fig. 7**; middle and bottom rows).

B

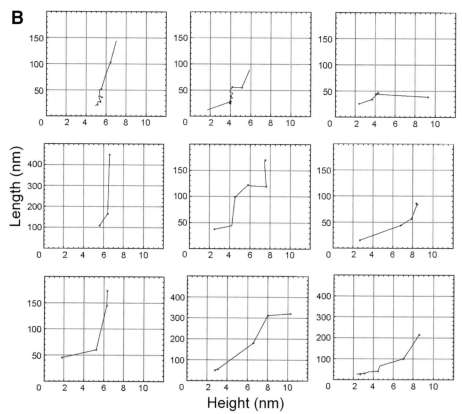

Fig. 5. **(B)** Examples of length vs height trajectories for individual hA oligomers.

3.3.4. Testing Inhibitors Using Time-Lapse AFM:
The Effect of Congo Red on hA Fibril Formation

We used time-lapse AFM to test the effect of Congo red on hA fibril formation (**Fig. 8**) *(3)*. The morphology and kinetics of hA fibrillar structures formed in the presence or absence of Congo red were compared (8 μ*M* hA; 240 μ*M* Congo red, 10 m*M* Tris-HCl, pH 7.4). Although the nucleation rate (appearance of oligomers) did not decrease in the presence of Congo red, the process of oligomer elongation in to fibrils was inhibited (**Fig. 8**). The height and length of hA oligomeric structures formed in the presence of Congo red were 3.74 nm and 50.6 nm respectively *(3)*. Mature fibrils formed in the absence of Congo red after 50 to 60 min of continuous scanning were 11.5 nm in height and 225 nm in length. These results complement other published data that has demonstrated that Congo red inhibits fibril formation *(22)*. Our data offers the additional insight that the inhibition is at the level of elongation rather than nucleation.

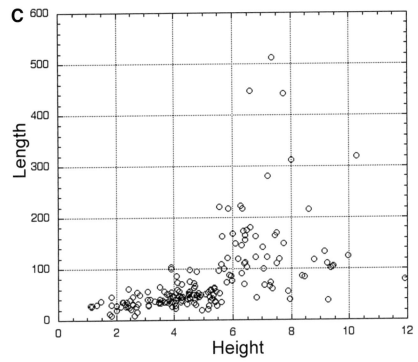

Fig. 5. **(C)** Length vs height summary plot for 49 hA oligomers that were pooled from four independent time-lapse AFM experiments. (Adapted with permission from **ref. 3**.)

3.3.5. Seeding Aβ Fibril Elongation With Pre-Assembled Fibrils

Aβ1-40 forms mature fibrils at a much slower rate compared to hA. In order to see Aβ1-40 fibril elongation by AFM within a reasonable time frame, it is necessary to seed the mica surface by adsorbing a few microliters of pre-poly-merized mature fibrils (*see* **Note 1**) (**Fig. 9**) *(15)*. Once a stable image is obtained of the pre-adsorbed mature fibrils, a fresh Aβ solution is injected into the fluid cell producing a final fresh Aβ1-40 concentration of 2.5 μ*M*. The surface is then scanned every 30 to 60 min over a period of 4 or 5 h to monitor the elon-gation of mature fibrils (**Fig. 9**). If the surface is not seeded before injecting the fresh solution of Aβ, only "oligomers" (containing approx 50 to 200 Aβ mono-mers) and occasionally very short "protofibrils" (not subunits of mature fibrils in this case) are seen over a time period of up to 10 h (*23*, Goldsbury et al., un-published). As a further application, potential inhibiting compounds could be co-injected with fresh peptide in order to measure their effect on fibril elonga-

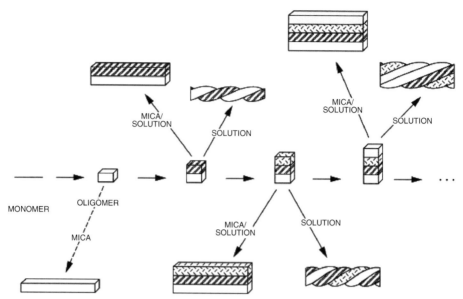

Fig. 6. Scheme for hA assembly. Oligomers initially grow in height, before elongating further into mature fibrils. The resulting fibrils can be composed of 2 or more protofibrils, depending to what extent the nuclei initially grew in height. Protofibrils are seldom observed in solution but can be stabilized when growth is constrained to a mica surface (*see* **Fig. 3**). In solution, mature fibrils often have distinctive fibril polymorphisms (*see* **Fig. 2A**), whereas on a mica surface, mature fibrils without any coiling are observed. (Adapted with permission from **ref. 3**.)

tion and morphology. Theoretically, such experiments could also be done to test for example whether one fibril type can seed elongation of fibrils of a different amyloid-protein type.

3.4. Effect of Amylin Peptides on Lipid Bilayers

Owing to the proposal that amyloid peptides are toxic through their membrane permeating ability, the effect of amylin peptides on a negatively charged lipid bilayer was investigated (**Fig. 10**) *(24–26)*. Lipid bilayers are deposited on to a mica surface using Langmuir Blodgett transfer as described in **Note 3**. After the original bilayer is scanned and any defects detected, 10 µ*M* of hA are injected into the fluid cell and the change in the lipid imaged (*see* **Note 9**) (**Fig. 10**). Upon injecting the peptide, a halo, intermediate in height in comparison to the bare mica and the lipid, is formed around the original defects. A similar decrease in height of the lipid is also seen in areas where no initial defect is detected (**Fig. 10**, asterisk). Continuous imaging of the same area of the lipid

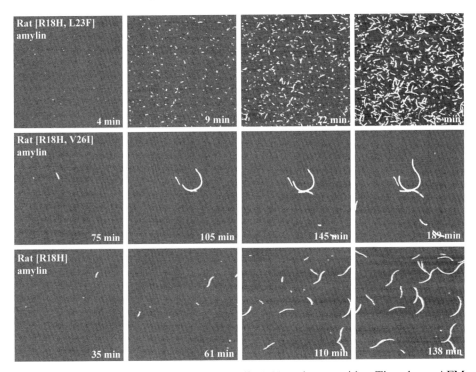

Fig. 7. Nucleation rates differ for rat amylin (rA) variant peptides. Time-lapse AFM for rA[R18H, L23F], rA [R18H, V26I], and rA [R18H] at a concentration of 40 μ*M* in PBS. Images are 2 μm² areas zoomed in from a 4 μm² image. (Adapted with permission from **ref. 2**.)

bilayer (5 μm²) reveals that the sections which exhibit a change in height of the lipid, grow with time in a concentric manner (**Fig. 10**). These results support the hypothesis that hA is cytotoxic through its interaction with lipid membranes *(27,28)*. This assay could theoretically be used in combination with time-lapse AFM on a mica surface, to test the ability of small molecules to inhibit fibril formation as well as lipid disruption.

4. Notes

1. Preparation of amyloid peptide solutions for AFM: All buffer solutions should be filtered through 0.2-μm filters to remove any particulate. Lyophilized Aβ is dissolved in DMSO at a stock concentration of 2 m*M* and stored at −20°C. For assembling fibrils, the stock solution is diluted in PBS (25 m*M* phosphate, 120 m*M* NaCl, pH 7.4) to a peptide concentration of 100 μ*M* (DMSO 10%). For imaging fibril elongation by AFM, this solution is further diluted 1:20 in PBS giving a peptide concentration of 5 μ*M*. This working solution is injected in to the fluid cell of

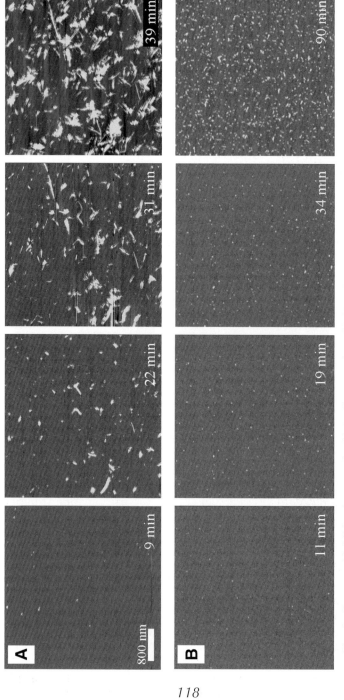

Fig. 8. Congo red inhibits the growth of hA oligomers into fibrils. Time-lapse AFM: 8 μM hA, 10 mM Tris-HCl, pH 7.5 in the absence (**A**) or presence (**B**) of 2.4 mM Congo red. (Adapted with permission from **ref. 3**.)

118

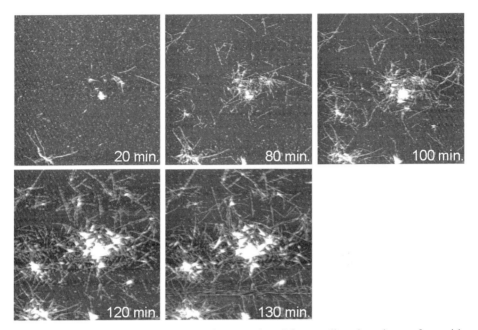

Fig. 9. Seeded Aβ1-40 fibril formation on mica. After seeding the mica surface with pre-assembled fibrils, a 5 μm² area was scanned at the times indicated after injection of a fresh solution of Aβ1-40 in PBS.

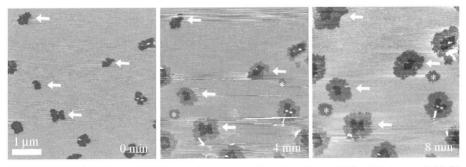

Fig. 10. A time-lapse AFM experiment showing the effect of 10 μ*M* hA on a POPC/ POPG (3:1) lipid bilayer (10 m*M* Tris-HCl, pH 7.4). The times after the injection of hA are displayed in each image. Arrows indicate defects present originally in the lipid bilayer. The asterisk indicates lipid disruption following the injection of peptide in an area where no defect was initially detected. Note that a color coding exists within the images: the darkest features are dips or holes within the lipid bilayer, whilst the lightest features represent structures lying above the lipid bilayer.

the AFM where it is further diluted 1:2 resulting in a final concentration of 2.5
μM. Before injecting the fresh peptide, the mica surface is "seeded" by the adsorption of pre-assembled fibrils (approx 2 µL of a 100 μM solution pre-incubated for
10 d at 37°C) and washed several times.

Amylin stock solutions are prepared by dissolving lyophilized peptide in Milli-
Q water or in 1,1,1,3,3,3-hexafluoro-2-isopropanol (HFIP). Stock solutions in
Milli-Q water should be <1 mM as fibril formation can occur spontaneously even
in Milli-Q water. Undissolved material from the lyophilised preparation may also
"seed" fibril formation, complicating or inhibiting the visualization of the early
assembly processes. Dissolving hA in HFIP (0.5–3.2 mM) produces a more homogeneous peptide solution circumventing this problem. Amylin HFIP stock solutions are either diluted straight into buffer, or the HFIP is evaporated off under a
nitrogen flow, before the peptide is dissolved in water prior to experimentation.
Final HFIP concentrations in the samples should be kept below 2% to avoid aber-
rations of the mica surface. Furthermore, it should be noted that the presence of
HFIP increases the rate of fibril formation *(29)*. For fibril assembly, stock solutions are diluted in 10 mM Tris-HCl, pH 7.4.

2. Preparation of the support and sample attachment: Amyloid proteins need to be
   adsorbed to a flat and stable support (*see* **Subheading 2.2.**). The quality of this support is one of the most critical factors for optimal results in AFM. Poorly prepared
   supports result in vibrations and poor image resolution. A good method for preparing AFM supports has been described by Schabert and Engel (1994) (*see* **Subheading 2.2.1.**) *(30)*. The amyloid solution is adsorbed to mica which is attached
   to hydrophobic Teflon. The Teflon repels the fluid droplet and thus contains it
   on the mica surface and protects the piezo from contact with the buffer solution
   (important since the piezo will be damaged if it gets wet—*see* **Note 9**). The key
   factor in support preparation to minimize subsequent vibration oscillations is ensuring that the surfaces are smoothly sealed, i.e., there are no bubbles or particulate
   material under the glued surfaces. Material with different properties—for example hydrophobic highly oriented pyrolytic graphite (HOPG) or glass, can accordingly replace mica as a support attached to the Teflon disk. For adsorption of the
   specimen, mica, HOPG or other surfaces should be as clean as possible. For this
   purpose, mica and HOPG can easily be cleaved to reveal a fresh and atomically
   flat surface. Glass can be washed, e.g., with concentrated HCl/HNO3 (3:1), and
   rinsed with ultra-pure water in a sonication bath in order to obtain a clean, flat surface *(31)*.

   Firm adsorption of a specimen to the support depends on the buffer conditions
   and the charge of the protein. Adsorption buffer conditions must be optimized for
   each new sample empirically. A good starting buffer could be Tris-HCl pH 7.3 adding increasing amounts of monovalent and divalent cations to maximize attachment and stability of the fibrils/oligomers. Different pHs can also be tested depending on the net charge of the protein. In the case of a partially hydrophobic sample,
   a hydrophobic support such as HOPG might be preferable to mica. In this case,
   hydrophobic interactions will mediate sample immobilization. However, features

of the HOPG surface itself can resemble filamentous structures (*see* **Note 4**, **Fig. 11A**). Once the sample is adsorbed, the buffer can be exchanged or kept the same for imaging.

3. Lipid bilayer support preparation: Non-covalent attachment of lipid layers to the support, can also facilitate adsorption of the protein of interest or enable amyloid protein interactions with lipids to be investigated (*see* **Subheading 3.4.**). The Langmuir-Blodgett technique is described later but lipids may also be adsorbed to mica using the vesicle fusion technique *(32)*. Langmuir-Blodgett transfer *(7,16–18)*: Lipid (approx 5–10 μL, 5 mg/mL in Chloroform) is added drop by drop on to the surface of a water filled trough until the pressure reaches 7 to 9 mN/m. A piece of mica, which has been freshly cleaved on both sides (100–500 μm thick, 1.5 m × 2.5 cm) is dipped underneath the water surface at maximum speed using a motor-controlled lift. The lipid is automatically compressed to a pressure of 40 to 42 mN/m with a barrier speed of zero. The mica is lifted up very slowly with a step motor at a speed of 1–2 mm/min resulting in a lipid monolayer surface (the mono-layer is stable in air for at least 1 h). The mica surface is cut in half and each half glued onto a circular metal disk (1-cm diameter) using insoluble epoxy glue. After waiting 5 min for the glue to set, the metal disk is turned so that the mica surface is facing downwards, and the excess mica is cut off, with care taken not to touch the monolayer surface. The metal disk/mica is placed in the lift holder and low-ered down through the monolayer surface on the trough at a speed of 1 to 2 mm/min, forming a stable bilayer on the mica surface. Once the metal disk/mica is immersed under the water in the trough, lowering is halted, and the lipid is sucked off the water surface in the trough using an aspiration system. The metal disk/mica is released from the lift holder into a small container (2 × 2 × 2 cm) inside the trough. In this container, the metal disk is removed while immersed under water. The disk is transferred into another container filled with buffer. A hydrophobic self-adhesive reinforcement ring (Herma, Germany) with an inner diameter of 6 mm and an outer diameter of 13 mm is carefully covered on the adhesive side with silicon sealant (732, Dow Corning, MI). The ring is placed with the silicon sealant facing down, onto the mica surface of the disk, which is still immersed under buf-fer. After 12 min the disk is lifted out of the buffer with tweezers. Finally, a water droplet will be over the center of the mica surface, but with some water over the hydrophobic reinforcement ring. To concentrate the water further into the center, toothpicks are used to clean off the excess silicon from the hydrophobic ring. Now the lipid layer is ready to be used as a substrate for imaging, but should be stable for at least 2 d in a moist chamber.

4. Specimen support artifacts: It is important to characterize the surface features of the substrate under the buffer used before the sample is adsorbed. With cleaved mica, the only effects that might be observed are layer edges, which usually exhibit very sharp edges, or large hill-like protrusions, probably originating from air inclu-sions between the mica layers. In contrast, HOPG is more prone to surface effects when imaged under buffer solution—for example, parallel stripes in various direc-tions (usually oriented in line with the 60 degree angles of the HOPG carbon lattice)

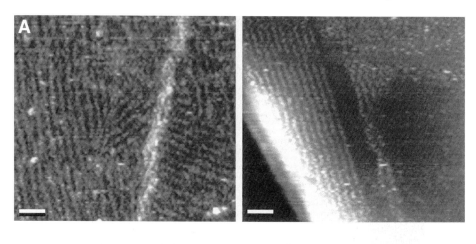

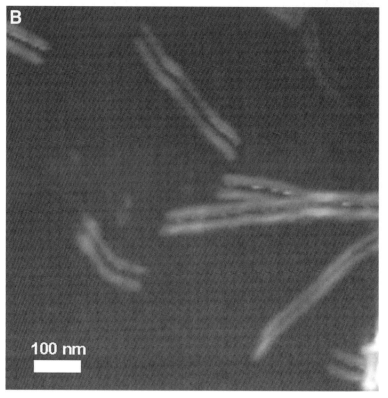

Fig. 11. Double tips and substrate artifacts. (**A**) Hydrophobic highly oriented pyro-lytic graphite (HOPG) imaged in tapping mode under buffer in the presence (left) or absence (right) of amyloid peptide. The same striated structural feature is seen—*see* **Note 4**. Scale bars = 20 nm. (**B**) Obvious doubling of the image indicates "double tip" artifact.

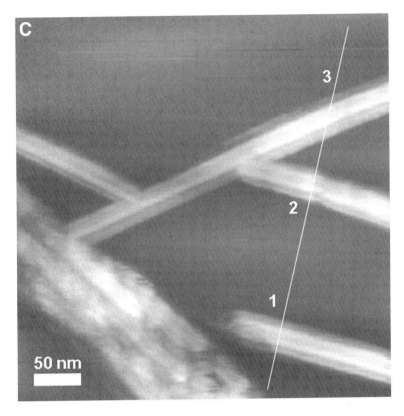

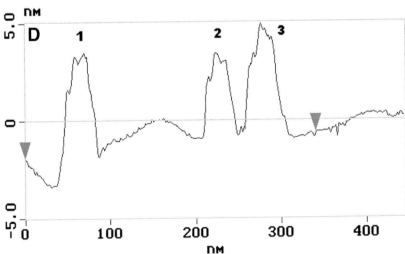

Fig. 11. (C) A more subtle tip shape effect. (D) Cross section line through fibrils 1, 2, and 3 marked in (C) show almost identical contours indicating features are probably representative of the overall tip shape and not fibril sub-structure.

are often observed with a laterally spacing of approx 5 to 10 nm and a height of 0.5 to 1 nm (**Fig. 11A**). These features pose a problem if they closely resemble features expected of the specimen, e.g., sheets of amyloid protofilaments.

5. The tip and cantilever: Tips attached to cantilevers with low spring constants (<1 N/m) are preferable since the lower the spring constant of the cantilever, the less force can be exerted between tip and sample reducing damage to soft protein complexes. Typically, silicon nitride cantilevers for biological applications are 100–200 μm in length and have spring constants of 0.06 to 0.6 N/m. For TappingMode™, cantilevers are oscillated close to their resonance peak—usually between 8 and 9 kHz. Oxide-sharpened, electron-beam deposited, or nano-tube attached tips provide sharper tips and thus higher resolution. Further information on tips—including modification of and cleaning of—can be found at the following website: http://www.veeco.com/appnotes/AN44_BiotipsApps.pdf (Veeco, CA). *See also* **Note 6**.

6. Tip shape: In the interpretation of AFM images, it is important to take into account that the shape of the scanning tip affects the shape of features in the image. The most common tip effect is convolution of the imaging tip's shape with the object's actual shape. In such convolutions, the height of the object (z-dimension) is not affected by tip geometry, but the width (x- y-dimension) becomes broadened owing to convolution with the interacting tips radius of curvature. Generally, if isolated objects on the surface appear to all have the same shape and orientation, the tip is probably being imaged and not the sample. Broadening of the width measurement is particularly predominant with isolated filamentous structures *(33,34)*. Some investigators have used gold particles to calibrate their tip so that accurate width measurements can be obtained *(35)*. Furthermore, the interaction between the surface and the tip very often involves more than one region of the tip, creating a double-image effect. This effect can be dramatic and easily recognized, or it may be more subtle (**Fig. 11 B,C**). If height cross-sections measured across different segments along a fibril exhibit almost identical contours, it is strongly indicative that the shape of the cross-section is dominated by, and represents the shape of the tip rather than sub-structure within the fibril (**Fig. 11 C,D**). In most cases, the tip must be replaced, but sometimes slightly changing the angle of the cantilever relative to the scanned support such that a slightly different region of the end of the tip contacts the sample helps reduce the problem.

Particular caution must be taken when investigating helical structures by AFM: an apparent existence of both right- and left-handed filaments in preparations of Alzheimer's disease paired helical filaments (PHFs) was shown by Markiewicz et al. *(36)*. (http://www.weizmann.ac.il/Chemical_Research_Support/surflab/peter/artifact/index.html) In computational simulations using different shaped tips, it was demonstrated that helical directionality was an artifact owing to a slightly oblong shaped tip. In the simulations, apparent beaded-structures often appeared, masking the filament sub-structure and, in some instances, further distortion occurred resulting in reversal of the apparent helical twist direction.

7. Tip contamination: Changes in tip shape may be owing to contamination resulting from material adsorbing to the very end of the stylus. This results in smears at

the left edge of the "trace-image" and at the right edge of the "retrace-image." The more material that is adsorbed on the end of the tip, the stronger the effect. To dislodge the contamination, a sample-free support region should be scanned with increased force for a few seconds, or if that does not work, then the tip should be scanned in a non-contact region above the surface for a few minutes. However, often the tip must be replaced. This issue is very important when performing time-lapse experiments where solutions are injected in to the fluid cell. Upon injection, the tip may become contaminated. During scanning by this tip, the surface will appear to be covered in aggregates when the real aggregates are actually adsorbed to the tip itself.

8. Imaging conditions: To reduce noise caused by mechanical vibrations, the AFM should be set up on a bungee system or damping table. For engaging the tip, the deflection voltage is adjusted to 0 ± 2 using the photodetector adjustment knobs. For initiating imaging, usually the scan area is set to zero. Once the tip is engaged, this enables the forces to be optimized before the sample is scanned and potentially damaged by the tip, or the tip is contaminated by scraping the sample. Once these conditions are achieved, the scan size can be increased for the sample to be viewed usually starting with a large scan area (10–20 μm^2) and finally zooming in to a region of interest (typically 0.5–5 μm^2).

In contact mode, reducing the force requires the set point (distance from the substrate) to be minimized as far as possible and the gains increased as much as possible without causing oscillations. During imaging, the set point, scan speed, and gains must be continually modulated to maintain low force and to optimize the image. Contact-mode imaging is restricted to samples that are firmly attached to the support and exhibit strong lateral stability. Fibrillar structures and isolated oligomers have a relatively small surface area for adhering to the support, and are often pushed away by the tip during scanning. Dynamic mode, also called "MAC Mode" or "Tapping-mode" (MAC Mode™, TappingMode™) was developed to address these issues (*see* **Subheading 1.1.**). In TappingMode™ the reduced contact time of the oscillating tip with the sample reduces lateral frictional forces compared to contact mode, and protein assemblies are therefore less likely to be pushed away. The forces can be further minimized by modulating the amplitude of oscillation and the distance of the tip from the sample. Once the image is optimized, scanning is generally more stable than in contact mode, and relatively few subsequent adjustments need to be made. A drawback though, is the slow scan rate required— typically a factor of four slower than contact mode. New innovations in microscope and tip design may circumvent this shortcoming and ultimately to allow higher temporal resolution.

9. Protection of the AFM piezo: We do not use the AFM fluid cell O-ring or tubing. Using the Teflon covered supports prevents the overflow of water to the piezo (*see* **Note 2** and **Subheading 2.2.1.**). However, it may be vigilant for any new AFM users to cover the piezo with a lubricant-free latex condom, cut to size, to avoid any water damage to the piezo. When performing experiments with prepared lipid bilayers it is advisable to always cover the piezo with a condom.

10. Measuring heights and periodicities: It should be noted that a color coding exists within the images: the darkest features are depressions below the surface, whilst the lightest features represent structures lying above the surface. The measured heights obtained by AFM of biomolecules, adsorbed to a solid support under buffer, are influenced by electrostatic interactions between the AFM tip and the sample *(37)*. These electrostatic forces, however, can be screened out by adjusting the electrolyte concentration and pH of the imaging buffer, allowing correct height measurements. It has been observed, that if the ionic strength of the imaging buffer is not strong enough, at low scanning forces, that height measurements can either be over- or underestimated, depending on the surface charge of the sample *(37)*. Appropriate conditions for any sample can be determined by measuring heights as a function of electrolyte concentration at forces applied to the tip of <0.2 nN, and adjusting the concentration of the electrolyte to a point where possible electrostatic repulsion is eliminated. Since the force on the sample cannot easily be determined in TappingMode™, to check accuracy of height data, it would be reasonable to image under buffers at neutral pH containing greater than 300 mM KCl or NaCl and/or low concentrations of divalent cations (e.g., 20 mM MgCl$_2$). Under these conditions, accurate height measurements of numerous biological samples by AFM have been obtained *(37)*.

   After the flattening of the acquired images, height, length, width, and periodicity information can be measured using the sectioning tool found in the microscope software. If particles have been imaged, the single particle analysis program can be used to automatically measure the heights of these structures. This data can be exported as a file compatible with Microsoft Excel. Width and length measurements can also be obtained for particles using this programme, however, to date this information has to be extracted manually. Note that width and length measurements are overestimated by AFM (*see* **Note 6**).

Acknowledgments

We acknowledge Professor Ueli Aebi for his advice and support. This work was performed at the M. E. Müller Institute for Structural Biology, Biozentrum, University of Basel, Switzerland. It was supported in part by an NCCR Program Grant on "Nanoscale Science" by the Swiss National Science Foundation and by the M. E. Müller Foundation of Switzerland.

References

1. Goldsbury, C., Kistler, J., Aebi, U., Arvinte, T., and Cooper, G. J. S. (1999) Watching amyloid fibrils grow by time-lapse atomic force microscopy. *J. Mol. Biol.* **285,** 33–39.
2. Green, J., Goldsbury, C., Mini, T., et al. (2003) Full-length rat amylin forms fibrils following substitution of single residues from human amylin. *J. Mol. Biol.* **326,** 1147–1156.

3. Green, J. D., Goldsbury, C., Kistler, J., Cooper, G. J. S., and Aebi, U. (2004) Human amylin oligomer growth and fibril elongation define two distinct phases in amyloid formation. *J. Biol. Chem.* **279,** 12206–12212.

4. Allison, D. P., Hinterdorfer, P., and Han, W. (2002) Biomolecular force measurements and the atomic force microscope. *Curr. Opin. Biotechnol.* **13,** 47–51.

5. Engel, A. and Muller, D. J. (2000) Observing single biomolecules at work with the atomic force microscope. *Nat. Struct. Biol.* **7,** 715–718.

6. Fisher, T. E., Marszalek, P. E., and Fernandez, J. M. (2000) Stretching single molecules into novel conformations using the atomic force microscope. *Nat. Struct. Biol.* **7,** 719–724.

7. Stolz, M., Stoffler, D., Aebi, U., and Goldsbury, C. (2000) Monitoring biomolecular interactions by time-lapse atomic force microscopy. *J. Struct. Biol.* **131,** 171–180.

8. Engel, A., Lyubchenko, Y., and Muller, D. (1999) Atomic force microscopy: a powerful tool to observe biomolecules at work. *Trends Cell Biol.* **9,** 77–80.

9. Engel, A., Gaub, H. E., and Muller, D. J. (1999) Atomic force microscopy: a forceful way with single molecules. *Curr. Biol.* **9,** R133–R136.

10. Heinz, W. F. and Hoh, J. H. (1999) Spatially resolved force spectroscopy of biological surfaces using the atomic force microscope. *Trends Biotechnol.* **17,** 143–150.

11. Hansma, H. G., Kim, K. J., Laney, D. E., et al. (1997) Properties of biomolecules measured from atomic force microscope images: a review. *J. Struct. Biol.* **119,** 99–108.

12. Bustamante, C., Rivetti, C., and Keller, D. J. (1997) Scanning force microscopy under aqueous solutions. *Curr. Opin. Struct. Biol.* **7,** 709–716.

13. Lal, R. and John, S. A. (1994) Biological applications of atomic force microscopy. *Am. J. Physiol.* **266,** C1–C21.

14. Clausen-Schaumann H., Seitz, M., Krautbauer, R., and Gaub, H. E. (2000) Force spectroscopy with single bio-molecules. *Curr. Opin. Chem. Biol.* **4(5),** 524–530.

15. Goldsbury, C., Aebi, U., and Frey, P. (2001) Visualizing the growth of Alzheimer's Aβ amyloid-like fibrils. *Trends Mol. Med.* **7,** 582.

16. Czajkowsky, D. M. and Shao, Z. (2002) Supported lipid bilayers as effective substrates for atomic force microscopy. *Methods Cell Biol.* **68,** 231–241.

17. Rinia, H. A., Kik, R. A., Demel, R. A., et al. (2000) Visualization of highly ordered striated domains induced by transmembrane peptides in supported phosphatidylcholine bilayers. *Biochemistry* **39,** 5852–5858.

18. Shi, D., Somlyo, A. V., Somlyo, A. P., and Shao, Z. (2001) Visualizing filamentous actin on lipid bilayers by atomic force microscopy in solution. *J. Microsc.* **201(Pt 3),** 377–382.

19. Goldsbury, C. S., Cooper, G. J., Goldie, K. N., et al. (1997) Polymorphic fibrillar assembly of human amylin. *J. Struct. Biol.* **119,** 17–27.

20. Kammerer, R. A., Kostrewa, D., Zurdo, J., et al. (2004) Exploring amyloid formation by a de novo design. *Proc. Natl. Acad. Sci. USA* **101,** 4435–4440.

21. Sharp, J. S., Forrest, J. A., and Jones, R. A. (2002) Surface denaturation and amyloid fibril formation of insulin at model lipid-water interfaces. *Biochemistry* **41,** 15810–15819.

22. Aitken, J. F., Loomes, K. M., Konarkowska, B., and Cooper, G. J. S. (2003) Suppression by polycyclic compounds of the conversion of human amylin into insoluble amyloid. *Biochem. J.* **374,** 779–784.

23. Goldsbury, C. S., Wirtz, S., Müller, S. A., et al. (2000) Studies on the in vitro assembly of Aβ1-40: implications for the search for Aβ fibril formation inhibitors. *J. Struct. Biol.* **130,** 217–231.

24. Green, J. D., Kreplak, L., Goldsbury, C. et al. (2004) Atomic force microscopy reveals defects within mica supported lipid bilayers induced by the amyloidogenic human amylin peptides. *J. Mol. Biol.,* in press.

25. Janson, J., Ashley, R. H., Harrison, D., McIntyre, S., and Butler, P. C. (1999) The mechanism of islet amyloid polypeptide toxicity is membrane disruption by intermediate-sized toxic amyloid particles. *Diabetes* **48,** 491–498.

26. Kagan, B. L., Hirakura, Y., Azimov, R., Azimova, R., and Lin, M. C. (2002) The channel hypothesis of Alzheimer's disease: current status. *Peptides* **23,** 1311–1315.

27. Anguiano, M., Nowak, R. J., and Lansbury, P. T. Jr. (2002) Protofibrillar islet amyloid polypeptide permeabilizes synthetic vesicles by a pore-like mechanism that may be relevant to type II diabetes. *Biochemistry* **41,** 11338–11343.

28. Harroun, T. A., Bradshaw, J. P., and Ashley, R. H. (2001) Inhibitors can arrest the membrane activity of human islet amyloid polypeptide independently of amyloid formation. *FEBS Lett.* **507,** 200–204.

29. Padrick, S. B. and Miranker, A. D. (2002) Islet amyloid: phase partitioning and secondary nucleation are central to the mechanism of fibrillogenesis. *Biochemistry* **41,** 4694–4703.

30. Schabert, F. A. and Engel, A. (1994) Reproducible acquisition of Escherichia coli porin surface topographs by atomic force microscopy. *Biophys. J.* **67,** 2394–2403.

31. Karrasch, S., Dolder, M., Schabert, F., Ramsden, J., and Engel, A. (1993) Covalent binding of biological samples to solid supports for scanning probe microscopy in buffer solution. *Biophys. J.* **65,** 2437–2446.

32. Reviakine, I. and Brisson, A. (2000) Formation of supported phospholipid bilayers from unilamellar vesicles investigated by atomic force microscopy. *Langmuir* **16,** 1806–1815.

33. Allen, M. J., Hud, N. V., Balooch, M., Tench, R. J., Siekhaus, W. J., and Balhorn, R. (1992) Tip-radius-induced artifacts in AFM images of protamine-complexed DNA fibers. *Ultramicroscopy* **42-44(Pt B),** 1095–1100.

34. Willemsen, O. H., Snel, M. M., van der Werf, K. O., et al. (1998) Simultaneous height and adhesion imaging of antibody-antigen interactions by atomic force microscopy. *Biophys. J.* **75(5),** 2220–2228.

35. Xu, S., Bevis, B., and Arnsdorf, M. F. (2001) The assembly of amyloidogenic yeast sup35 as assessed by scanning (atomic) force microscopy: an analogy to linear colloidal aggregation? *Biophys. J.* **81,** 446–454.

36. Markiewicz, P. (1988) Orientational Dependency of Atomic Force Microscopic Images Revealed by Alzheimer Paired Helical Filaments. In: Doctoral Thesis, Department of Chemistry, University of Toronto.

37. Muller, D. J. and Engel, A. (1997) The height of biomolecules measured with the atomic force microscope depends on electrostatic interactions. *Biophys. J.* **73,** 1633–1644.

9

Fourier Transform Infrared and Circular Dichroism Spectroscopies for Amyloid Studies

Miguel Calero and María Gasset

Summary

Amyloids, found as extracellular protein deposits in a diverse group of human and animal disorders, are characterized by a basic scaffold consisting of cross β-sheet structure. Both far-UV circular dichroism and Fourier transform infrared spectroscopy (FTIR) spectroscopies are the most commonly used techniques for determining the secondary structure of proteins and peptides that either have not been or cannot be studied by nuclear magnetic resonance or X-ray crystallography. Both techniques are complementary and preferentially used depending on the physical state of the analyte and the major secondary structure element. Although there are special setups for working with films, circular dichroism is best suited for diluted solutions of polypeptides exhibiting α-helix as major structural element. On the other hand, FTIR works best with concentrated solutions, solids, and films and resolves with accuracy the β-sheet composition. Both spectroscopies need a small amount of protein for analysis, are non-destructive and can monitor very accurately relative changes owing to the influence of environment of the sample, though display interferences with some widely used chemicals. Within the amyloid field, conjunction of both spectroscopies has provided the first filter step for amyloid detection and has contributed to decipher the structural aspects of the amyloid formation mechanism.

Key Words: Amyloids; circular dichroism; Fourier transform infrared spectroscopy; secondary structure determination; β-sheet content.

1. Introduction

Amyloids are a class of ordered non-crystalline protein aggregates that were initially defined in terms of their staining properties *(1,2)*. Such staining properties consisted of anisotropic binding of Congo red and later, that of thioflavin T *(2–5)*. Anisotropy in probe binding relies on a common basic structure, the intermolecular cross β-sheet, that concurs with a parallel gain of insolubility. Presence of the cross β-sheet motive is singularly asserted by the detection of the 4 to 5 Å equatorial and 10 to 12 Å axial reflections by X-ray fiber diffraction

From: *Methods in Molecular Biology, vol. 299: Amyloid Proteins: Methods and Protocols*
Edited by: E. M. Sigurdsson © Humana Press Inc., Totowa, NJ

that correspond to the inter-chain distance and to the face-to-face separation of β-sheets, respectively *(6,7)*.

To better understand the structure of amyloid-forming polypeptides (short peptide fragments or proteins), the mechanisms by which polypeptides assemble into these specific polymers would be invaluable. Amyloids are generally the final state of an either nucleated- or nucleated-conformational polymerization process that macroscopically correlates with the formation of insoluble fibrillar structures *(8–10)*. Such a mechanism indicates the existence of (i) a critical concentration above which polymerization occurs, (ii) a size polydispersity on the polymerization pathway, and (iii) insolubility. The appearance of the β-sheets and their assembly into cross-scaffolds can be synchronic with the nuclei formation or with its polymerization *(8–12)*. The presence or enrichment in β-sheet structure does not necessarily correlate with the acquisition of amyloid properties, but can also reflect the collapse of polypeptide chains into amorphous or ribbon-like aggregates *(13–15)*.

A number of spectroscopic techniques exist for assessing protein structure (**Table 1**). The choice of the appropriate technique for determining the overall structure or conformation of a protein depends on several factors, which include structural information required, protein availability, solubility trend, concentration requirements, access to instrumentation and expertise. A brief guideline for deciding on the most adequate technique was recently given by Colón *(16)*. The presence and relative abundance of β-sheet structure in peptides and proteins is usually assessed by both circular dichroism (CD) and Fourier transformed-infrared (FTIR) spectroscopies.

Circular dichroism spectroscopy is a form of light absorption spectroscopy that measures the difference in absorbance of right- and left-circularly polarized light (rather than the commonly used absorbance of isotropic light) as a function of the wavelength of a substance. Far-UV CD spectroscopy (180–240 nm) provides the most convenient method to monitor protein conformation in solution. The same chromophores responsible for the absorption bands in proteins (peptidic bond, aromatic amino acids, and disulfide bridges) also give rise to the CD signals. These signals depend on the environment of the peptide bond (which absorbs below 230 nm) and are also determined by the secondary structure of the protein. Knowing the base spectra of the α-helix, β-sheet, β-turns, and the disordered conformation, the secondary structure content of a protein can be estimated by a mathematical component analysis of the CD spectrum. It should be kept in mind that aromatic side chains of Trp, Tyr and Phe can also be active in this spectral region generating the so-called non-amide chiral contributions *(17)*. In the cases of amyloid-forming peptides, the strength of this spectroscopic technique relies upon the common characteristic mechanism of conformational transition. Hence, although different amyloidogenic proteins

Table 1
Spectroscopic Methods for Probing Protein Structure

Method	Structure	Concentration (mM)	Advantages	Disadvantages
UV absorption	Tertiary	0.01–0.1	Inexpensive and easy	Very low resolution
CD	Secondary (far UV)	0.01–0.1	Easy, ideal method for studying the secondary structure and the effect of solvent composition on secondary and tertiary structure	Low resolution, interference by aromatic residues, expensive
	Tertiary (near UV)	0.05–0.5		
Fluorescence	Tertiary	0.0001–0.01	High sensitivity, relatively inexpensive and easy to use, information on compactness and dynamics	Limited to the study of semi-buried aromatic residues
IR absorption	Secondary	0.5–2	Relatively inexpensive and easy, applicable to solid samples	Water interference, high protein concentration
Raman	Secondary and Tertiary	>0.5	Minimal interference by water, applicable to solid samples	Slow, not generally available
NMR	Secondary and Tertiary	1–10	Highest resolution, provides information on protein dynamics	Expensive, difficult, high protein concentration limited to small proteins

Adapted and modified from **ref. *16***.

lack any significant primary sequence identity, the formation of their aggregates consistently involves an increase in β-sheet content, which is associated with fibrillar morphology, relative insolubility, and protease resistance. It has been shown that CD spectra ideally between 240 and approx 180 nm can be analyzed for the different secondary structural types achieving accuracies of 0.97 for α-helices, 0.75 for β-sheet, 0.50 for turns, and 0.89 for other structure types *(18–20)*.

FTIR spectroscopy is an absorption spectroscopy in which the transitions detected are those arising from vibration modes of bonds involving heteroatoms *(21,22)*. Among the different bands of the spectrum, that occurring in the 1700 to 1600 cm^{-1} range (known as amide I band) is the region of choice for analysis. Within this region, about 80% of its intensity arises from the stretching of the amide carbonyl group, whose vibration frequency depends on its interaction with hydrogen bonds *(23)*. Infrared (IR) spectroscopy is one of the most versatile techniques available, and methodologies exist to acquire spectra from proteins in any physical state, including crystals, powder, thin film, and aqueous solutions, as well as membrane-bound proteins. The major practical problem in FTIR arises from the universal solvent H_2O, which is not only active but the dominant signal in IR. To overcome this limitation different strategies have been developed: (i) mathematic algorithms for water signal subtraction, (ii) replacement of H_2O with D_2O and the consideration of isotopical exchange effects, and (iii) manipulation of the analyte as a thin or hydrated film or solid. Similarly to CD, the analysis of the amide I band (or I' after HxD isotopic exchange), assumes that the experimental spectrum is the result of the convolution of the spectral signatures of pure secondary structure elements with a noise or broadening function *(24,25)*. The conjunction of the normal mode analysis and the resolution enhancement methods have allowed identification of the vibration signatures of different secondary structure elements *(25–31)*. Interestingly, the spectroscopical components of the β-sheet structure appear as very well-resolved elementary bands at the highest and lowest frequency ranges of the 1700 to 1600 cm^{-1} *(25,31)*. In general, amyloid staining properties concur with the appearance in the spectrum of a doublet with maxima at 1682 and 1625 cm^{-1}, which indicate the presence of β-sheet structures of intermolecular nature *(11,13,15)*.

2. Materials

1. Amyloid source (*see* **Notes 1** and **2**).
2. Non-interfering buffers for CD and FTIR measurements.
3. Polarimetric certified quartz cuvets for CD.
4. CaF$_2$ or BaF$_2$ windows, Teflon spacers, and universal transmission cell holder, for FTIR trasmission.

5. Ge, ZnSe, or diamond crystals and ATR setup (mirrors for beam focusing and crystal holder) for ATR-FTIR.
6. Spectropolarimeter instrumentation for CD analysis. The main suppliers of spectropolarimeters are JASCO, AVIV, APL Ltd., OLIS, and Jovin Yvon. The APL instrument is specifically designed for stopped-flow solution kinetic CD measurements. Normally, the instruments are composed of a whole optics design and piezo-mechanic modulator, and are equipped with an air cooled 150W Xenon lamp, a temperature controller (*see* **Note 3**), and an oxygen-free nitrogen supply (*see* **Note 4**).
7. FTIR instrumentation. The main suppliers of FTIR spectrometers are Bruker, Thermo Nicolet, Perkin-Elmer, Hitachi, and Jasco, among others. Conventionally, they are provided with DTGS, MCT (requires liquid N_2 cooling), or both detectors (*see* **Note 5**), and require to be continually purged with high-quality dry air (*see* **Note 6**).
8. Software for CD data analysis. The software for CD data analysis can be used online or downloaded from different web pages such as:
   http://www2.umdnj.edu/cdrwjweb/
   http://www.cryst.bbk.ac.uk/cdweb/html/home.html
   http://www-structure.llnl.gov/cd/cdtutorial.htm
   http://www.med.unc.edu/wrkunits/2depts/biochem/MACINFAC/cd.html
9. Software for FTIR data analysis. The software for spectral analysis may be supplied with the equipment, acquired from GALACTIC (Grams versions) or obtained from specialists in the field.

3. Methods

Spectroscopical analysis of amyloid formation by both FTIR and CD focus the interest on the determination of the presence of β-sheet elements, which have specific spectral signatures. For an accurate attribution, there are several precautions that must be taken into account and will be detailed.

3.1. Analysis of Amyloids by Circular Dichroism in the Far-UV Region (190–250 nm)

3.1.1. Sample Preparation

3.1.1.1. QUARTZ CUVETS

Conventional quartz cuvets for CD measurements are those having a 0.1-cm optical path and exposing to the beam a spherical surface. These cuvets, when sealed, offer the lowest birefringence and the best quality spectra, but have limitations in titrations, parallel absorption measurements, and recovery. To overcome them, the rectangular quartz cuvets with 0.1-cm optical path and Teflon lid are recommended. For far-UV CD, cells with a width of 0.1 to 2 mm are used and the choice of width will be determined by the need to obtain an optimum total absorbance of the sample at the desired protein concentration.

Larger cells (5–10 mm) can be used if a low protein concentration is needed to avoid aggregation. The type of cell used for measurements depends on a number of factors, mainly: protein concentration, amount of protein available, buffer used, and requirements for temperature control. Quartz cuvets are available from different manufacturers such as Hellma GmbH & Co KG (www.hellma-worldwide.de) and Spectrocell (www.spectrocell.com).

For filling the short path cuvets, it may be necessary to use a syringe and needle to slowly inject the sample into the cell. For cleaning, wash the cells thoroughly with detergent (e.g., Hellmanex from Hellma), rinse them with 95% ethanol, and finally with water, blowing the cells completely dry each time.

3.1.1.2. Buffer Selection

The choice of buffer used for CD measurements is critical. Generally speaking the buffer used for CD measurement should be as "optically transparent" as possible in the far-UV region to avoid masking of the protein CD signals. Potassium phosphate (10 mM) buffers best with a wide pH range but can only be used within wavelengths of 250 to 195 nm. The advantage of borate (20 mM), Tris (25 mM), sodium cacodylate (20 mM), potassium acetate (20 mM) and Hepes (25 mM) is that these buffers can be used down to 190 nm. To achieve the desired pH, prepare the proper ratios of the mono- and dibasic forms of potassium phosphate. Do not use HCl to adjust pH as chloride ions will interfere with CD in the lower UV. If a counter ion is needed, do not use NaCl. Use Na$_2$SO$_4$, KF (*see* **Note 7**) or NaF as counter ions. To accurately reach short wavelengths (below 200 nm) needed for secondary structure estimation, it is recommended to avoid or keep at minimum the concentration of additives (*see* **Note 8**) in the protein solution. Similarly, protein unfolding experiments using urea or guanidine-HCl cannot be followed by CD below 205 nm, due to the strong absorbance of these reagents.

3.1.1.3. Protein/Peptide Solutions

If possible, protein and peptide solutions at about 1 to 5 mg/mL should be equilibrated in the working buffer by dialysis. This will ensure the elimination of scavengers and other chemicals from synthetic approaches (*see* **Note 9**), salt removal in the case of large-scale protein purification protocols, and the existence of the desired pH. As any additional protein or peptide will contribute to the CD signal, the protein to be studied should be as pure as possible. Given these factors, it follows that the protein solution should only contain those chemicals required to maintain protein stability/solubility at the lowest concentrations possible. In order to improve the signal-to-noise ratio, unfolded proteins and particulate matter should be removed from the solution by filtering through a 0.2- or 0.45-µm filter or by centrifugation.

3.1.1.4. Protein/Peptide Concentration and Cuvets Path Length

Getting the exact concentration of the protein solution is a primary requirement for CD analysis, especially when quantitative information about secondary structure is needed (*see* **Subheading 3.1.3.**). Moreover, by knowing the correct concentration of the protein, one can optimize the signal-to-noise ratio and avoid artifacts owing to improper measurement conditions. Protein concentration should be accurate determined by UV-absorption readings (if chromophores are present), or preferably by quantitative amino acid analysis before and after spectra acquisition.

If the absorbance of the sample rises above 1.0 at any point of the spectra, accurate CD measurements are simply not possible. As reference for getting the sample into the right range, it should be considered that the starting concentration (mg/mL) needed for a 10 mm cell is approx 0.015 mg/mL. Multiply this value by 10 if a 1 mm cell is used or by 100 for 0.1 mm cells. From these values, the concentration may have to be increased or decreased by a factor of 2 to optimize signal-to-noise ratio *(32)*. It is worthwhile to run first an UV absorption scan over the full range at which CD measurements will be done to determine if solvent-plus-sample absorbance will be a problem. For far-UV CD measurements on proteins, shorter pathlength (0.1 mm or 0.05 mm) cells decrease solvent absorbance significantly and permit scanning down to shorter wavelengths. These cells will require higher protein concentrations (~1 mg/mL). It should be kept in mind that amyloidogenic proteins are prone to aggregate at high concentrations. Even in the absence of turbidity, soluble aggregates may cause light scattering and spectrum distortion, and therefore aggregation should be ruled out by performing the experiments at different concentrations of by using size exclusion chromatography before analysis.

3.1.2. Parameters for CD Spectra Acquisition

Spectra acquisition should be performed according to the manufacturer's instructions, paying special attention to the calibration state of the instrument as well as other parameters that directly affect the spectrum by altering the signal-to-noise ratio and have to be modified by the user before measurements. A basic understanding of these parameters will aid in the selection of the right conditions for obtaining a high-quality spectrum.

3.1.2.1. Instrument Wavelength and Magnitude Calibrations

Spectropolarimeters must be calibrated on a regular basis using a standard sample such as (+)-10-camphorsulfonic acid (CSA). For this purpose, a 1 mg/mL solution (*see* **Note 10**) should be prepared in water, which should have an absorbance of 0.149 at 285 nm ($\varepsilon = 34.5\ M^{-1}cm^{-1}$) in a 1-cm cell, and run a CD scan of the sample in a 1-mm cell between 180 and 320 nm. The CD of CSA displays a maximum at 290.5 nm with a $\Delta\varepsilon$ of ~2.36 $M^{-1}cm^{-1}$ and a minimum at 192.5 nm of $\Delta\varepsilon$ ~ –4.9 to –4.7 $M^{-1}cm^{-1}$. The ratio $\Delta\varepsilon$ 192.5/$\Delta\varepsilon$ 290.5 should always be greater than 2 *(33,34)*.

Spectral range of measurement: The ideal working wavelength for far-UV CD analysis is in the 260 to 190 nm range, but often this has to be restricted to 260 to 200 nm owing to interference from buffer components.

Spectral bandwidth (or slit bandwidth): Slit bandwidth should be as large as possible for optimal signal-to-noise ratio, however too large a slit bandwidth will distort the spectrum.

Data resolution: How often a data point is collected throughout the spectrum is indicated. Resolutions from 0.2 to 1 nm are a good starting point for general measurements.

Instrument response time (or time constant): The time over which the instrument averages a data point is indicated. Signal-to-noise ratio increases with the square root of the time constant and should be kept as large as possible.

Scan speed (speed of measurement): The number of nanometers scanned per unit of time (nm/min) is usually indicated. This parameter is largely determined by the time constant and the data resolution.

Sensitivity: Sensitivity of the instrument needs to be selected according to the intensity of the signal from the sample.

Accumulations: Obtaining a good quality spectrum usually requires more than 1 hour, to avoid spectral drift it is better to run faster spectra and take an average value. The accumulation refers to the number of spectra averaged.

3.1.2.2. BASELINE CORRECTION

The baseline is usually not straight in a CD experiment. Therefore, it is important to perform a baseline correction by collecting the CD spectrum of the solvent and subtracting it from that of the sample.

Detailed below is a typical experimental setup for far-UV CD analysis, that can be used as a guideline for choosing the appropriate conditions and parameters:

Protein concentration: 0.15 mg/mL
Buffer: 10 m*M* potassium phosphate, pH 7.4
Cell path: 1 mm
Temperature: 10–25°C
Range: 260–190 nm
Bandwidth: 1 nm
Resolution: 1 nm
Time constant: 1 s
Scan speed: 50 nm/min
Accumulations:10
Sensitivity: 50 mdegrees

3.1.3. Spectra Analysis

After the CD data have been acquired and baseline subtraction performed, the data can be analyzed (*see* **Note 11**). All analytical methods pursuing the determination of the secondary structure composition of a polypeptide from its far-UV CD spectrum (260–190 nm) consider it as a linear combination of the spectra of the pure secondary structure elements and a noise function (*19,35*). In general, the selection of the method of analysis depends on the aim of the study as follows:

- For the determination of the protein conformation in solution: SELCON, CDNN or K2D.
- For determination of polypeptide conformation having a suitable polypeptide set of references: LINCOMB.
- For the determination of relative changes (mutations, ligands, perturbants): LINCOMB.
- For evaluating the number of folding states giving rise to a set of spectra: the CCA algorithm or SVD.

It is very important to determine accurately the protein concentration since all the methods described for analysis rely on the comparison to reference structures on a residue molar basis. Therefore, the analyst must know precisely the concentration, molecular weight, and number of residues of the protein analyzed. Incorrect assessment of the protein concentration will lead to erroneous calculation of the percentage of secondary conformers (*see* **Note 12**).

Regarding the specific field of amyloids, the presence or appearance of β-sheet elements can be assessed by these analytical methods, but they have several inherent limitations and offer not very high accuracy (0.97 for helices, 0.75 for beta-sheet, 0.50 for turns, and 0.89 for other structure types [18–20]). These methods, however, are very reliable for monitoring changes in the conformation of proteins under different conditions such as denaturation studies, unfolding experiments, helix induction by TFA, etc (*see* **Note 13**).

3.1.4. Basic Protocol for CD Analysis

A basic protocol for the acquisition and processing of a CD spectrum of a protein solution at constant temperature follows:

1. Prepare the sample by choosing adequate protein concentration, buffer and cell path (*see* **Subheading 3.1.2.**).
2. Allow the system to be purged with nitrogen for at least 5 min before switching on the lamp. Leave the lamp on for 15 min before measurement.
3. Set the acquisition parameters (*see* **Subheading 3.1.2.** for guidance).
4. Acquire buffer spectrum and check for interferences in the far-UV region.
5. Acquire sample spectrum.
6. Correct baseline by calculating sample-buffer difference spectrum.

7. Transform CD data (millidegrees) in mean residue ellipticity (degrees·cm^2·dmol^{-1}) (*see* **Note 11**).
8. Use the appropriate software package for secondary structure estimation (*see* **Subheading 3.1.3.**).

3.2. Analysis of Amyloids by FTIR

3.2.1 Sample Preparation

There are two main modes for FTIR spectra acquisition: plain transmission and attenuated total reflection (ATR). Each mode requires a specific setup and its usage depends on specific aims.

3.2.1.1. OPTICAL WINDOWS FOR TRANSMISSION

The most widespread method for acquiring FTIR spectra of proteins in solution is by standard transmission method, where the sample is placed between two optical windows separated by a defined path length formed by a spacer of known thickness. These windows must be transparent in the region of interest, inert, and insoluble in the experimental conditions. For protein studies, the best material is calcium fluorite or barium fluorite depending upon the pH (*see* **Note 14**). The choice of the correct type of window will minimize spectral artifacts and damage to the windows.

3.2.1.2. CRYSTALS FOR ATR

FTIR spectra of proteins may also be obtained by ATR, using infrared transparent crystal such as Ge, ZnSe, or diamond crystals (*see* **Note 15**). This technique is based on single or multiple reflections of the infrared light at the crystal–sample interface (*see* **Note 16**) and it is applicable to protein or peptide samples in solid state, in solution or as thin or hydrated films.

3.2.1.3. BUFFER SELECTION

The most physiological method for studying proteins is in solution. Proteins may be dissolved in a range of solvents, including TFE, DMSO, H$_2$O, and D$_2$O, but it must be noted that organic solvents cause significant structural perturbations in proteins and peptides. TFE mimics membranes and induces α-helix conformation, while DMSO renders the protein completely unstructured. Other buffers containing amide (i.e., formamide) or carboxylate groups (i.e., formic and acetic acid) should be avoided because these give strong signals in the IR region. In addition and specifically for synthetic peptides, removal of TFA traces (*see* **Note 17**) from cleavage steps is highly recommended since it gives a characteristic peak at 1672 cm^{-1} that overlaps with the vibration of the β-turn structure.

Water exhibits a series of extremely intense absorption bands in the infrared region. The most important is the O-H bending absorption band, centered at 1640 cm^{-1}, which is much more intense that the amide I absorption band of proteins. A valid approach for solving this problem is the use of deuterium oxide (D_2O) as solvent for proteins to be studied by FTIR, since the O-D bending absorption band is shifted to lower wavenumbers, leaving the amide I region free from interferences. The use of D_2O offers some others advantages for the analysis of the spectra that will be discussed ahead.

3.2.1.4. PROTEIN/PEPTIDE SOLUTIONS AND PATH LENGTHS

As discussed in the previous subheading dedicated to CD analysis, any additional protein or peptide will contribute to the spectroscopic signal, and consequently the protein to be studied should be as pure as possible. Unlike CD spectroscopy, FTIR does not require an exact estimation of the protein concentration for quantitative analysis of the secondary structure elements. Moreover, the strength of this technique is that it is especially useful for the study of membrane-associated proteins, aggregates, or proteins not in solution, where it is difficult or impossible to estimate the protein concentration.

For transmission setups, protein and peptide solutions at about 1 to 5 mg/mL should be dissolved in the working buffer, ideally D_2O, free from interfering substances. Under these conditions, paths of 25 to 50 μm are used. However, if H_2O is used as solvent, the concentration should be increased to 20 to 50 mg/mL and the path should be reduced to 5 to 10 μm to minimize interferences in the amide I region.

Alternatively, for transmission and ATR setups, the sample can be applied between the windows, onto the crystal as a solid or as a thin or hydrated protein film. These methods are applicable to insoluble samples such as fibrils or aggregate and have the advantage of diminishing problems owing to the interference of buffers; however they may offer lower sensitivity.

As a general requirement, transmission and ATR FTIR modes call for stock solutions as concentrated as possible (about 1 to 2 mg/mL is the lower limit) and have similar buffer restrictions. The samples can be prepared in various ways as described in the following. These three sample preparation methods assume deposition of a small aliquot of the sample (approx 10 μL) onto a surface (*see* **Note 18**) that for transmission FTIR can be either a calcium or barium fluorite window. For ATR-FTIR, a Ge, ZnSe, or diamond crystal is appropriate.

1. Dissolve peptide in HFIP or TFE solution, then dry under N_2 current an aliquot corresponding to about 50 to 100 μg, hydrate it in D_2O-buffer (*see* **Note 19**), and subsequently seal the sample in a universal holder.
2. Dry an aliquot of a protein/peptide solution in H_2O-buffer, hydrate it in equal D_2O volume, and then seal the sample in the holder.

3. Equilibrate by dialysis the protein/peptide solution in D_2O buffer and then seal as previous.

3.2.2. Spectral Acquisition

Generally, FTIR spectrometers are supplied with several default parameter sets for each type of substance and experimental conditions (nominal resolution, apodization, wavenumber range, number of scans, etc.). The resulting data are plotted as an interferogram or absorption spectra (*see* **Note 20**).

Before obtaining the spectra of the sample, it is necessary to acquire the background interferogram or spectrum using the same parameters that will be used for the samples. Background interferogram is measured without sample in the chamber, and is stored automatically by the spectrometer. Subsequently, the transmittance or absorbance spectra of samples are obtained by rationing their interferograms with the background to compensate for the energy peaks from the lamp, and the CO_2 and H_2O bands.

For samples prepared in H_2O buffers, the spectrum of the sample will be dominated by the very strong absorption bands of the solvent at 3400 cm^{-1} (H-O stretching), at 2125 cm^{-1} (water association band), and at 1640 cm^{-1} (H-O-H bending), masking the region of interest. To subtract the water spectra (in this case, the buffer spectrum), two automatic routines have been developed, which consist either of the zeroing of the 2125 cm^{-1} band (*37*) or the flattening of the 1900 to 1740 cm^{-1} region (*38*).

For samples prepared in D_2O-buffers the most significant problem is the subtraction of water vapor that appears as spikes in the amide I band region and that will affect the mathematical analysis if it is not properly eliminated. Compensation for the interference of the water vapor (sharp bands in the 1800 to 1200 cm^{-1} region) requires the acquisition of a water vapor spectrum collected without sample at incomplete purging conditions.

3.2.3. Spectral Analysis

The peptidic bond gives up to nine characteristic bands named amide A, B, I–VII. The amide A and B bands, at about 3500 and 3100 cm^{-1} respectively, originate from the Fermi resonance between the first overtone of the amide II and the N-H stretching vibration. The amide I (1700–1600 cm^{-1}) is 80% made up by the amide C=O stretching, whereas about 50% of the amide II (1600–1500 cm^{-1}) results from the N-H bending, and about 30% from the C-N stretching vibrations. The amide I and II (I' and II', after HxD exchange) are the major bands for backbone conformation analysis as the frequency of bands is strongly dependent upon the strength of any hydrogen bond formed. Amide III and IV bands contain a complex mixture of coordinate displacements, whereas amide V, VI, and VII bands contain out-of-plane motions (*23,31,39*).

The methods for the analysis of the amide I (I') band in terms of secondary structure composition are based on a first step of Fourier self-deconvolution or resolution enhancement, a second curve fitting step and a final band assignation process.

3.2.3.1. Fourier Self-Deconvolution

The concept of Fourier self-deconvolution is based on the assumption that a spectrum is composed of several single bands (each narrow band is characteristic of a secondary structure), which are broadened in the liquid or solid state. Therefore, the bands overlap and cannot be distinguished in the amide envelope. In the pioneer work by Susi and Byler (1986) *(40)*, the amide I (I') was deconvoluted with a Lorentzian line shape function and a resolution enhancement factor of 2.4 was applied. Similar protocols are generally used to generate resolution-enhanced spectra to allow the identification of the signals from the various secondary structures present in the protein by curve fitting.

3.2.3.2. Curve Fitting

The deconvoluted spectrum is then fitted to Lorentzian/Gaussian band shapes by an iterative curve fitting procedure. For such a task, the input parameter set should contain: number of bands, their frequency position, band widths, band narrowing factor, and the absorption intensity. Two alternative strategies have been employed that differ in the choice of input parameters. One of them is used for fitting all bands detected after resolution enhancement. The other one utilizes the intensities at given frequencies and full width at half height values obtained as average from model proteins (*see* **Note 21**). As a final step, the outcome of the fit is then used to fit the experimental band resulting in band frequency and area pair of values.

3.2.3.3. Band Assignment

Table 2 summarizes the frequency of the band maximum observed experimentally and predicted by theoretical mode analysis of the different secondary structure elements, both for the H-forms and for the D-forms. The proportion of each secondary structure element is calculated as its area percentage referenced to the sum of the band areas. It should be kept in mind that even though the results of the secondary structure composition are in good agreement with the X-ray crystallographic structures, FTIR is still a relative approach and works better for establishing changes compared to a reference system (ligand binding, perturbation, etc.).

- Some precautions in the band assignment process should be taken:
- Some amino acid side chains are active in the amide I (I') region (**Table 3**). In an average polypeptide sequence, their contribution can be considered negligible, but

Table 2
Assignment of the Different Elementary Components
of the Amide I Band Obtained After Fourier Self-Deconvolution
and Curve Fitting to Secondary Structure Elements

Secondary structure	H-form wavenumber range (cm^{-1})	D-form wavenumber range (cm^{-1})
β-sheet	1695–1674	1689–1682
Turns		1682–1662
α-helix none exchangeable loops	1657–1648	1662–1645
Unordered exchangeable loops	1662–1645	1645–1637
β-sheet extended structures	1641–1623	1637–1627
Low frequency β-sheet	1627–1613	1627–1613

in short peptides or for certain repetitive sequences, such as glutamine repeats, the contribution of the side chain can be significant *(29)*. In these types of samples, the band assignment process can be misleading.

- Registration of the IR spectra of the polypeptide sample both in H_2O and in D_2O (or in H- and in D-hydration forms) is needed for more accurate determination of unordered regions (fast HxD exchange) and helical structure (very slow exchange rate) to get insight into the degree of solvent exposure. Inspection of HxD exchange and extent requires the consideration of the 1600 to 1400 cm^{-1} region and can be used as an index of changes in tertiary structure *(31)*.

3.2.4. Basic Protocol for FTIR Analysis

A basic protocol for the acquisition and processing of a FTIR spectrum from a protein or peptide sample is as follows:

1. Prepare the sample by choosing appropriate protein concentration, buffer and mode (transmission or ATR) (*see* **Subheading 3.2.1.**).
2. Allow the system to be purged with dry air (*see* **Note 22**) for at least 15 min (*see* **Note 23**).
3. Set the acquisition parameters as follows: nominal resolution: 0.5 cm^{-1}, wavenumber range: 4000–1000 cm^{-1}, number of scans: 256.
4. Acquire background spectrum (*see* **Note 24**).
5. Acquire spectra of (i) sample (*see* **Note 25**), (ii) buffer, and (iii) water vapor and convert to absorbance mode (*see* **Subheading 3.2.2.**).
6. Calculate sample-buffer difference spectrum (*see* **Note 26**).
7. Correct the sample-buffer difference spectrum by subtracting the water vapor contribution (*see* **Note 27** and **Subheading 3.2.2.**).

Table 3
Amino Acid Side Chain Vibrations

Amino acid	Vibrating group	Wavenumber (cm^{-1})	A_0 ($M^{-1} \cdot$cm^{-1})
Asp	$-COO- / -COOH$	1574 / 1716	380 / 280
Glu	$-COO- / -COOH$	1560 / 1712	470 / 220
Arg	$-CN_3H_5^+$	1673 / 1633	420 / 300
Lys	$-NH3^+$	1629 / 1526	130 / 100
Asn	$-C = O / -NH_2$	1678 / 1622	310 / 160
Gln	$-C = O / -NH_2$	1670 / 1610	360 / 220
Tyr	ring-OH / ring-O$^-$	1518 / 1602, 1498	430 / 160, 700
His	ring	1596	70
Phe	ring	1494	80
C-terminal	$-COO- / -COOH$	1598 / 1740	240 / 170

The contribution of the side chain vibrations in the 1800–1400 cm^{-1} region has been thoroughly investigated by Venyaminov and Kalin *(29)* and probed to be essential in cases in which the protein/peptide under study is highly enriched in any of amino acids with IR active side chains. It should be noted that vibration properties of side chains are susceptible to the environment, and thereby to the interaction with ligands.

8. For secondary structure estimation proceed to self-deconvolution of the spectrum, curve fitting and band assignment (*see* **Subheading 3.2.3.**) using the available software.

3.3. Example

Secondary structure analysis of wild-type and variant cystatin C by Circular Dichroism and Attenuated Total Reflection-Fourier Transformed Infrared Spectroscopy (Adapted and modified from reference 41).

Figure 1A shows the far-UV CD spectra of wild-type and variant cystatin C proteins. Analysis of the data indicates that both proteins contain a mixture of secondary structure motifs with a clear predominance of β-sheets. The percentage of secondary structures of wild-type cystatin C calculated by the Estima algorithm are 11.5% α-helix, 43.5% β-sheet, 8.5% β-turns, and 36.5% random coil. However, the secondary structure of the variant cystatin C differs from that of the wild-type protein. The CD spectrum of the variant cystatin C indicated a lower content of α-helix and higher content of random coil: 1% α-helix, 46% β-sheet, 7% β-turns, and 46% random coil. As depicted in **Fig. 1B**, similar results were obtained by ATR-FTIR spectroscopy. Those samples were subjected to deuteration for 1 h to allow the discrimination between the relative contributions of random coil and α-helix conformers. Self-deconvolution and

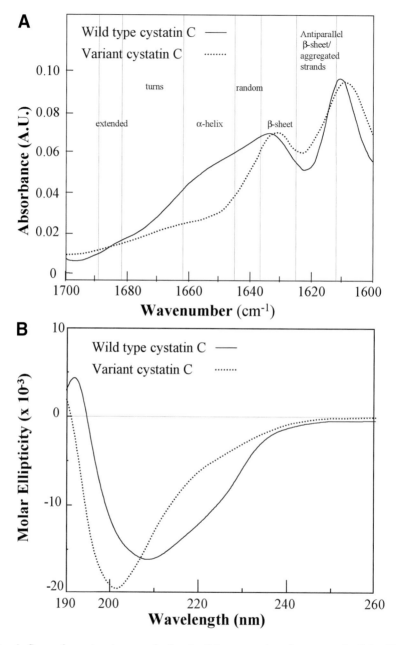

Fig. 1. Secondary structure analysis of wild-type and variant cystatin C by CD and ATR-FTIR. (**A**) The circular dichroic spectra of wild-type (solid line) and variant cystatin C (dotted line) were recorded from 190 to 260 nm at 24°C with a Jasco J-720 spectropolarimeter at a protein concentration of 0.15 mg/mL. The percentages of the different secondary structure motifs based on the CD data were calculated by the CD Estima

curve fitting of the spectra in the amide I' region showed maxima centered at wavenumbers characteristic of β-turns, α-helix, random, β-structure, and antiparallel β-sheet/aggregated strands. The two major differences between both proteins observed in the CD experiments were also detected by FTIR, namely a lower α-helix content and more unordered structures for the variant cystatin C compared to the wild-type protein. The high contribution to the spectra of the band at 1614 cm^{-1} (more prominent in the variant protein spectra), characteristic of antiparallel β-sheet structures and/or aggregated strands, likely reflects the tendency of cystatin C to polymerize and aggregate.

This study, together with insights on the 3D structure by intrinsic fluorescence studies and molecular modeling, indicates that the secondary structure of the variant cystatin C is different from that of the wild-type protein, and differences in the 3D structure may account for the tendency of the variant protein to form fibrils. These data suggest that the L68Q substitution affects cystatin C in a similar fashion to that caused by amino acid replacements in other amyloid forming proteins, such as the E22Q variant of Aβ deposited in cerebral vasculature of patients with HCHWA, Dutch type, immunoglobulin light-chain associated with light-chain amyloidosis, transthyretin variants found in familial amyloid polyneuropathy, and the amyloid formation in lysozyme amyloidosis.

4. Conclusions

CD and FTIR spectroscopy are the two most common techniques used for fast secondary structure analysis of proteins and peptides, and its relative changes induced by environmental perturbations. Being complementary, the choice between them depends on several factors, including equipment availability, expertise, protein solubility, and concentration requirements. When possible, both CD and FTIR, as well as other spectroscopic techniques should be used in conjunction with structural analysis of amyloid precursors and fibrils in order to achieve more informative and reliable results. These techniques allow the researcher to quantitate the amyloid load, explore the mechanism and kinetics of

---

Fig. 1. (*Continued*) algorithm (42). (**B**) ATR-FTIR spectra of wild-type (solid line) and variant cystatin-C (dotted line) were recorded at nominal resolution of 0.5 cm^{-1} and 256 scans accumulations on a BioRad FTS6000 infrared spectrophotometer equipped with a Golden Gate diamond ATR cell (Specac, Smyrna, GA). Fourier self-deconvolution of the spectra in the amide I region were performed using a Bessel apodization function with a resolution enhancement factor of k = 2 and a peak half-width of 12 cm^{-1}, in the Win-IR Pro system (BioRad). Assignment of the different components of amide I after Fourier self-deconvolution to secondary structure was performed as described *(23,31,39)*.

fibril formation, and to study the binding of ligands. These methods are also powerful tools to explore new strategies for therapy and diagnosis of amyloid diseases.

As an example of the application of CD and FTIR, we present a comparative study of the structure of wild-type cystatin C and a mutant variant (L68Q), which is found deposited in the brain vessel walls of patients suffering from an autosomal dominant form of cerebral amyloid angiopathy (hereditary cerebral hemorrhage with amyloidosis, Icelandic type [HCHWA-I]). This example has been adapted from *(41)*.

5. Notes

1. Structural studies by CD or FTIR require peptides or proteins of the highest purity. When isolated from natural sources (tissue, cell culture, etc.) special precautions should be taken to avoid contamination from other peptides or proteins that will give rise to CD and FTIR signals. Although purity is not usually a major concern for studies with synthetic peptides, residual organic contaminants (e.g., TFA) may alter the properties of the peptide or interfere with the spectroscopic analysis (*see also* **Notes 9** and **17**).

2. For CD analysis, the sample must always be in solution. Samples for FTIR analysis can be in solution, in suspension, in solid state (fibrils, aggregates, lyophilized powder, etc.), or as a thin or hydrated film.

3. The CD spectrum of proteins is temperature-dependent, and therefore it is necessary to measure the CD at constant temperature.

4. In order to prevent damage to the CD optics, caused by the ozone generated during the measurement, it is necessary to purge the instrument by flushing continuously with nitrogen at a flow rate of 15 to 20 L/min. Not taking this simple precaution will lead to a deficient performance of the instrument below 200 nm. The use of nitrogen that is 99.999% pure is recommended.

5. DTGS is a pyroelectric detector that offers a constant sensitivity in a wide infrared region. MCT is a semiconductor detector that operates a 77°K (liquid nitrogen temperature) and offers higher sensitivity than DTGS. MCT detector has become the standard detector used in FTIR for handling weak light.

6. In contrast to CD spectropolarimeters, where it is essential to completely remove the oxygen from the instrumentation, FTIR spectrometers require high quality dry air or dry oxygen free nitrogen to avoid interference from water vapor.

7. For salt enrichment KF and NaF are recommended for pH above 5.0. Fluoride salts should never be used below pH 5.0, as HF traces will damage the quartz cuvets.

8. Citrate, EDTA, or imidazole should not be used for CD analysis below 210 nm. Other buffers or additives have to be tested before they are used for sample preparation. If the protein is dissolved in unsuitable buffer, it should be dialyzed against 10 mM potassium phosphate or other appropriate buffer.

9. Note that synthetic peptides can carry traces of TFA that are detectable by FTIR (but not by CD) and appear as a neat band at 1672 cm^{-1}. For efficient removal of

TFA, several steps of lyophilization followed by HCl washes are required (*see also* **Note 17**).

10. CSA solution for CD calibration should be prepared freshly by weighing a small amount (~1 mg) of the product and dissolving it in 1 mL of distilled water. The exact concentration of the solution should be assessed by measuring the absorbance in spectrophotometer using a 1 cm cell. Under these conditions, a 1 mg/mL solution has an absorbance of 0.149 at 285 nm (ε_{285nm}= 34.5 M^{-1} cm^{-1}). CSA can be purchased as an ammonium salt from different manufactures such as Sigma-Aldrich (no. A1142, FW: 249.33).

11. CD is measured as *observed ellipticity* (θ), whose units are *millidegrees*. For secondary structure assessment, the observed ellipticity must be transformed into a quantity called *mean residue ellipticity* ($[\theta]_{MRE}$), whose units are *degrees·cm²·dmol⁻¹*, that is referenced to the molar concentration of peptide bonds (or amino acid units). Mean residue ellipticity can be calculated using the formula $[\Theta]_{MRE} = \theta/(10\ N\ c\ l)$, where θ is the observed ellipticity, N is the number of amino acids of the protein, c is the molar concentration of the protein, and l is the cell path in cm. All commercial spectropolarimeters perform this conversion on request, provided concentration, molecular weight (or number of residues) and cell path.

12. Recently, Goormaghtigh and collaborators have devised a mathematical method to estimate secondary structure motifs from the CD data, which is relatively independent of concentration *(36)*. However, at present this method has not been widely tested.

13. Shifts in conformation involving as few as 10 amino acids may be readily detectable by CD spectroscopy and conformer analysis.

14. Barium fluorite is soluble in strong acids and care should be exercised if studies are performed at low pH.

15. ZnSe is a good general purpose material, although it has a limited use with strong acids and alkalis, as well as with complexing agents such as ammonia or EDTA. Ge has a higher refractive index and is used for samples that produce strong absorption bands. Diamond crystals are very resistant and suited for a wide range of samples including acids, bases and oxidizing agents, although these crystals have an intrinsic absorption from 2300 to 1800 cm^{-1}.

16. Although ATR is a powerful tool, it must be stressed that protein adsorption onto the crystal may represent a problem if it occurs with an unexpected structural change. In addition, the penetration of the light into the sample is wavelength dependent, and may cause artifacts owing to differential absorption between the high and low wavenumber sides. Most FTIR spectrophotometers perform an ATR conversion to correct these artifacts. This option is only applicable for spectra measured by the ATR method; when applied to transmission mode it will give meaningless results.

17. To remove TFA, cycles of lyophilization and resuspension in 10 mM HCl give satisfactory results. If performed in this manner, the pH of the final solution should be acidic and therefore a high concentrated buffer or an additional step for buffer exchange is required (*see* **Note 9**).

18. Samples should be applied with care to avoid bubble formation that will interfere with FTIR signal.
19. Note that pD = pH + 0.4.
20. Infrared spectra are typically represented as absorption spectra. In contrast to UV absorption spectra, in IR spectroscopy, the absorbance is traditionally plotted against the wavenumber, which is expressed in cm^{-1} (1 cm^{-1} = 0.03 THz). Provided we are working with a FTIR absorption spectrum, there is no need for conversion of units prior to data analysis.
21. For curve fitting of the amide I region bands, a half-width of 13 to 15 cm^{-1} and a band narrowing factor of 1.5 to 1.8 are convenient as initial deconvolution parameters.
22. For purging, the FTIR spectrometer good quality dry air or oxygen-free nitrogen should be used.
23. Correct purging of the sample compartment may be assessed either by leaving the system on purge for an extended period of time, or by continuously acquiring and monitoring the spectra until no sharp negative bands are observed in the region 1800 to 1200 cm^{-1}.
24. Water vapor and carbon dioxide in air, as well as dust on mirrors of the spectrometer can generate intense infrared absorption bands. Purging the system with good quality dry air and the acquisition of a background spectrum are required to compensate efficiently for these spectral interferences.
25. Acquiring the spectrum of the sample involves opening the sample compartment. Therefore it is important to allow the re-purging of the compartment for 15 min before acquiring the spectrum.
26. The subtraction scaling factor may be obtained interactively by zeroing the 1850 to 1740 cm^{-1} slopes arising mainly from residual water or deuterium oxide.
27. Zoom in on the 1800 to 1600 cm^{-1} region and estimate the subtraction factor interactively by zeroing the water vapor peaks at 1772 and 1717 cm^{-1}.

References

1. Virchow, R. (1854) Zur cellulosefrage. *Virchows Arch. Pathol. Anat. Physiol.* **6,** 416–426.
2. Glenner, G. G. and Page, D. L. (1976) Amyloids, amyloidosis and amyloidogenesis. *Int. Rev. Exp. Pathol.* **15,** 2–92.
3. Vassar, P. S. and Culling, C. F. (1959) Fluorescent stains, with special reference to amyloid and connective tissues. *Arch. Pathol.* **68,** 487–498.
4. Kelenyi, G. (1967) Thioflavin S fluorescent and Congo red anisotropic stainings in the histologic demonstration of amyloid. *Acta Neuropathol. (Berl.)* **7,** 336–348.
5. Vassar, P. S. and Culling, C. F. (1962) Fluorescent amyloid staining of casts in myeloma nephrosis. *Arch. Pathol.* **73,** 59–63.
6. Inouye, H., Fraser, P. E., and Kirschner, D. A. (1993) Structure of beta-crystallite assemblies formed by Alzheimer beta-amyloid protein analogues: analysis by x-ray diffraction. *Biophys J.* **64,** 502–519.

7. Sunde, M., Serpell, L. C., Bartlam, M., Fraser, P. E., Pepys, M. B., and Blake, C. C. (1997) Common core structure of amyloid fibrils by synchrotron X-ray diffraction. *J. Mol. Biol.* **273,** 729–739.

8. Jarrett, J. T. and Lansbury, P. T. (1993) Seeding "one-dimensional crystallization" of amyloid: a pathogenic mechanism in Alzheimer's disease and scrapie? *Cell* **73,** 1055–1058.

9. Janek, K., Behlke, J., Zipper, J., et al. (1999) Water-soluble beta-sheet models which self-assemble into fibrillar structures. *Biochemistry* **38,** 8246–8252.

10. Fandrich, M., Forge, V., Buder, K., Kittler, M., Dobson, C. M., and Diekmann, S. (2003) Myoglobin forms amyloid fibrils by association of unfolded polypeptide segments. *Proc. Natl. Acad. Sci. USA* **100,** 15463–15468.

11. Fraser, P. E., Nguyen, J. T., Surewicz, W. K., and Kirschner, D. A. (1991) pH-dependent structural transitions of Alzheimer amyloid peptides. *Biophys. J.* **60,** 1190–1201.

12. Munishkina, L. A., Phelan, C., Uversky, V. N., and Fink, A. L. (2003) Conformational behavior and aggregation of alpha-synuclein in organic solvents: modeling the effects of membranes. *Biochemistry* **42,** 2720–2730.

13. Gasset, M., Baldwin, M. A., Lloyd, D. H., et al. (1992) Predicted alpha-helical regions of the prion protein when synthesized as peptides form amyloid. *Proc. Natl. Acad. Sci. USA* **89,** 10940–10944.

14. Wille, H., Zhang, G. F., Baldwin, M. A., Cohen, F. E., and Prusiner, S. B. (1996) Separation of scrapie prion infectivity from PrP amyloid polymers. *J. Mol. Biol.* **259,** 608–621.

15. Lopez De La Paz, M., Goldie, K., Zurdo, J., et al. (2002) De novo designed peptide-based amyloid fibrils. *Proc. Natl. Acad. Sci. USA* **99,** 16052–16057.

16. Colon, W. (1999) Analysis of protein structure by solution optical spectroscopy. *Meth. Enzymol.* **309,** 605–632.

17. Woody, R. W. (1972) The circular dichroism of aromatic polypeptides: theoretical studies of poly-L-phenylalanine and some para-substituted derivatives. *Biopolymers* **11,** 1149–1171.

18. Woody, R. W. (1995) Circular dichroism. *Meth. Enzymol.* **246,** 34–71.

19. Johnson, W. C. (1990) Protein secondary structure and circular dichroism: a practical guide. *Proteins* **7,** 205–214.

20. Manavalan, P. and Johnson, W. C. Jr. (1987) Variable selection method improves the prediction of protein secondary structure from circular dichroism spectra. *Anal. Biochem.* **167,** 76–85.

21. Smith, B. C. (ed.) (1995) *Fundamentals of Fourier Transform Infrared Spectroscopy.* CRC Press, Boca Raton, Florida.

22. Stuart, B. (ed.) (1997) *Biological applications of infrared spectroscopy.* David J. Ando Ed., ACOL series, John Wiley & Sons, New York.

23. Goormaghtigh, E., Cabiaux, V., and Ruysschaert, J. M. (1994) Determination of soluble and membrane protein structure by Fourier Transform Infrared Spectroscopy: I. Assignments and model compounds. *Subcell. Biochem.* **23,** 329–362.

24. Kauppinen, J. K., Moffatt, D. J., Mantsch, H. H., and Cameron, D. G. (1986) Fourier self-deconvolution: a method for resolving intrinsically overlapped bands. *Appl. Spectrosc.* **35,** 271–276.

25. Byler, D. M. and Susi, H. (1986) Examination of the secondary structure of proteins by deconvolved FTIR spectra. *Biopolymers* **25,** 469–487.

26. Yang, W. J., Griffiths, P. R., Byler, D. M., and Susi, H. (1985) Protein conformation by infrared spectroscopy: resolution enhancement by fourier self deconvolution. *Appl. Spectrosc.* **39,** 282–287.

27. Krimm, S. and Bandekar, J. (1986) Vibrational spectroscopy and conformation of peptides, polypeptides, and proteins. *Adv. Protein Chem.* **38,** 181–364.

28. Dousseau, F. and Pezolet, M. (1990) Determination of the secondary structure content of proteins in aqueous solutions from their amide I and amide II infrared bands. Comparison between classical and partial least- squares methods. *Biochemistry* **29,** 8771–8779.

29. Venyaminov, S. Y. and Kalnin, N. N. (1990) Quantitative IR spectrophotometry of peptide compounds in water (H$_2$O) solutions. I. Spectral parameters of amino acid residue absorption bands. *Biopolymers* **30,** 1243–1257.

30. Sarver, R. W. and Krueger, W. C. (1991) Protein secondary structure from fourier transform infrared spectroscopy: a data base analysis. *Anal. Biochem.* **194,** 89–100.

31. Goormaghtigh, E., Cabiaux, V., and Ruysschaert, J. M. (1994) Determination of soluble and membrane protein structure by Fourier Transform Infrared Spectroscopy: III. Secondary structures. *Subcell. Biochem.* **23,** 405–450.

32. Copeland, R. A. (ed.) (1994) *Methods for protein analysis: a practical guide to laboratory protocols.* Chapman and Hall, New York, NY.

33. Tuzimura, K., Cono, T., Meguro, H., Hatano, M., and Murakami, T. (1977) A critical study of the measurement and calibration of circular dichroism. *Anal. Biochem.* **81,** 167–174.

34. Hennessey, J. P. Jr. and Johnson, W. C. Jr. (1982) Experimental errors and their effect on analyzing circular dichroism spectra of proteins. *Anal. Biochem.* **125,** 177–188.

35. Greenfield, N. and Fasman, G. D. (1996) Computed circular dichroism spectra for the evaluation of protein conformation. *Biochemistry* **8,** 4108–4116.

36. Raussens, V., Ruysschaert, J. M., and Goormaghtigh, E. (1993) Protein concentration is not an absolute prerequisite for the determination of secondary structure from circular dichroism spectra: a new scaling method. *Anal. Biochem.* **319,** 114–121.

37. Dousseau, F., Therrien, M., and Pezolet, M. (1989) On the spectral subtraction of water from the FTIR spectra of aqueous solutions of proteins. *Appl. Spectrosc.* **43,** 538–542.

38. Powell, J. R., Beals, J. M., and Castellino, F. J. (1986) Secondary structure predictions of human plasminogen and the bovine prothrombin kringle loops. *Arch. Biochem. Biophys.* **248,** 390–400.

39. Goormaghtigh, E., Cabiaux, V., and Ruysschaert, J. M. (1994) Determination of soluble and membrane protein structure by Fourier Transform Infrared Spectroscopy: II. Experimental aspects, side chain structure and H/D exchange. *Subcell. Biochem.* **23,** 363–403.
40. Susi, H. and Byler, D. M. (1986) Resolution-enhanced Fourier transform infrared spectroscopy of enzymes. *Methods Enzymol.* **30,** 290–311.
41. Calero, M., Pawlik, M., Soto, C., et al. (2001) Distinct properties of wild-type and the amyloidogenic human cystatin C variant of hereditary cerebral hemorrhage with amyloidosis, Icelandic type. *J. Neurochem.* **77,** 628–637.
42. Yang, J. T., Wu, C. S., and Martinez, H. M. (1986) Calculation of protein conformation from circular dichroism. *Methods Enzymol.* **130,** 208–269.

10

Quasielastic Light Scattering for Protein Assembly Studies

Aleksey Lomakin, David B. Teplow, and George B. Benedek

Summary

Quasielastic light scattering (QLS) spectroscopy is an optical method for the determination of diffusion coefficients of particles in solution. In this chapter, we discuss the principles and practice of QLS with respect to protein assembly reactions. Particles undergoing Brownian motion produce fluctuations in scattered light intensity. We describe how the temporal correlation function of these fluctuations can be measured and how this correlation function provides information about the distribution of diffusion coefficients of the particles in solution. We discuss the intricacies of deconvolution of the correlation function and the assumptions incorporated into data analysis procedures. We explain how the Stokes–Einstein relationship can be used to convert distributions of diffusion coefficients into distributions of particle size. Noninvasive observation of the temporal evolution of particles sizes provides a powerful tool for studying protein aggregation and self-assembly. We use examples from studies of Aβ fibrillogenesis to illustrate QLS application for understanding the molecular mechanisms of the nucleation and growth of amyloid fibrils.

Key Words: Dynamic light scattering; diffusion; size distribution; self-assembly; aggregation; Alzheimer's disease, amyloid; fibrillogenesis.

1. Introduction

Quasielastic light scattering (QLS), also known as dynamic light scattering (DLS), is an optical method well-suited for the determination of diffusion coefficients of particles undergoing Brownian motion in solution *(1,2)*. Diffusion coefficients are determined by particle size, shape, and flexibility, as well as by inter-particle interactions. All these parameters provide important information about the kinetics and structural transitions within systems of particles in solution and can be studied by QLS. The QLS method is rapid, sensitive, non-invasive, and quantitative. QLS instruments may be constructed relatively easily and are also available commercially.

From: *Methods in Molecular Biology, vol. 299: Amyloid Proteins: Methods and Protocols*
Edited by: E. M. Sigurdsson © Humana Press Inc., Totowa, NJ

QLS is a useful tool for studying particle aggregation and, in particular, for monitoring protein assembly. The processes of protein assembly are ubiquitous. They are required to produce multi-subunit structures, including enzymes, enzyme complexes, ribosomes, ion channels, and viral capsids. Protein assembly reactions also are associated with a number of diseases, including Alzheimer's *(3)*, prion (4), and other neurodegenerative diseases *(5)*, sickle cell anemia *(6)*, and cataract disease *(7)*. In each case, proteins that exist normally in a soluble, disaggregated state assemble into oligomeric and polymeric structures which cause cell and tissue injury, disrupting normal organismal function. To study the factors controlling both normal and pathological protein assembly processes, a method is required to monitor these processes with high sensitivity and resolution. The subject of this chapter, QLS, is such a method *(8)*.

In its essence, QLS simply measures the fluctuations in light intensity emanating from a sample irradiated by a laser. These fluctuations contain information about the physical nature of the particles in the sample solution. However, for this information to be interpreted accurately, the theoretical and practical underpinnings of the method must be understood and incorporated into the execution of the QLS experiment. To assist the reader in this endeavor, this chapter is organized in a tripartite manner. The introduction to **Subheading 2.** provides a general overview of the entire QLS method. The remainder of **Subheading 2.** reviews general principles of light scattering, including the critical mathematical formulations upon which the method is based. **Subheading 3.** discusses the QLS hardware, practical aspects of its use, and how the raw data are analyzed to produce a physical picture of the particles in the solution under study. Accompanying these sections are illustrative examples of QLS in practice.

2. QLS Theory

Coherent light generated by a laser is an electromagnetic wave. Particles irradiated by this wave produce secondary waves, i.e., scattered light. This scattered light has a number of characteristics, which make its analysis of experimental interest. The amplitude of the scattered wave depends on scatterer mass and refractive index. The phase of the scattered wave depends on the position of the scattering particle. The intensity of the scattered light is the square of the sum of amplitudes of all scattering waves with their phases taken into account (*see* **Subheading 2.1.**). As the scattering particles move, the phases of their scattered electromagnetic waves change, causing fluctuations in the intensity of light registered by a photodetector. The essence of the QLS technique is the measurement of the temporal correlations in the fluctuations in the scattered light intensity (*see* **Subheading 2.2.**), and from analysis of these data, the reconstruction the dynamics of scatterer motion.

The QLS method probes dynamics of the scattering system on a time scale ranging from tens of nanoseconds to seconds. The dynamics, which is of a particular interest for studying protein aggregation, is the diffusive motion of the particles in solution. The measurement of the intensity correlation function allows evaluation of the diffusion coefficients of the scattering particles (*see* **Subheading 2.3.**). In studies of protein aggregation, the sample usually contains particles of different sizes. In these polydisperse systems, "regularization" procedures provide the means to reconstruct smooth distributions of the scattering intensity over the scatterers' diffusion coefficient (*see* **Subheading 3.3.**).

The diffusion coefficient depends on particle size and shape, as well as on the ambient temperature and solution viscosity. By converting a diffusion coefficient into a hydrodynamic radius (the radius of a hard sphere with the same diffusion coefficient as the scatterer), temperature and viscosity are factored out (*see* **Subheading 3.4.1.**). The intensity of scattering by an individual particle is proportional to its mass squared (*see* **Subheading 2.1.2.**). This makes the QLS method particularly sensitive to large scatterers. If the relationship between particle mass and size is known, the distribution of the relative concentrations of particles over their size can be reconstructed (*see* **Subheading 3.4.3.**).

A QLS instrument consists of a laser with beam delivery optics, a cuvet containing the sample, a photodetector with light-collecting optics, and a correlator with data analysis software (*see* **Subheading 3.1.**). The degree to which the user has control over instrument settings varies from system to system. The more expensive systems usually allow more user control and require more expertise. Here, we focus on key general features of QLS method, features whose proper use and optimization are critical for the successful implementation of the QLS method.

2.1. General Principles of Light Scattering

2.1.1. Scattering Vector

Relative to the phase of a wave scattered at the origin, the phase of a wave scattered at a point with radius vector $\mathbf{r}$ is $\mathbf{q} \cdot \mathbf{r}$ (**Fig. 1**). The vector $\mathbf{q}$ is called the "scattering vector" and is a fundamental characteristic of any scattering process. Its length is $q \equiv |\mathbf{q}| = 4\pi / \lambda \sin \theta / 2$, where θ is the scattering angle and λ is the light wavelength in the scattering medium, $\lambda = \lambda_0 / n$. Here, n is the refractive index of the medium and λ_0 is the wavelength of the incident light in vacuum. The summary electromagnetic field resulting from scattering by many particles is

$$E = \sum_k \mathbf{E}_k \exp(i\mathbf{q} \cdot \mathbf{r}_k + i\varphi), \qquad (1)$$

where $\mathbf{E}_k$ is the amplitude of the wave scattered by the k-th particle located at position $\mathbf{r}_k$, and φ is the common phase shift equal to the phase of the putative

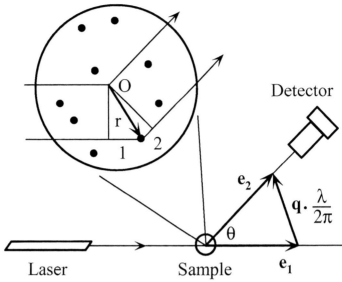

Fig. 1. The scattering vector **q**. The path traveled by a wave scattered at the point with radius vector **r** differs from the path passing through the reference point O by two segments, 1 and 2, with lengths l_1 and l_2, respectively. The phase difference is $\Delta\phi = 2\pi$ $(l_1 + l_2) / \lambda$. The segment l_1 is a projection of **r** on a unit vector $\mathbf{e}_1$ in the direction of the incident beam, i.e., $l_1 = \mathbf{r} \cdot \mathbf{e}_1$. Similarly, $l_2 = \mathbf{r} \cdot \mathbf{e}_2$, where $\mathbf{e}_2$ is a unit vector in the direction of scattering. Thus $\Delta\phi = (2\pi / \lambda)(\mathbf{e}_1 - \mathbf{e}_2) \cdot \mathbf{r}$. The vector $\mathbf{q} = (2\pi / \lambda)(\mathbf{e}_1 - \mathbf{e}_2)$ is called the scattering vector.

wave scattered at the origin, **r** = 0. The intensity I of the scattered light per unit area is proportional to the square of the amplitude of the electromagnetic field, that is

$$I \propto |\mathbf{E}|^2 = \left| \sum_k \mathbf{E}_k \exp(i\mathbf{q} \cdot \mathbf{r}_k) \right|^2 \tag{2}$$

2.1.2. Intensity of Scattering by a Small Particle

Consider an aggregate composed of m monomers, each producing scattered electromagnetic wave $\mathbf{E}_0$. Let the aggregate center of mass be at point **R**. If the size of the aggregate is small, so that $\mathbf{q} \cdot (\mathbf{r}_k - \mathbf{R}) \ll 1$ for all monomers, Eq. 1 reduces to $\mathbf{E} \approx m\mathbf{E}_0 \exp(i\mathbf{q} \cdot \mathbf{R} + i\phi)$. Thus the intensity of the light scattered by the aggregate is proportional to the aggregation number squared, $I = m^2 I_0$, where I_0 is the intensity of scattering by a monomer. The quadratic dependency of scattering intensity on the mass of the scatterer is the basis for optical determination of the molecular weight of macromolecules (*9*), for various turbidimetry and nephelometry techniques, and for understanding many natural phenomena, from appearance of clouds to cataractogenesis in the eye lens (*10*).

2.1.3. Intensity of Scattering by a Large Particle

If an aggregate particle is not small, the destructive interference of waves scattered from different points in the particle must be taken into account. This destructive interference reduces the intensity of light scattering by a factor of $|\alpha|^2$, where $\alpha(\mathbf{q})$ is an averaged value of the phase factors $\exp(i\mathbf{q} \cdot \mathbf{r}_k)$ or all monomers (*see* **Note 1**). The factor $|\alpha|^2$ should be averaged over all possible orientations of the particle. The result of this averaging yields the structure factor. A table of expressions for the structure factors for particles of various shapes can be found elsewhere *(11)*.

2.1.4. Intensity of Light Scattered by Many Particles

In a typical experiment, light scattered by N individual molecules at random locations $\mathbf{r}_k$ within the scattering volume is detected. The size of the scattering volume generally is large, thus $\mathbf{q} \cdot \mathbf{r}_k \gg 1$ and the phase factors $\exp(i\mathbf{q} \cdot \mathbf{r}_k)$ in Eq. 1 vary randomly. As a result, the mean square amplitude of the scattered wave is proportional to $\sqrt{N}$, and the average intensity of the scattered light is simply N times the intensity scattered by an individual particle, as expected. The local intensity, however, fluctuates from one point to another around its average value, as described by Eq. 2. The pattern of these fluctuations in light intensity, called an interference pattern or "speckles," is determined by the positions of the scattering particles. The characteristic size of the speckles is called coherence length.

2.2. Correlation Function

2.2.1. Correlation Function of Intensity Fluctuations

In QLS, the photodetector registers the fluctuations of the intensity of the scattered light, which are in fact random. Information is contained only in the temporal correlations in this signal. The correlation function of the intensity $I(t)$ is defined as:

$$G^{(2)}(\tau) = < I(t)\, I(t+\tau) >. \tag{3}$$

In the above formula, the angular brackets denote an average over time t. This time averaging, an inherent feature of the QLS method, is necessary to extract information from the random fluctuations in the intensity of the scattered light.

2.2.2. Connection to Field Correlation Function

The notation $G^{(2)}(\tau)$ had been introduced to distinguish the correlation function of the intensity $I(t)$ from the correlation function of the electromagnetic field

$$G^{(1)}(\tau) = < \mathbf{E}(t)\, \mathbf{E}^*(t+\tau) >, \tag{4}$$

which is the Fourier transform of the light spectrum. In the majority of practical applications of QLS, the scattered light is a sum of waves scattered by many independent particles and therefore displays Gaussian statistics. This being the case, there is a relation between the intensity correlation function $G^{(2)}(\tau)$ and the field correlation function $G^{(1)}(\tau)$:

$$G^{(2)}(\tau) = I_0^2 (1 + \gamma g^{(1)}(\tau)^2). \qquad (5)$$

Here I_0 is the average intensity of the detected light and γ is the efficiency factor. When scattered light is collected from a small area, $\gamma \approx 1$. If light is collected from area large compared to the coherence area, fluctuations in light intensity are averaged out and $\gamma \ll 1$. The key element in Eq. 5 is $g^{(1)}(\tau) = |G^{(1)}(\tau)/G^{(1)}(0)|$, the absolute value of the normalized field correlation function. This is the instrument independent function that is measured and analyzed in the QLS method.

2.3. Dynamics of the Scattering System

2.3.1. Brownian Motion

The dynamics which is of particular interest for studying protein aggregation is Brownian motion of the particles. Brownian motion is responsible for the diffusion of the solute and is quantitatively characterized by the diffusion coefficient D. According to Eq. 1, electromagnetic waves scattered by a pair of individual particles have, at the observation point, a phase difference of $\mathbf{q} \cdot \Delta\mathbf{r}$, where $\Delta\mathbf{r}$ is the vector distance between particles. As the scattering particles move over distance $\Delta x \approx q^{-1}$, the phases for all pairs of particles change significantly and the intensity of the scattered light becomes completely independent of its initial value. Thus the correlation time of the intensity fluctuations, τ_c, is the time required for a Brownian particle to move a distance q^{-1}. The laws of diffusive motion stipulate that the mean square displacement of a Brownian particle over time Δt is characterized by the relationship $\Delta x^2 = 2D\Delta t$. Thus for $\Delta x \approx q^{-1}$, $\tau_c \approx 1/Dq^2$. Rigorous mathematical analysis of the process of light scattering by noninteracting small Brownian particles leads to the following expression for the normalized field correlation function in Eq. 5:

$$g^{(1)}(\tau) = \exp(-Dq^2\tau). \qquad (6)$$

2.3.2. Polydispersity

Equation 6 represents the simplest single exponential form of the correlation function for a solution of small, isotropic, non-interacting identical particles. In polydisperse systems, i.e., where particles of different shape or size are present in the solution, Eq. 6 has to be generalized as follows:

$$g^{(1)}(\tau) = \sum_i I_i \exp(-D_i q^2\tau). \qquad (7)$$

Here, I_i is the normalized ($\Sigma\ I_i = 1$) contribution to scattering from particles with diffusion coefficients D_i. Reconstruction of the distribution of scattering particles over their diffusion coefficients from experimentally measured correlation function Eq. 7 is discussed in **Subheading 3.3.**

2.3.3. Other Types of Dynamic Behavior

Orientational and conformational dynamics of large (comparable to wavelength of light) particles as well as inter-particle interactions all lead to complex, non-single-exponential correlation functions, even for solutions of identical particles *(12,13)*. These effects are usually insignificant for scattering by particles small compared to the length of the inverse scattering vector q^{-1}, but become important, and often overwhelming, for larger particles. In those cases, QLS probes not the pure diffusive Brownian motion of the scatterers, but also other types of dynamic fluctuation in the solution. Fortunately, the relaxation times of these other types of fluctuations rarely depend on the scattering vector as Dq^2, which is characteristic for the diffusion process. Thus, in principle, measurement of the correlation function at several different angles of scattering, and therefore at several different q, allows polydispersity to be distinguished from multimodal relaxation of a non-diffusive nature.

3. The QLS System

A QLS system consists of five elements: a source of light (laser), a sample, light collecting optics, a correlator, and the data analysis software (**Fig. 2**). To obtain satisfactory results, all five elements must meet certain criteria. The laser should operate in single mode and be stable. The correlator should work efficiently for sample times significantly shorter than the correlation times characteristic of the molecules under investigation. The optical setup should collect at least one photon per correlation time per coherence area. These requirements determine whether a particular system can be studied by QLS at all. If instrument performance is suitable for the study, the quality and reliability of the results will be determined by the other two elements of the QLS system, the sample and the data analysis.

QLS instruments are available commercially. Among the suppliers of QLS systems are ALV (Germany), Brookhaven Instruments (USA), Malvern Instruments (UK), and Precision Detectors (USA). We use a custom-built optical system with an Innova 90+ Argon Ion (wavelengths 488 nm and 514 nm, power up to 0.5 W in single mode, single frequency regime) or a He-Ne (wavelength 633 nm power 50 mW) laser as the light source. Both lasers are from Coherent (USA). The photodetector is a Hamamatsu (Japan) R4220P PTM (photomultiplier) coupled with a 144-channel Langley-Ford (Amherst, MA) correlator and custom developed deconvolution software. Alternatively, a He-Ne laser and

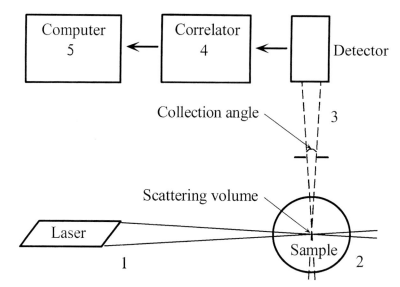

Fig 2. A block diagram of a QLS instrument. The source of light (**1**) consists of a laser and the focusing optics that delivers incident light into the sample. This part of the QLS system determines wavelength and intensity of the incident beam as well as the geometry of the illuminated volume. The sample cuvet (**2**) is the source of stray scattering both at the entry and at the exit points of the incident beam. A larger cuvet is better, but requires more sample. Collecting optics and the photodetector (**3**) register the intensity of the light scattered in a certain direction. Important characteristics of the collecting optics are collection volume and solid angle from which the light is collected. The intersection of the illuminated volume and the collection volume defines the scattering volume. Light intensity data obtained by the photodetector are continually fed to the correlator (**4**) that computes the correlation function of the intensity fluctuations. The correlation function is then analyzed (**5**) to determine the size and the distribution of the scattering particles.

APT (avalanche photo diode) built in the Precision Detectors 256 channel correlator is used.

3.1. Hardware

3.1.1. Laser

Intensity. The key to quality measurements in QLS is to have sufficient intensity of light scattering. If a count rate of 3 to 5 photocounts per correlation time can be achieved when light is collected from within one coherence area (*see* **Subheading 3.1.2.**), the laser intensity is sufficient.

Single mode. The laser must operate in the single mode regime, i.e., it should generate a single transverse mode (called TEM_{00}). Different transverse modes have different cross-sectional intensity profiles and very close frequencies. The mode TEM_{00} produces a beam with a nearly Gaussian intensity profile. If the laser generates several transverse modes simultaneously, the beam will have an irregular intensity profile. More importantly, optical "beating" between the modes would be registered by the photodetector. This may result in a distortion of the correlation function.

Stability. The fluctuations in laser intensity make the factor I_0^2 in Eq. 5 dependent on delay time τ. From Eq. 5, it is clear that for a low efficiency factor γ, even small fluctuations in laser intensity can have significant effects on the determination of $g^{(1)}(\tau)$.

Polarization. At a given scattering angle, maximum scattering occurs in a plane perpendicular to the incident light polarization. Therefore, the incident beam should be polarized perpendicular to the plane formed by light source, sample and photodetector.

Overheating. High laser intensity may lead to increased temperature in the scattering volume. Because the diffusion coefficient is temperature-dependent, heating within the scattering volume can cause errors. Significant heating is more likely in concentrated and highly absorbing samples. To check this possibility, it is a good idea to do measurements at several intensities of the incident beam, especially when a precise absolute measurement of scattering particles size is desired.

Focusing. Generally, the narrower the beam diameter, the better. Compared to a poorly focused incident beam, a well-focused beam produces the same scattering intensity from a smaller scattering volume. This increases the coherence angle, which is an important parameter affecting the efficiency factor γ in Eq. 5 (*see* **Subheading 3.1.2.**)

3.1.2. Light Collection

Scattering volume. Collecting optics ensures that light reaching the photodetector originates from (passes through) only a small "collection volume" within the sample. This minimizes stray light entering the photodetector. The intersection of the collection volume and the volume illuminated by the incident beam is termed the "scattering volume." The scattering volume does not have sharp boundaries. It is a qualitative characterization of the region within the sample where the majority of the single scattering events registered by the photodetector occur.

Particle number fluctuations. The number of particles in the scattering volume should be large. If this is not the case, intensity fluctuations associated with fluctuations in the number of particles within the scattering volume will contribute significantly to the measured correlation function. This contribution is dependent on the geometry of the scattering and is difficult to account for quantitatively. For this reason, a narrowly focused incident beam may not be suitable for studying large, strongly scattering objects (e.g., bacteria).

Multiple scattering. Equation 6 describes the correlation function of light that is scattered once. However, photons that are scattered several times also reach the detector. In QLS, multiple scattering should be minimized since it complicates data analysis significantly. The first scattering can only occur within the illuminated volume. The last scattering before the photon reaches the photodetector occurs within the collection volume, by definition. All intermediate scatterings can occur anywhere in the sample. Thus multiple scattering is collected from a much larger volume than is single scattering, which only comes from within the scattering volume. For this reason, multiple scattering may be a problem even in a sample with turbidity barely visible to a naked eye. To minimize effects of multiple scattering, the collection volume, the illuminated volume, and the total sample volume should be made as small as possible.

Coherence angle. Scattered light is optically collected from the scattering volume within a particular solid angle. The larger this angle, the more intensity is registered. However, if this angle becomes larger than the coherence angle, the efficiency γ in Eq. 5 drops. The coherence angle is the angular size of the speckles in the interference pattern. The fluctuations in light intensity collected from different speckles are statistically independent. Increasing the light-collecting aperture in a QLS spectrometer beyond the coherence angle does not lead to improvement in the signal-to-noise ratio because the temporal fluctuations in the intensity are averaged out. The coherence angle is approximately $\lambda / 2\pi L$, where L is the cross-sectional size of the scattered volume perpendicular to the direction of the scattering. The coherence angle is fairly small, about 1° for a 10-μm wide scattering volume.

Efficiency. Parameter γ in Eq. 5 is 1 when the scattered light is collected within the coherence angle. Otherwise the efficiency is the ratio of the coherence solid angle to the collection solid angle. If the collection angle is large compared the coherence angle, its further increase does not improve the signal-to-noise ratio in the correlation function but only magnifies the effects of instability of the incident beam. A collection angle of more then 2 to 3 coherence angles is useful only when the intensity of the scattering light falls below the intensity of stray light (*see* **Note 2**). Efficiency can be determined experimentally by examining the correlation function of a sample (*see* **Fig. 3**).

Angular dependency. Some commercial instruments allow measurements at different scattering angles. The scattering angle determines the scattering vector **q** (*see* **Fig. 1**). Equation 6 specifies how the correlation function for pure

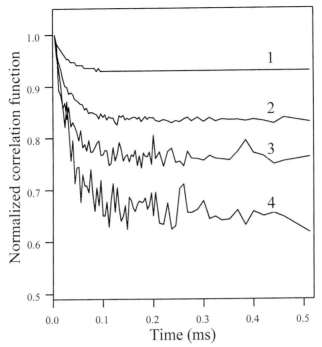

Fig 3. Correlation functions of detergent micelles (0.1% Triton X-100) measured with different collection angles. Curves 1 to 4 represent estimated collection angles of approx 10, 3, 1, and 0.5 times the coherence solid angle, respectively. Corresponding efficiency factors γ are 0.06, 0.19, 0.30, and 0.55. Efficiency does not reach 1 at small collection apertures because of the contribution of stray scattering into the baseline of the correlation function. Curve 2 represents the best choice of the collection angle for this sample. Note that these are short (3 s each) measurements made for demonstration purposes. In practice, the measurement duration of this sample would be two orders of magnitude longer and the noise in the data would be 10 times less than in the figure.

Brownian motion depends on the absolute value of this vector. Other types of dynamic behavior (*see* **Subheading 2.3.3.**) will result in deviations in this Dq^2-type dependency of the inverse relaxation time. Measurements of angular dependency thus could verify the diffusive nature of the system dynamics. Large particles, which scatter disproportionally more light, scatter even stronger at small angles (*see* **Subheadings 2.1.2.** and **2.1.3.**). Unless particles >100 nm are being investigated, small angle scattering should be avoided (*see* **Note 3**). In typical protein aggregation studies, there is little reason to do measurements at angles other then 90°. At this angle, the stray scattering from the cuvet (*see* **Note 4**) is usually at a minimum.

3.1.3. Correlator

The correlator digitally computes the correlation function of the photocurrent $G^{(2)}$ (τ). In the most common approach, the number of photons registered by the photodetector within a short time interval, the "sample time" (denoted Δt), is stored in the correlator. This number represents an instantaneous value of the scattered light intensity. The correlator keeps many, usually several thousands, of these consecutive intensity measurements in its memory. According to Eq. 3, to obtain the correlation function $G^{(2)}$ (τ) at $\tau = n\Delta t$, the average product of counts separated by n sample times should be determined. The number n is referred to as a channel number. Clearly, the shortest delay time at which the correlation function is measured by the procedure described above is Δt (channel 1). Modern commercial correlators can simultaneously process several hundred channels without loss of information at sample times as low as a fraction of a microsecond.

3.1.4. Cuvet

Where the incident beam hits the outer and inner surfaces of cuvet walls are sources of stray light that might enter the photodetector. This stray light, and the light scattered by the sample, originate from the same source and are partially coherent, producing interference patterns in addition to the interference pattern derived solely from the particles in the sample. This may result in systematic distortion of the correlation function and should be minimized. Appropriate design and good alignment of the optics, as well as careful cleaning of the cuvet, serves this goal. It is desirable to know the intensity of the background stray scattering from pure buffer and to ensure that the scattering from the sample significantly exceeds this intensity level (*see* **Note 4**).

3.2. Sample

3.2.1. Role of Large Particles

Samples monitored by QLS must be optically pure. This concept is quite different from the concept of chemical purity. Even a small weight fraction of chemically inert dust particles can completely dominate light scattering from a large population of relatively small proteins. The presence of a few large particles is often a factor in QLS experiments, especially when monitoring protein assembly. As these particles drift into and out of the scattering volume, the total scattering intensity fluctuates significantly. This can render QLS measurements unusable.

3.2.2. Cleaning the Sample

Avoiding dust. Dust in the air has a strong tendency to adsorb electrostatically to charged groups on the surfaces of the empty cuvet. These adsorbed particles will be suspended in the sample solution. If the cuvet is not to be filled completely, it is desirable to introduce the sample into the bottom of the cuvet to avoid washing dust off the cuvet walls. Do not vortex the partially filled cuvet.

Washing cuvet. Do not wash disposable cuvets or test tubes—they are usually made in a dust-free environment and filled with clean air. When washing cuvets, remember that the dust gets inside the cuvet when you empty it of washing solvent. Empty and dry cuvets in a dust-free atmosphere, e.g., in a glove box filled with filtered argon. After a cuvet is filled with sample and sealed, clean it from the outside using lens paper and lens cleaning solution or methanol.

Filtration. The easiest way to remove large impurities from solution is by filtration. Standard 0.22-µm filters generally are too porous to be of use. We have found that 20-nm Anatop filters are satisfactory in most studies of protein aggregation. When aggregates of interest are too big to pass through the 20-nm filter, they usually scatter much more than dust and do not require filtering at all.

Centrifugation is another effective way to remove large impurities from the solution, provided that the sample is spun in the same sealed cuvet in which the QLS measurements will be done. Transferring the sample into another cuvet after centrifugation defeats the purpose of the procedure. Typical airborne dust can be pelleted in 30 min at 5000g. However, there are always very "flaky" dust particles, which will not sediment by this procedure.

Flow-through filtering. Successful measurement of low molecular weight molecules is absolutely dependent on the optical purity of the sample. In studies of the amyloid β-protein (Aβ) *(14)*, we have used the intrinsic fractionation and filtering potential of size exclusion chromatography (SEC), and a continuous flow procedure for washing the QLS cuvet, to produce optically pure Aβ samples. We attached one end of a micropipet to the SEC UV detector outflow line and placed the other end at the very bottom of a cuvet (a standard disposable 5 × 50 mm, 0.5-mL, round bottom, glass test tube). In this way, "filtered" buffer is constantly washing the interior of the cuvet from the bottom up and spilling over the top. When the peak of interest is detected and fills the cuvet, the micropipet is removed from the cuvet and the cuvet is sealed and washed from the outside. This procedure, although somewhat cumbersome, provided excellent dust-free samples. We recommend it for any QLS study of peptides and small proteins.

3.2.3. Digital Filtration

In protein solutions, large protein aggregates may form that cannot be removed. When such aggregates pass through the scattering volume, they cause spikes of intensity, which distort the measured correlation function. Some correlators allow suppression of data acquisition during spikes of intensity. These algorithms are called "software dust filters" and involve establishing certain cut-off levels for the intensity of the scattered light. When software dust filtering is employed, it may be beneficial to focus the laser beam in order to minimize the scattering volume and increase the coherence angle. Dust particles or aggregates pass through a small scattering volume less frequently and in fewer numbers than through larger volumes (*see* discussion of laser focusing in **Subheading 3.1.1.**) Spikes in intensity associated with dust particles then become larger in intensity, shorter in time and less frequent. That allows for better discrimination between these spikes and the regular intensity fluctuations in the interference pattern.

3.3. Data Analysis

3.3.1. Polydispersity

The term "polydispersity" is used to describe the presence of non-identical particles in a sample. Polydispersity can be an inherent property of the sample, for instance, when polymer solutions or protein aggregation are studied, or it can be a consequence of impurities or deterioration of the sample. In the first case, the polydispersity itself is an object of interest, whereas in the second case it is an obstacle. In both instances, polydispersity significantly complicates data analysis.

For a continuous distribution of scattering particle size, Eq. 7 is generalized as follows:

$$g^{(1)}(\tau) = \frac{1}{I_0} \int I(D) \exp(-Dq^2\tau)dD. \tag{8}$$

Here $I(D)dD = N(D)I_0(D)dD$ is the intensity of light scattered by particles having their diffusion coefficient in interval $[D, D + dD]$, $N(D)dD$ is the number of these particles in the scattering volume, and $I_0(D)$ is the intensity of light scattered by each of them. The goal of the mathematical analysis of QLS data is to reconstruct as precisely as possible the distribution function $I(D)$ (or $N(D)$) from the experimentally measured correlation function.

Unfortunately, the corresponding mathematical minimization problem is "ill-posed" *(15)*, meaning that dramatically different distributions $I(D)$ lead to nearly identical correlation functions of the scattered light and therefore are equally acceptable fits to the experimental data. We discuss below three approaches for dealing with this ill-posed problem.

3.3.2. The Direct Fit Method

The simplest approach is the direct fit method. Here the functional form of $I(D)$ is assumed *a priori* (single modal, bimodal, Gaussian, etc.). The parameters of the assumed function that lead to the best fit to the experimental data are then determined. This method is only as good as the original guess of the functional form of $I(D)$. Moreover, the method can be misleading because it tends to "confirm" any *a priori* assumption made. It is also important to note that the more parameters there are in the assumed functional form of $I(D)$, the better the experimental data can be fit but the less meaningful the values of the fitting parameters become. In practice, typical QLS data allow reliable determination of about three independent parameters of the size distribution of the scattering particles.

3.3.3. The Method of Cumulants

The cumulant method is free from bias introduced by *a priori* assumptions about the shape of $I(D)$. In this approach, the focus is not on the shape of the distribution but instead on the moments of the distribution, or closely related quantities called cumulants *(16)*. Cumulants are stable characteristics insensitive to spikes in the distribution. The first cumulant (moment) of the distribution $I(D)$, the average diffusion coefficient $\overline{D}$, can be determined from the initial slope of the field correlation function. Indeed, using Eq. 8, it is straightforward to show that:

$$\frac{d}{d\tau} \ln | g^{(1)}(\tau) | = \frac{1}{I_0} \int I(D)Dq^2 dD \equiv \overline{D}q^2 \qquad (9)$$

The second cumulant (moment) of the distribution can be obtained from the curvature (second derivative) of the initial part of the correlation function. As in the direct fit method, the accuracy of the real QLS experiment allows determination of at most three moments of the distribution. The first moment, $\overline{D}$, can be determined with better than ±1% accuracy (*see* **Note 5**). The second moment, the width of the distribution, can be determined with an accuracy of ±5 to 10%. The third moment, which characterizes the asymmetry of the distribution, usually can be estimated with an accuracy of only about ±100%.

3.3.4. Regularization

The regularization approach combines the best features of both of the previous methods. It assumes that the distribution $I(D)$ is an arbitrary, but smooth function, and seeks a non-negative distribution producing the best fit to the experimental data. The regularization requirement of smoothness precludes spikes in the distribution, allowing unique solutions to the minimization problem. There are several regularization algorithms that differ in the specific mathematical implementation of the smoothness condition. One popular program is called CONTIN *(17)*. We have developed and use our own algorithm *(18)*, which is also utilized in PrecisionDeconvolve software supplied with QLS instruments produced by Precision Detectors (USA).

Smoothness parameter. The choice of the smoothness parameter is one of the most difficult and important parts of the regularization method. If the smoothing is too strong, the distribution will be very stable but will lack details. If the smoothing is too weak, false spikes can appear in the distribution. The "rule of thumb" is that the smoothing parameter should be just sufficient to provide stable, reproducible results in repetitive measurements of the same correlation function. Two facts are helpful for choosing the appropriate smoothing parameter. First, the lower the statistical errors of the measurements, the smaller the smoothing parameter can be without loss of stability. This will yield finer resolution in the reconstructed distribution $I(D)$. Second, narrow distributions generally require much less smoothing and can be reconstructed much better than can wide distributions. This is because oscillations in narrow distributions are effectively suppressed by non-negativity conditions.

Resolution of the regularization procedure. In a typical QLS experiment, regularization analysis can resolve a bimodal distribution with two narrow peaks of equal intensity if the diffusion coefficients corresponding to these peaks differ by more than a factor of approx 2.5. The moments of the distribution reconstructed by the regularization procedure with properly chosen smoothness parameter coincide closely with those obtained by the cumulant method. In **Fig. 4**, we present the bimodal distribution function of oligomers observed in a solution of Aβ *(19)* computed with three different smoothing parameters. The distribution in **Fig. 4A** was computed without smoothing. This spiky distribution is the best fit to the measured correlation function, but it is not stable and varies from one experiment to another. The same data with an appropriately chosen smoothing parameter is shown in **Fig. 4B**. This distribution fits the measured data only marginally worse (about 0.5% larger average deviation) than the distribution in **Fig. 4A**, but is stable from one experiment to another. In **Fig. 4C**, we show an excessively smoothed distribution. This distribution is stable, but lacks details, and it fits the measured data noticeably worse than do the distributions in **Fig. 4A,B** (about 3% larger average deviation). Note that the average diffusion coefficient and the overall width of the reconstructed distribution are affected little by the choice of smoothing parameter.

3.4. Interpretation of Distribution of Scattering Intensity I(D)

3.4.1. Determination of the Sizes of Particles in Solution

Measurement of the intensity correlation function allows evaluation of the diffusion coefficients of the scattering particles. For spherical particles, the relation between the radius R_h and its diffusion coefficient D is given by the Stokes-Einstein equation *(20)*:

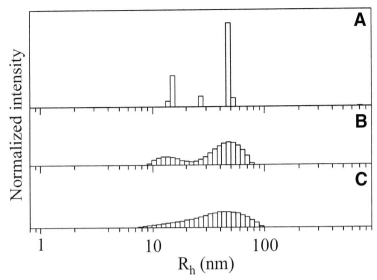

Fig 4. Oligomer size distribution of Aβ. **(A)** Distribution computed with insufficient smoothing. Peak positions in this distribution are not reliable, even though they provide the best fit to the experimental data. **(B)** Properly chosen regularization parameters allow observation of two fractions—oligomers with hydrodynamic radius in the range 10 to 20 nm and their aggregates with average radius of 60 nm. This distribution is stable and would not vary much from one measurement to another. The average deviation from the experimental data is only 0.5% worse then in **(A)**. **(C)** An excessively smoothed distribution does not show separate oligomer and aggregate populations. This distribution is stable, and it fits the experimental data well, with average deviation only 3% more than in **(A)**, but important details are not resolved.

$$D = \frac{k_B T}{6\pi\eta R_h} \tag{10}$$

Here k_B is the Boltzmann constant, T is the absolute temperature, and η is solution viscosity. For non-spherical particles, Eq. 10 defines the effective hydrodynamic radius of the particle. It is customary to introduce the apparent hydrodynamic radius R_h^{app}, defined as:

$$R_h^{app} = \frac{k_B T}{6\pi\eta D^{app}} \tag{11}$$

where D^{app} is the average (apparent) diffusion coefficient measured in the QLS experiment. The apparent hydrodynamic radius is calculated numerically, and in some cases analytically, for a variety of particles shapes *(21)* (*see* **Note 6**).

3.4.2. Effect of Concentration

At finite concentrations, two additional factors affect the diffusion of particles, viscosity and inter-particle interactions. Viscosity generally increases with the concentration of macromolecular solute. According to Eq. 10, this leads to a lower diffusion coefficient and, therefore, to an increase in the apparent hydrodynamic radius. Interactions between particles can act in either direction. If the effective interaction is repulsive, which is usually the case for soluble molecules (otherwise they would not be soluble), local fluctuations in concentration tend to dissipate faster, meaning higher apparent diffusion coefficients and lower apparent hydrodynamic radii. If the interaction is attractive, fluctuations in concentration dissipate more slowly and the apparent diffusion coefficients are lower. Thus, depending on whether the effect of repulsion between particles is strong enough to overcome the effect of increased viscosity, both increasing and decreasing types of concentration dependence of the hydrodynamic radius are observed *(22)*.

3.4.3. Determination of the Relative Concentrations of Particles

The regularization procedure reconstructs the distribution of the scattering intensity over diffusion coefficient, $I(D) = N(D)I_0(D)$, as defined by Eq. 8. Using Eq. 10, this distribution can be converted readily into a distribution over hydrodynamic radius, R_h. However, to deduce the relative concentration of particles with a particular diffusion coefficient one must know the scattering intensity from these particles, $I_0(D)$. As discussed in **Subheading 2.1.2.**, the scattering intensity is proportional to the mass of an aggregate squared. The question thus reduces to the connection between mass of an aggregate and its diffusion coefficient, or equivalently, R_h. For solid spherical particles, $M \propto R_h^3$. For discs and ideal polymer coils, $M \propto R_h^2$. For long rigid rods with diameter d the connection is $M \propto R_h \ln(R_h / d)$. With an appropriately postulated dependency $M(R_h)$, the relative molar concentration of the scattering particles is given by $N(R_h) = I(R_h) / M^2$. The relative weight concentration is $C(R_h) = MN(Rh) = I(Rh) / M$.

In **Fig. 5**, we show several alternative representations of two distributions of Aβ aggregates (data adapted from *[23]*). Panel A shows the distribution of scattering intensity over diffusion coefficient of micelle-like structures observed immediately after dissolution of of Aβ in 0.1 N HCl. Panel B shows the same type of distribution observed in this sample 30 h later, when Aβ fibrils were formed. These same data converted into distributions of scattering intensity over hydrodynamic radius are shown in panels C and D, respectively. In these panels, the fraction of average R_h ~7 nm represents micelles. In panel D, we also observe the contribution from Aβ fibrils, which have R_h within the 15 to 40 nm range in this sample. The distribution of the scattering intensity over diffusion

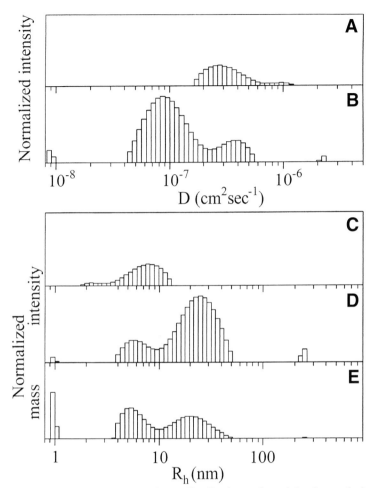

Fig. 5. Different representations of two distributions of particles in a solution of Aβ. **(A)** The distribution of scattering intensity over diffusion coefficient of micelle-like structures observed immediately after dissolution of Aβ(1-40) at a concentration of 1 mg/mL in 0.1 N HCl. **(B)** The same distribution 30 h later, when Aβ fibrils were formed. The distributions are normalized to the total intensity of scattering, which increased approx fourfold during the incubation. Therefore, the total area of the distribution in panel B is about 4 times the area of the distribution in panel A. **(C and D)** The data in **(A)** and **(B)**, respectively, converted into distributions of scattering intensity over hydrodynamic radius. **(E)** The distribution in **(D)** converted into the mass distribution, assuming that the mass of a scatterer is proportional to R_h.

coefficient is the primary information extracted directly from the correlation function. The average (apparent) diffusion coefficient D^{app} is a stable characteristic and is well-reproducible. The distribution over hydrodynamic radius is computed using Eq. 4 and in a logarithmic scale is an appropriately shifted reflection of the distribution over the diffusion coefficient. Distribution over R_h is a popular representation of the data, since it factors out uninteresting temperature and viscosity dependence.

Panel D in **Fig. 5** shows the distribution of the intensity $I(R_h) = N(R_h) \cdot I_0(R_h)$ (*see* Eq. 7) coming from particles of a particular size. This distribution is renormalized in panel E to show the actual particle mass concentration, $N(R_h) \cdot M(R_h)$, where $M(R_h)$ is the mass of the particle with a given R_h. Conversion is done under assumption that the scattering intensity is proportional to $M(R_h) \sim R_h$. This assumption is an approximation roughly appropriate for fibrillar scatterers. The distribution in panel E indicates that about half of the material remains in micellar form after incubation, even though large fibrils dominate the scattering. Large aggregates thus may contribute significantly to scattering even though their concentration is small. The high sensitivity of QLS to large particles makes the method particularly valuable for studying aggregation processes.

4. Notes

1. This analysis is applicable only if the probability of multiple scattering within one particle is negligible. Analysis of scattering intensity from strongly scattering and/or absorbing particles (metallic, for example) requires a more complicated approach *(24)*.

2. Increasing the collection aperture beyond the coherence angle increases the number of photocounts available for computation of the correlation function. That makes the correlation function look "smoother." However, it does not improve the accuracy with which the diffusion coefficient can be deduced.

3. If visible light is used, and the beam is unblocked (note that this is a safety hazard), the visual inspection of the speckle pattern observed at small angles may be informative. A bright static pattern indicates the dirt on the walls of the cuvet, whereas flashes of intensity indicate large particles crossing the beam. A pattern of speckles moving in a vertical direction is indicative of convection caused by heating.

4. Measurement of the pure buffer purified in the same way as the sample will show the level of stray scattering and the level of dust contamination. Do not use an empty cuvet to judge stray scattering. Empty cuvets produce much more scattering from inner surfaces than do filled cuvets, because of larger differences in refractive index between glass and air than between glass and water.

5. The average diffusion coefficient $\overline{D}$ is a stable characteristic of the distribution. That is not so for the average size of the particles (hydrodynamic radius defined in **Subheading 3.4.1.**) $\overline{D}$ is proportional to $\overline{1/R_h}$, which is not the same as $1/\overline{R_h}$. As a consequence, an average hydrodynamic radius $\overline{R_h}$ for a wide distribution may vary dramatically from one measurement to another. A much better characteristic of the typical particle size is R_h^{app} computed from D^{app} using Eq. 11.

6. For non-spherical particles, the diffusion coefficient is actually a tensor—the rate of particle diffusion in a certain direction depends on the particle orientation relative to this direction. Small particles, as they diffuse over a distance q^{-1}, change their orientation many times. QLS measures the average (apparent) diffusion coefficient for these particles. Particles of a size comparable to, or larger than, q^{-1} essentially preserve their orientation as they travel a distance smaller than their size. For these particles, the single exponential expression of Eq. 6 for the field correlation function is not strictly applicable. This fact is particularly important in QLS applications designed for studying long fibrils *(8)*.

Acknowledgments

This work was supported by the National Institutes of Health Grants AG14366, AG18921, NS38328 (to D.B.T.), and IY05127 (to G.B.B), by the Foundation for Neurological Diseases (to D.B.T), and by the National Aeronautics and Space Administration Grant NAG8-1659 (to G.B.B.).

References

1. Pecora, R. (1985) Dynamic Light Scattering: Applications of Photon Correlation Spectroscopy. Plenum Press, New York.
2. Schmitz, K. S. (1990) An Introduction to Dynamic Light Scattering by Macromolecules. Academic Press, Boston.
3. Selkoe, D. J. (1994) Alzheimer's disease—A central role for amyloid. *J. Neuropath. Exp. Neurol.* **53,** 438–447.
4. Prusiner, S. B. (1998) Prions. *Proc. Natl. Acad. Sci. USA* **95,** 13363–13383.
5. Kirkitadze, M. D., Bitan, G., and Teplow, D. B. (2002) Paradigm shifts in Alzheimer's disease and other neurodegenerative disorders: the emerging role of oligomeric assemblies. *J. Neurosci. Res.* **69,** 567–577.
6. Eaton, W. A. and Hofrichter, J. (1990) Sickle cell hemoglobin polymerization. *Adv. Prot. Chem.* **40,** 63–279.
7. Benedek, G. B. (1997) Cataract as a protein condensation disease. *Invest. Ophtalmol. Vis. Sci.* **38,** 1911–1921.
8. Lomakin, A., Benedek, G. B., and Teplow, D. B. (1999) Monitoring protein assembly using quasielastic light scattering spectroscopy. *Meth. Enzymol.* **309,** 429–459.
9. Debye, P. (1947) Molecular-weight determination by light scattering. *J. Phys. Col. Chem.* **51,** 18–32.
10. Benedek, G. B. (1971) Theory of transparency of eye. *Appl. Optics.* **10,** 459–473.
11. Kerker, M. (1969) The scattering of light and other electromagnetic radiation. Academic Press, New York.
12. Chichoki, B. and Felderhof, B. U. (1993) Dynamic scattering function of a semidiluted suspension of hard spheres. *J. Chem. Phys.* **98,** 8186–8193.
13. Balabonov, S. M., Ivanova, M. A., Klenin, S. I., Lomakin, A., Molotkov, V. A., and Noskin, V. A. (1988) Quasielastic light scattering study of linear macromolecules dynamics. *Macromolecules* **21,** 2528–2535.

14. Walsh, D. M., Lomakin, A., Benedek, G. B., Condron, M. M., and Teplow, D. B. (1997) Amyloid β-protein fibrillogenesis. Detection of protofibrillar intermediate. *J. Biol. Chem.* **272,** 22364–22372.

15. Tikhonov, A. N. and Arsenin, V. Y. (1977) *Solution of Ill-Posed Problems.* Halsted Press, Washington.

16. Koppel, D. E. (1972) Analysis of macromolecular polydispersity in intensity correlation spectroscopy. The method of cumulants. *J. Chem. Phys.* **57,** 4814–4820.

17. Provencher, S. W. (1982) A constrained regularization method for inverting data represented by linear algebraic or integral equations. *Comput. Phys. Commun.* **27,** 213–227.

18. Braginskaya, T. G., Dobitchin, P. D., Ivanova, M. A., et al. (1983) Analysis of the polydispersity by photon correlation spectroscopy: regularization procedure. *Physica Scripta* **28,** 73–79.

19. Bitan, G., Kirkitadze, M. D., Lomakin, A., Vollers, S. S., Benedek, G. B., and Teplow, D. B. (2003) Amyloid β-protein (Aβ) assembly: Aβ40 and Aβ42 oligomerize through distinct pathways. *Proc. Natl. Acad. Sci. USA* **100,** 330–335.

20. Einsten, A. (1905) Über die von der molekularkinetischen Theorie der Wärme geforderte Bewegung von in ruhenden Flüssigkeiten suspendierten Teilchen. *Annalen der Physik und Chemie* **17,** 549–560.

21. de la Torre, J. G. and Bloomfield, V. A. (1981) Hydrodynamic properties of complex, rigid, biological macromolecules: theory and applications. *Quarterly Reviews of Biophysics* **14,** 81–139.

22. Muschol, M. and Rosenberger, F. (1995) Interactions in undersaturated and supersaturated lysozyme solutions: static and dynamic light scattering results. *J. Chem. Phys.* **103,** 10424–10432.

23. Yong, W., Lomakin, A., Kirkitadze, M. D., Teplow, D. B., Chen, S.-H., and Benedek, G. B. (2001) Structure determination of micelle-like intermediates in amyloid β-protein assembly by using small angle neutron scattering. *Proc. Natl. Acad. Sci. USA* **99,** 150–154.

24. Van de Hulst, H. C. (1981) Light Scattering by Small Particles. Dover, New York.

11

Intrinsic Fluorescent Detection of Tau Conformation and Aggregation

Martin von Bergen, Li Li, and Eckhard Mandelkow

Summary

The polymerization of the microtubule-associated protein tau into paired helical filaments (PHFs) is one of the hallmarks of Alzheimer's disease. Insights into the prerequisites and kinetics of the polymerization was obtained by the in vitro analysis of this process. In the past, fluorescent dyes were used to stain amyloidogenic material in histology and later on similar dyes were used in in vitro studies as well. To circumvent the flaws of extragenous dyes, namely the alteration of the polymerization kinetic or incompatibility with other chemical compounds needed for stability analysis, we applied tryptophan fluorescence to the in vitro analysis of PHF formation. Single tryptophans were introduced into the hexapeptide PHF6 within the third repeat, which was shown to be involved in β sheet formation and scattered around the whole microtubule binding domain. Tryptophan fluorescence was then used to scan the microtubule binding domain for accessibility to quenching reagent in the soluble and the aggregated state and the fluorescence resonance energy transfer (FRET) between tryptophan and tyrosine 310. Furthermore, this approach enables the analysis of stability of PHFs in the presence of Guanidinium hydrochloride. The examples given here could be applied in modified ways to other amyloidogenic proteins.

Key Words: Alzheimer's disease; paired helical filament; tau protein; self-assembly; tryptophan fluorescence, conformational stability.

1. Introduction

Up to 18 amyloidogenic peptides and proteins are known that are associated with a human disease. Nearly the half of them cause neurodegenerative disorders (*1*). One of them is Alzheimer's disease (AD), in which two proteins aggregate in an abnormal fashion, the Aβ peptide which builds the plaques in the extracellular space and the microtubule-associated protein tau which polymerizes into the intracellular paired helical filaments (PHF) (*2,3*). Tau normally functions as a microtubule-associated protein that is involved in microtubule stabilization and neurite outgrowth (*4*). Tau exists in six isoforms in the adult

From: *Methods in Molecular Biology, vol. 299: Amyloid Proteins: Methods and Protocols*
Edited by: E. M. Sigurdsson © Humana Press Inc., Totowa, NJ

brain that are encoded by one gene on chromosome 17 and are formed by alternative splicing *(5)*. Recently, familial forms of frontotemporal dementias have been associated with mutations in the tau gene (for review *see* **ref.** *6)*. One group of mutations are intronic mutations which cause a shift in the ratio of isoforms, others are exonic mutations and cause single missense mutations. Up to now it is uncertain how the shift in the ratio of isoforms could cause neurotoxic effects. The same is true for the missense mutations where both a loss of physiological function (stabilization of microtubules) or a gain of pathological function (facilitated filament formation) are discussed *(7)*. Here we will demonstrate how intrinsic markers can be used to analyze the aggregation process and reveal differences in stability of filaments between wild-type and mutated tau protein.

The evaluation of kinetic parameters as well as accelerators and inhibitors of amyloid formation requires a reliable quantitative analysis in solution. A proven approach is the use of fluorescent markers which are highly sensitive to the local environment, and thus to the conformation and assembly state of the protein. Several fluorescent dyes have been applied successfully to the analysis of tissue slices, such as Congo red *(8)*, thioflavin-S (ThS) *(9)*, and thioflavin-T (ThT) *(10)*. Especially the recognition by Congo red was claimed to be a characteristic feature of amyloid aggregates *(11)*. A more detailed analysis showed that the disease related amyloids share the structural element of the cross-β structure *(12)*. However, assays of filament formation in solution revealed several drawbacks of the use of Congo red and thioflavin-S. For example, the addition of an exogenous dye with unknown binding properties may alter the course of polymerization, and higher concentrations of Congo red are known to inhibit filament formation of the Aβ peptide and Huntington protein *(13,14)*. The binding of dyes can influence the stability of amyloid aggregates in both directions, and the interplay with other components (for example, amyloid inhibitors to be tested) is unpredictable. Additionally there exists a great variability among the different amyloids regarding the binding of Congo red, thioflavin-S, and thioflavin-T. In the case of tau aggregation, Congo red does not bind efficiently to the tau-containing neurofibrillary tangles in the brain *(15)*, and thioflavin-T is sensitive to cofactors (e.g., polyanions) used to accelerate tau polymerization *(16)*. Furthermore, the binding of the reporter molecule may have an influence on the interactions within the fiber. These disadvantages provided the rationale for developing intrinsic markers of amyloid polymerization. In order to avoid labeling with detector molecules (radioactive or fluorescent) or laborious techniques like FTIR, CD, EM, STEM, or AFM we developed a tryptophan-based assay, which makes use of a single mutation in the tau protein. The naturally occuring tau sequence does not contain any tryptophan and only five tyrosines, most of which lie outside the repeat domain which forms the core of Alzheimer PHFs (residues 18, 29, 197, 310, 394, **Fig. 1A**). We therefore introduced a con-

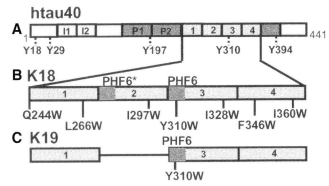

Fig. 1. Tau isoforms and constructs. (**A**) htau40, the longest isoform of tau contains four microtubule binding repeats in the C-terminal half and two inserts near the *N*-terminus. The five naturally occuring tyrosines are marked (Y18, Y29, Y197, Y310, Y394). (**B**) Construct K18 comprising the four repeats in the microtubule binding domain. (**C**) Construct K19 containing three repeats. In **B** and **C**, the hexapeptide motifs PHF6 (third repeat) and PHF6* (second repeat) that promote the formation of β-structure are highlighted. The positions of the point mutations for the tryptophan scanning are indicated.

servative exchange from tyrosine to tryptophan within the microtubule binding region of tau (residue 310). In earlier studies we had identified two hexapeptide motifs (PHF6: $V_{306}QIVYK_{311}$ and PHF6*: $V_{275}QIINK_{280}$) in the microtubule-binding domain of tau which are involved in the formation of cross-β structure during the pathological aggregation *(17)* (recommendations for the sequence of choice in **Note 1**). In order to simplify the analysis of aggregation, we used tau constructs which contain three or four microtubule-binding repeats (termed K19 and K18, *see* **Fig. 1B,C**), because their aggregates have been shown to resemble the morphology and structural composition of Alzheimer PHFs and polymerize in a reasonably short time *(18)*.

2. Materials

2.1. Chemicals and Proteins

Heparin (average M_r of 3000) and ThS were obtained from Sigma. Human tau isoforms and constructs were expressed in *Escherichia coli* as described earlier *(19)*. For purification of recombinantly expressed tau protein please *see also* Chapter 4. Homogenity of protein fractions was analyzed by 17% SDS-PAGE. Protein concentrations were determined by the Bradford assay. Acrylamide was obtained from Serva as a 40% (w/v) solution in water and diluted into the cuvet to the final concentrations. Guanidine hydrochloride was purchased from Sigma and prepared as a stock solution in PBS at 8 *M*. The mutations of K18 (Q244W,

L266W, ΔK280, ΔK280-Y310W, I297W, Y310W, I328W, and I360W), K19 (Y310W), were cloned by site-directed mutagenesis according to the recommendations of the supplier of the Quick Change kit (Stratagene, Germany) on the plasmid pNG2 *(20)*. Plasmids were sequenced on both strands.

3. Methods

3.1. PHF Assembly

Formation of PHFs was achieved by incubating varying concentrations of tau isoforms or tau constructs (typically in the range of 50–100 μ*M*) in volumes of 20–500 μL at 37°C in PBS, pH 7.4 (recommendations for buffers in **Note 2**) in the presence of heparin. The ratio of heparin:protein was adjusted to a ratio of 1:4 as described in the Chapter 4. Incubation time varied between minutes up to several days. The formation of aggregates was measured by ThS fluorescence and electron microscopy as described in Chapter 4.

3.2. Tyrosine and Tryptophan Fluorescence Spectroscopy

The fluorescence of tryptophan is highly sensitive to the local environment *(21)*, which changes during aggregation, whereas the fluorescence of tyrosine is less affected by conformational changes. These features are also observed during the polymerization of tau. The fluorescence of the endogenous tyrosine 310 shows little change (**Fig. 2A,B**), the emission maximum is unaffected (**Fig. 2B**), and the excitation maximum is slightly redshifted by about 1 to 2 nm (**Fig. 2A**). This small change makes tyrosine unattractive as an intrinsic marker. However, after exchanging the residue into tryptophan the emission maximum becomes much more intense, and it is strongly affected by tau polymerization. In the soluble state, the emission maximum (**Fig. 2D**) at 352 nm indicates that the Trp310 is completely exposed to solvent. This is consistent with the "natively unfolded" structure of tau in solution. Upon aggregation, the emission maximum exhibits a pronounced blueshift towards 340 nm, while change in the excitation maximum is small (only 1–2 nm).

Every change of the protein by mutation or modification requires the proof that the important biochemical and biophysical properties are not fundamentally altered. Therefore the aggregation behavior of different tau constructs and mutations was checked by other methods, such as the ThS fluorescence assay (based on the binding of exogenous ThS, Friedhoff et al., 1998), and the conformation of the soluble and aggregated samples were measured by spectroscopy (CD, FTIR). Furthermore the morphology of PHFs was checked by electron microscopy. Neither the course of aggregation nor the secondary structure or morphology of PHFs was noticeably affected by the point mutations *(22)* (general comments on point mutations in **Note 3**).

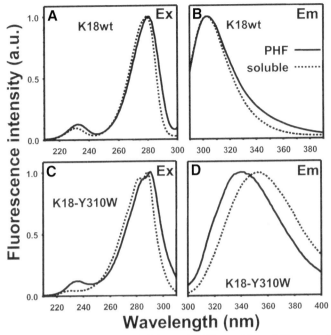

Fig. 2. Tryptophan fluorescence exhibits a significant blue shift upon PHF formation of tau construct K18. In (**A**) and (**B**) the excitation (*Ex*) and emission (*Em*) spectra of construct K18wt in the soluble state (dashed line) and after PHF formation (solid curve) are shown. The spectra are normalized to the same height (peaks at ~280 and ~305 nm, respectively). The tyrosine excitation scan exhibits a slight red shift upon aggregation of about 1 to 2 nm. The tryptophan-containing mutant K18/Y310W shows only a slight conformation dependency of the excitation scan in (**C**), but a remarkable blue shift of the maximum of the emission scan in (**D**). In the excitation scan the emission wavelanth was fixed at 350 nm for the emission scan, the excitation was fixed at 280 nm. The emission maximum of the soluble protein (dashed line) of 354 nm is shifted toward 339 nm in the aggregated state (solid curve).

In order to cover different areas of the tau sequence by fluorescent markers, we performed tryptophan scanning mutagenesis over the microtubule binding domain to determine the core of tau aggregates. The soluble proteins exhibit similar emission maxima at about 352 nm, whereas the aggregated proteins yield clear blueshifts of varying magnitude. This indicates the degree to which the reporter tryptophan is buried in a more hydrophobic environment in the core of the PHFs. From these data, a correlation between the sequence position and the emission maximum of the tryptophan within the polymerized protein can be drawn (**Fig. 3A**).

All fluorescence experiments were performed on a Spex Fluoromax spectrophotometer (Polytec, Waldbronn, Germany), using 3 × 3-mm quartz microcuvets

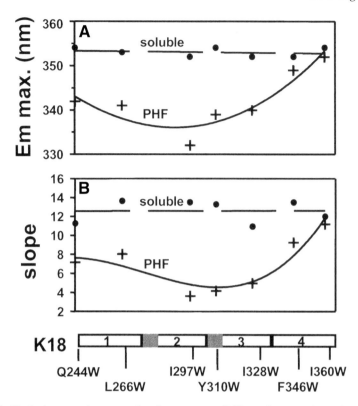

Fig. 3. Emission maximum and solvent accessibility of tryptophans in the repeat domain of tau in the soluble state and after PHF aggregation is sequence-dependent. In the tryptophan scanning approach covering the repeat domain (construct K18), seven residues were mutated into tryptophans and analyzed in regard to their exposure to the solvent (Gln_{244}, Leu_{266}, Ile_{297}, Tyr_{310}, Ile_{328}, Phe_{346}, and Ile_{360}) before and after aggregation. Emission scans with excitation fixed at 290 nm were measured from 300 to 420 nm and the maxima were taken from the normalized spectra. **(A)** The emission maximum wavelength is plotted versus the position of the mutation within the sequence. For the soluble proteins (filled circles, dashed line), the maxima are around 350 to 355 nm, indicating a hydrophilic environment. After PHF assembly (+ symbols, solid line), the emission maximum shows a pronounced blue shift. As in **(B)** there is a U-shaped dependence on the position of Trp in the sequence, arguing that the residues in the middle are more shielded than those at the edges of the repeat domain. **(B)** Fluorescence quenching by acrylamide of samples are represented as Stern-Volmer plot $F_o/F_c = 1 + Kc$ (c = acrylamide concentration). A high slope K indicates high solvent accessibility. The dashed curves and filled circles represent the mutants of K18 in the soluble state, with high slopes indicating nearly full solvent accessibility. The significantly lower slopes of the polymerized mutants (crosses and solid line) hint to a sequence-dependent lower accessibility to the solvent, which causes a U-shaped curve, illustrating that the residues near the ends of the sequence (in R1 and R4) are more exposed than those in the middle (in R2 and R3).

from Hellma (Mühlheim, Germany) with 20 µL of sample volume, the protein concentrations was adjusted to be about 10 µ*M*. For tyrosine excitation spectra, scans ranged from 250 to 300 nm at a fixed emission wavelength of 310 nm, and for emission spectra, scans ranged from 290 to 450 nm at fixed excitation wavelength of 275 nm. For tryptophan excitation spectra, scans ranged from 210 to 310 nm at a fixed emission wavelength of 350 nm, and for emission spectra, scans ranged from 300 to 400 nm at fixed excitation wavelength of 290 nm. In all cases, the slit widths were 5 nm and integration time was 0.25 s; photomultiplier voltage was 950 V (comments on technical spectra in **Note 4**).

3.3. Fluorescence Quenching Experiments

In the polymerized state the emission maxima vary from about 350 nm for mutations near the end of the fourth repeat to 340 nm for mutations within the second and the third repeat, and to an intermediate value of about 344 nm for a tryptophan located in the first repeat. This indicates a tighter molecular packing of the second and the third repeat within the PHFs. A quencher accessibility study using acrylamide (**Fig. 3B**) supports this interpretation. In this approach, the protein is first polymerized, then incubated with acrylamide of increasing concentrations. The slope of the Stern-Volmer plot determines the accessibiliy of the analyzed tryptophan mutation to the solvent. The slope of the soluble mutations is nearly equal for all Trp mutations at the different sequence positions, whereas the slope of the polymerized mutations follows a U-shape with the minimum of accessibility located in the second and third repeat. These results are in excellent agreement with the position-dependence of the blue-shift (**Fig. 3A**) and indicate that repeats R2 and R3 form the least accessible, most deeply buried core of the PHFs.

Steady-state fluorescence quenching experiments were performed on either soluble or aggregated proteins. Aliquots of the stock quenching solutions (5 *M*) were added into the cuvet containing 10 µ*M* protein in PBS, pH 7.4. Quenching experiments were performed with excitation at 280 nm. Stock quenching solutions were freshly prepared at concentrations of 5 *M*. Quenching data were fitted to the Stern-Volmer equation, $F_o/F_c = 1 + K_{SV}[Q]$, where F_o and F_c are the fluorescence intensity in the absence and in the presence of quencher, $[Q]$, at concentration c, respectively, and K_{SV} is the Stern-Volmer quenching constant.

3.4. Denaturation and Disassembly of PHFs by GuHCl

A further unique application of the tryptophan assay is the measurement of the stability of the aggregated tau protein. This can be determined by exposing PHFs to denaturing agents like urea and guanidine hydrochloride (GuHCl). When the emission maximum of the aggregated protein is plotted versus increasing concentrations of the denaturant, it changes gradually from its blue-

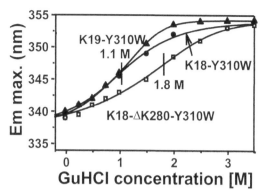

Fig. 4. Denaturation of PHFs measured by tryptophan fluorescence. Preformed fila-ments from K19Y310W, K18Y310W, and K18-ΔK280Y310W were incubated at increas-ing concentrations of GuHCL and analyzed by tryptophan emission scans with the excitation fixed at 290 nm. The emission maximum is plotted versus increasing con-centrations of GuHCl. The emission maxima increased with higher GuHCl concentra-tions, from about 339 nm in the absence of GuHCl to about 354 nm in the presence of 3 M GuHCl, indicating the disassembly kinetics of the polymerized samples. PHFs from K19 and K18 show nearly identical stabilities (half-point of denaturation at ~1.1 M GuHCl, filled symbols), arguing that stability of PHFs from K19 and K18 is not influenced by the number of repeats. By contrast, PHFs made from the mutant K18-ΔK280Y310W (carrying one of the FTDP-17 mutations) are much more stable (half-point ~1.8 M GuHCl, lower curve, open square), consistent with the higher propensity for β-structure.

shifted position to that of the soluble protein, and the half-maximal value is a measure of the stability of the aggregates (**Fig. 4**). The analysis of PHFs made from the three repeat-containing construct K19 and the four repeat construct K18 revealed a surprisingly low half-maximal concentration of GuHCl of about 1.1 M. By comparison, well-folded globular proteins exhibit typical half-maxi-mal concentrations for unfolding from 2 up to 5 M GuHCl. Interestingly, the half-maximal concentration of GuHCl for the disassembly of the FTDP-mutant ΔK280 is about 0.7 M higher, revealing the greater stability of PHFs afforded by this mutant. For this mutant, an increased tendency for beta sheet formation is reported *(23)*. In an extension of this approach, the tryptophan assay can be used to screen for the inhibition of PHF aggregation or promotion of PHF disassembly by small molecular compounds.

Prior to the disassembly experiments, pre-formed PHFs were separated from soluble protein by centrifugation at 100,000g for 30 min at room temperature (table top centrifuge TL-100, Beckman). The PHFs in the pellet were resus-pended in PBS. Ten micromoles of this PHF solution were diluted with GuHCl (stock solution of 8 M) to various concentrations, incubated at 37°C overnight, and measured the next day. The emission spectra were normalized to their maximum fluorescence.

4. Notes

1. In order to investigate the conformation, solvent accessibility, stability, and other parameters of interest, one should engineer tryptophan mutations in sequence positions that are known or suspected to undergo change upon aggregation.
2. Avoid thiol-containing buffers for fluorescence analysis in order to avoid quenching effects.
3. Make sure the fluorimeter yields correct spectra, calibrate the instrument for correction of the technical spectrum. The Tecan safire fluorimeter, for example, requires such a correction owing to the wavelength-dependent efficiency of the optical filters.
4. Control the effect of the mutation by known methods and parameters (e.g., make sure by independent methods that the mutation as such does not affect conformations and aggregation characteristics).

Acknowledgments

We thank Drs. S. Barghorn, J. Biernat, and E.-M. Mandelkow for stimulating discussions. This work was supported in part by grants from Deutsche Forschungsgemeinschaft.

References

1. Taylor, J. P., Hardy, J., and Fischbeck, K. H. (2002) Toxic proteins in neurodegenerative disease. *Science* **296,** 1991–1995.
2. Selkoe, D. J. (2003) Folding proteins in fatal ways. *Nature* **426,** 900–904.
3. Johnson, G. V. and Bailey, C. D. (2002) Tau, where are we now? *J. Alzheimers Dis.* **4,** 375–398.
4. Garcia, M. L. and Cleveland, D. W. (2001) Going new places using an old MAP: tau, microtubules and human neurodegenerative disease. *Curr. Opin. Cell Biol.* **13,** 41–48.
5. Goedert, M., Wischik, C. M., Crowther, R. A., Walker, J. E., and Klug, A. (1988) Cloning and sequencing of the cDNA encoding a core protein of the paired helical filament of Alzheimer disease: identification as the microtubule-associated protein tau. *Proc. Natl. Acad. Sci. USA* **85,** 4051–4055.
6. Hutton, M. (2000) Molecular genetics of chromosome 17 tauopathies. *Ann. NY Acad. Sci.* **920,** 63–73.
7. Goedert, M., Ghetti, B., and Spillantini, M. G. (2000) Tau gene mutations in frontotemporal dementia and parkinsonism linked to chromosome 17 (FTDP-17). Their relevance for understanding the neurogenerative process. *Ann. NY Acad. Sci.* **920,** 74–83.
8. Glenner, G. G., Eanes, E. D., and Page, D. L. (1972) The relation of the properties of Congo red-stained amyloid fibrils to the β-conformation. *J. Histochem. Cytochem.* **20,** 821–826.

9. Vallet, P. G., Guntern, R., Hof, P. R., et al. (1992) A comparative study of histological and immunohistochemical methods for neurofibrillary tangles and senile plaques in Alzheimer's disease. *Acta Neuropathol. (Berl)* **83**, 170–178.

10. LeVine, H. (1993) Thioflavine T interaction with synthetic Alzheimer's disease beta-amyloid peptides: detection of amyloid aggregation in solution. *Protein Sci.* **2**, 404–410.

11. Sipe, J. D. and Cohen, A. S. (2000) Review: history of the amyloid fibril. *J. Struct. Biol.* **130**, 88–98.

12. Kirschner, D. A., Teplow, D. B., and Damas, A. M. (2000) Twist and sheet: variations on the theme of amyloid. *J. Struct. Biol.* **130**, 87.

13. Findeis, M. A. (2000) Approaches to discovery and characterization of inhibitors of amyloid beta-peptide polymerization. *Biochim. Biophys. Acta* **1502**, 76–84.

14. Kuner, P., Bohrmann, B., Tjernberg, L. O., et al. (2000) Controlling polymerization of beta-amyloid and prion-derived peptides with synthetic small molecule ligands. *J. Biol. Chem.* **275**, 1673–1678.

15. Braak, H., Braak, E., Ohm, T., and Bohl, J. (1989) Alzheimer's disease: mismatch between amyloid plaques and neuritic plaques. *Neurosci. Lett.* **103**, 24–28.

16. Friedhoff, P., Schneider, A., Mandelkow, E. M., and Mandelkow, E. (1998) Rapid assembly of Alzheimer-like paired helical filaments from microtubule-associated protein tau monitored by fluorescence in solution. *Biochemistry* **37**, 10223–10230.

17. von Bergen, M., Friedhoff, P., Biernat, J., Heberle, J., Mandelkow, E. M., and Mandelkow, E. (2000) Assembly of tau protein into Alzheimer paired helical filaments depends on a local sequence motif ((306)VQIVYK(311)) forming beta structure. *Proc. Natl. Acad. Sci. USA* **97**, 5129–5134.

18. Friedhoff, P., von Bergen, M., Mandelkow, E. M., and Mandelkow, E. (1998) A nucleated assembly mechanism of Alzheimer paired helical filaments. *Natl. Acad. Sci. USA* **95**, 15712–15717.

19. Biernat, J., Mandelkow, E. M., Schröter, C., et al. (1992) The switch of tau protein to an Alzheimer-like state includes the phosphorylation of two serine-proline motifs upstream of the microtubule binding region. *EMBO J.* **11**, 1593–1597.

20. Barghorn, S., Zheng-Fischhofer, Q., Ackmann, M., Biernat, J., von Bergen, M., and Mandelkow, E. (2000) Structure, microtubule interactions, and paired helical filament aggregation by tau mutants of frontotemporal dementias. *Biochemistry* **39**, 11714–11721.

21. Eftink, M. R. (1994) The use of fluorescence methods to monitor unfolding transitions in proteins. *Biophys. J.* **66**, 482–501.

22. Li, L., von Bergen, M., Mandelkow, E. M., and Mandelkow, E. (2002) Structure, stability, and aggregation of paired helical filaments from tau protein and FTDP-17 mutants probed by tryptophan scanning mutagenesis. *J. Biol. Chem.* **277**, 41390–41400.

23. von Bergen, M., Barghorn, S., Li, L., Marx, A., Biernat, J., Mandelkow, E. M., and Mandelkow, E. (2001) Mutations of tau protein in frontotemporal dementia promote aggregation of paired helical filaments by enhancing local beta-structure. *J. Biol. Chem.* **276**, 48165–48174.

12

Quantitative Measurement of Fibrillogenesis by Mass Spectrometry

Andrew D. Miranker

Summary

In this chapter, a method for the quantitative determination of amyloid conversion by electrospray mass spectrometry is presented. Mass spectrometry is typically used for the purpose of measuring the mass of unknowns. However, the judicious selection of an internal standard permits the quantitative determination of protein concentration. For amyloid formation, this is particularly useful in circumstances where either the protein under study is in limited abundance, or separation of the precursor from other protein factors is impractical or undesirable. For the measurement of amyloid formation, internal standards are typically mass distinct variants of the amyloid precursor. In addition, the extreme stability of amyloid fibers permits assessment of residual precursor concentration with or without separation of fibers from unreacted precursor. Lastly, by using internal standards which are not amyloidogenic and do not interfere with fiber formation, electrospray mass spectrometry permits quantitative measurement of fiber formation in real-time and with small (pmoles) quantities of protein.

Key Words: Amyloid; fibrillogenesis; IAPP; amylin; β-2microglobulin mass spectrometry; electrospray.

1. Introduction

The kinetics of amyloid fiber formation are both fascinating and challenging as a result of their nucleation dependence *(1,2)*. Fiber formation reactions are generally characterized by a lag phase in which no detectable fibers are formed. This is then followed by an explosive elongation phase in which fiber is formed over a period of time that is often shorter that the lag phase itself. The paradigm for understanding any chemical reaction is the identification of intermediate and transition states on the reaction pathway. For protein folding and other unimolecular reactions, the existence of intermediate states may be

From: *Methods in Molecular Biology, vol. 299: Amyloid Proteins: Methods and Protocols*
Edited by: E. M. Sigurdsson © Humana Press Inc., Totowa, NJ

inferred from reaction kinetics alone. Amyloid formation is more complex as both conformational and oligomeric changes occur during the reaction. As oligomerization is highly heterogeneous, quantitative measurement of fiber formation is often elusive. Here we describe a method for using mass spectrometry quantitatively in the context of fibrillogenesis reactions *(3)*. By capitalizing on the intrinsic stability of amyloid fibers, and through the introduction of internal standards, a molar value for the incorporation of precursor into fibril may be determined. The advantage of this approach is that it is rapid, uses small quantities of protein, and can be performed in a heterogeneous environment, such as the inclusion of other proteins, lipids, etc.

2. Materials

1. Mass spectrometer.
2. C-18 ZipTip™ (Millipore Inc., Billerica, MA).
3. HPLC grade acetonitrile (ACN).
4. Hexafluoroisopropanol (HFIP), ideally redistilled.
5. Dimethyl sulfoxide (DMSO).
6. Human islet amyloid precursor polypeptide (IAPP) (BACHEM).
7. Rodent IAPP (BACHEM, Torrance, CA).
8. Syringe filters, e.g., HT Tuffryn® 13 mm, 0.2 µm (Gelman Inc., Ann Arbor, MI).
9. Freeze dryer.
10. Buffer A: 7 *M* GuHCl, 10% DMSO, 0.1% TFA.
11. Buffer B: 10% ACN, 0.1% TFA.
12. Buffer C: 0.2% formic acid.
13. Buffer D: 50% ACN, 0.2% formic acid.
14. Buffer E: 150 m*M* ammonium acetate, 1 m*M* Tris pH 7.4 (pH with acetic acid/ammonia, not HCl/NaOH).

3. Methods

The methods described here may be applied to any amyloidogenic protein. Some of the physical properties of amyloid fibers, most notably stability and size, are essential to the success of these protocols. These considerations are described in detail, with particular emphasis given for IAPP *(4,5)*.

3.1. Quenched Measurement

A typical fibrillogenesis reaction will be conducted over a period of time in which aliquots can be removed and assessed. Mass spectrometry then serves as direct measurement of incorporation of precursor during the fibrillogenesis reaction. Quantitative mass spectrometry for fibrillogenesis typically requires that fibers be separable from unreacted precursor. The extreme disparity in size and stability of fiber from its precursor makes this straightforward. Separations may be achieved by centrifugation, filtration, size exclusion chromatography, or irreversible binding to disposable reverse-phase columns. In some cases, the extreme stability of fibers eliminates the requirement for separation (*see* **Subheading 3.2.**). In the protocol described later, desalting for electrospray ionization, concentration of the protein, and separation of fiber from precursor are performed in a single step.

Mass spectrometry is generally treated as a tool for measuring mass. However, it has long been used quantitatively on systems of known mass provided an internal standard is introduced to account for variations in ionization efficiency. In this protocol, an internal standard is introduced directly to the reaction aliquot prior to any separation or desalting steps. In this case, the internal standard is not only present for the purposes of normalizing ionization efficiency, but also for any variations associated with sample manipulation. This is a powerful approach which has enabled subtle comparisons to be made in hydrogen exchange measurements of protein folding (*6,7*), ligand binding in mutagenesis studies (*8*), and proteome-wide differential expression analysis (*9,10*).

Mass spectrometric analysis is complementary to other methods of detection. Amyloid formation kinetics can often be assessed indirectly using intrinsic (e.g., far UV-CD) and extrinsic (e.g., dye binding using thioflavin-T [ThT] *[11]*) spectroscopic approaches. These approaches are particularly powerful when reaction monitoring can be done in real-time. However, subsequent analysis of the kinetic profiles requires direct measurement of protein incorporation into amyloid fibers. Solution factors can also greatly affect the response of dyes such as ThT. For example, the presence of lipid greatly enhances the fluorescence of ThT (**Fig. 1A**) without actually affecting the extent of reaction (**Fig. 1B**) Thus, mass spectrometry can serve to relate order parameter measurement of reaction kinetics with the molar consumption of precursor.

IAPP fibrillogenesis is typically initiated by >20-fold dilution from a 1 m*M* stock solution in denaturant into physiological media (*12*). The sensitivity of the IAPP fiber formation reactions to seed makes care in the preparation of IAPP stock solutions of critical importance. Reverse-phase purified IAPP should be lyophilized immediately, as it turns into fibers on the days timescale even in approx 30% acetonitrile, 0.1% TFA. In addition, ambient moisture slowly gives rise to the formation of fibers within the lyophilized powder. Stock solutions made by direct dissolution into DMSO, or HFIP are therefore capable of self-seeding. As an alternative, we use reverse-phase spin columns for the preparation of stocks. In the case of IAPP, fibers bind irreversibly to C18 reverse-phase media. We treat this as a reliable approach for any given lot of C18 spin columns. For added assurance, an additional filtration step is applied prior to the reverse-phase treatment.

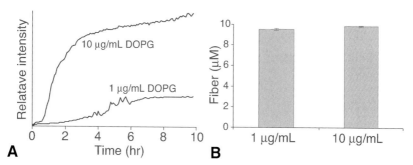

Fig. 1. Kinetic assay of IAPP fiber formation monitored by ThT fluorescence. Extent of reaction determined at the end of the reaction by quantitative mass spectrometry. (**A**) Fiber formation reaction was initiated by dilution of IAPP stock solution in DMSO to 10 μM in aqueous buffer containing 50 μM ThT and either 1 μg/mL or 10 μg/mL liposomes prepared from DOPG. Fluorescence was measured by exciting at 450 nm and collecting emitted light at 485 nm. (**B**) After each 12-h time course, rat IAPP was added to the reaction mixture at a concentration of 2 μM. Samples were analyzed by ESI-MS and final monomer concentration was determined by comparing peak areas from rat and human IAPP.

3.1.1. Preparation of IAPP Stock Solutions

IAPP stock solutions are prepared from lyophilized powder, which has been stored at –20°C and allowed to warm to room temperature prior to opening the vessel.

1. One to two milligrams of protein are dissolved in approx 1 mL buffer A. Fiber formation does not take place in this buffer on the week to month timescale. Importantly, seed fibers are not resolubilized by this step.
2. Protein solution is filtered at 0.2 μm using a 13 mm syringe filter.
3. Following manufacturer's instructions, RP-spin columns are first activated using 400 μL 100% HPLC grade ACN. Spin washes are performed in a benchtop centrifuge. 2000g × 30 s.
4. Column is then washed twice with buffer B.
5. Protein solution in A is applied to the spin column.
6. Column is washed twice with buffer B.
7. Column is washed twice with milliQ water.
8. Top and bottom of column interior are dabbed dry with laboratory tissue.
9. Protein is eluted two times with 200 μL HFIP or DMSO. Solvent is applied directly to the spin column using a P200 pipetman. Eluent is collected without centrifugation, and the two fractions combined.

Concentration of the stock solution is determined by freeze drying 1 to 10 μL aliquots, resuspending in buffer A and assessing UV absorbance. $\varepsilon_{280} = 1400/M$/cm.

3.1.2. Clean Up for Mass Spectral Analysis

Nanoelectrospray is our method of choice in analyzing these samples. Protein concentrations on the order of 1 to 10 μM are ideal, with manual sample handling requiring the use of 0.5 to 5 μL. These restrictions reflect the capabilities of our LC-TOF (Micromass). The use of fused silica-based reverse-phase columns in place of ZipTips (*see* **Note 1**) can significantly reduce material consumption. In principle, matrix-assisted laser desorption ionization (MALDI) should work equally well for this application.

1. Remove aliquot of fibrillogenesis reaction. Aliquot size should nominally correspond to 10 to 100 pmoles in precursor units.
2. Quench reaction. For IAPP this is typically performed by dilution with buffer A. Alternatively, reaction can be quenched by addition of acetic acid to 1%.
3. Pipet a known quantity of protein standard. For IAPP, this is rodent IAPP suspended in H_2O.
4. C18 ZipTips are used, as per manufacturer's instructions.
5. Tips are activated by pipetting with 100% ACN.
6. Tips are then washed three times with buffer C. All washes for ZipTips are performed three times by arraying three tubes with buffer. Tips are washed by repipetting in successive tubes from left to right.
7. Samples is loaded by transferring sample using ZipTip from sample tube to empty tube.
8. Tip is washed three times with buffer C (as in **step 5**).
9. Sample is eluted by repipetting 10 μL of buffer D.

The ZipTip is essentially a small reverse phase column and can therefore be run dry and rehydrated over the course of this procedure. This can be problematic for subsequent mass analysis particularly if the quenched sample is high in salt. In this case, ZipTip steps should be performed while visually inspecting the tip to avoid running air into the matrix. Excess liquid at the top of the column can be carefully removed by capillary action touching laboratory tissue to the end of the tip.

3.1.3. Considerations for the Internal Standard

Quantitative mass spectrometry requires the use of an internal standard which possesses an alternative mass compared to what is being measured (**Fig. 2**). Its use here is comparable to that of proteomic investigations *(9,10)*. Firstly, the greatest accuracy is achieved by introducing the internal standard before both chromatographic and mass spectrometric analyses. Small differences in the standards' interaction with the C18 material of the columns is not of great importance here since chromatography is simply being used as a means of buffer exchange. A step is used to elute the protein eliminating differences in adsorption to the column. The most important consideration is that the standard has

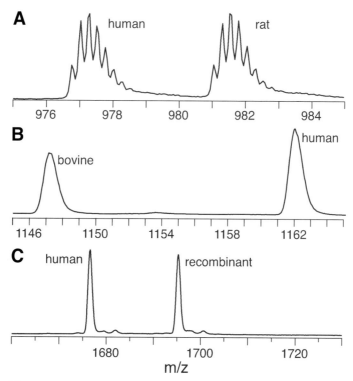

Fig. 2. Electrospray mass spectra of a selected set of human amyloidogenic proteins doped with equimolar internal standards for quantitation. (**A**) Human IAPP using
rat IAPP as standard (reprinted from **ref**. *3* with permission). (**B**) Human insulin using
bovine insulin as standard (reprinted from **ref**. *16* with permission). (**C**) Human derived
β2-microglobulin using recombinant β2 microglobulin as standard (reprinted from **ref**.
8). The mass of recombinant protein is larger due to presence of N-terminal methionine residue.

comparable ionization efficiency to the protein being measured (*see* **Note 2**).
This is best achieved by using metabolic labeling. The use of ^{15}N or ^{13}C provides a chemically identical protein with an increased mass corresponding to
the number of nitrogen and/or carbon atoms in the sample respectively.

The use of isotopically labeled protein is ideal, however for many systems
this approach is prohibitive. In this case, a closely related protein may be used.
For IAPP, we use the sequence variant from rodents. Rat IAPP possesses 6
mutations compared to human. None of these alterations affect the net charge
of the protein under acidic conditions. This likely accounts for the nearly identical ionization efficiency which we have reported for rat vs human IAPP *(3)*.

Interactions of the standard with the reactant are irrelevant, provided the standard is added after the fibrillogenesis reaction is quenched. For many amyloid systems, obvious internal standards exist. e.g., in Alzheimer's disease, Aβ residues 1–42 could be used as standard for fibrillogenesis studies of residues 1–40.

3.2. Real-Time Measurement of Fiber Formation

Quantitative measurement of fiber formation may also be performed in real-time. There are several requirements for this to be possible.

1. Fibrillogenesis must be possible under low salt conditions (or in the presence of volatile salts), without agitation and in a concentration range of 1–50 μ*M*.
2. The internal standard must not be fibrillogenic and not interfere with fibrillogenesis.
3. The fibers must not dissociate upon ionization.

Real-time measurement of fibrillogenesis must be performed using electrospray ionization techniques. This is performed using nanoelectrospray because of its tolerance to near physiological solution conditions *(13,14)*. There is no requirement for denaturing levels of heat, acid or organic solvent. Furthermore, small quantities (<5 m*M*) of involatile buffer salts can often be included with only modest degradation of signal and nonspecific adduct formation. These properties of nanoelectrospray are ideal for biomedically derived amyloids. Model systems for the study of amyloid formation, often require elevated temperatures, high organic solvent content, or acidic pH for amyloid formation to take place. Notably, all three of these conditions enhance the ionization process.

Nanoelectrospray is typically achieved using 1-mm borosilicate glass capillaries which have been drawn into needles using an automated pipette puller. These are then sputter coated with gold and the tips broken manually to an approx 1 μm orifice under a stereomicroscope using fine tweezers. Needles are also commercially available with well-defined orifice sizes, e.g., through New Objective Inc. The initial droplet size of nanoelectrospray is of the order of 10^{-11} μL and the flow rate is approx 10 nL/min. It is the small initial droplet size of nanoelectrospray which permits ionization to take place from aqueous solutions at neutral pH and room temperature. Acid plays a dual role in electrospray ionization. It serves to protonate the analyte and to increase the conductivity of the solution for effective electrospray. We find the former to be irrelevant to the quality of electrospray spectra of macromolecules. We routinely use positive ion electrospray to ionize proteins from solution conditions in which they are negatively charged. In contrast, it is typically essential to compensate for the loss of conductivity caused by performing reactions at neutral pH. This is achieved by using volatile buffer salts, e.g., 150 m*M* ammonium acetate. Increased conductivity results in a consistent electrospray without any pneu-

matic assistance to achieve liquid flow (*see* **Note 3**). As a result, a few microliters of a fibrillogenesis reaction may be analyzed in real-time for hours while
requiring as little as 1 to 2 s for a spectrum to be taken.

The requirement for a closely related, nonfibrillogenic internal standard for
ionization can be orthogonal to the requirement that the internal standard not
interfere with fibrillogenesis. For IAPP, we identified that the nonfibrillogenic
variant of IAPP from rodents *(15)* does not interfere with fibrillogenesis. For
other amyloid systems, this may be more challenging. An alternative approach
is to forgo the utility of absolute quantitation of the mass spec signal when con-
ducting a real time experiment. The intensity of precursor may be renormalized
to other signals in the spectrum such as buffer salts (e.g., Tris). The ionization
efficiency of an involatile buffer is unrelated to that of the protein, however,
for a given reaction in a given needle, the relative ionization efficiency of the
buffer and protein may be assumed to be a constant. Kinetic profile information
can therefore be determined by renormalizing the precursor signal with the buffer signal.

The focus of quantitative mass spectrometry in this protocol is on the analysis of unreacted precursor. As amyloids are typically noncovalent assemblages,
it is a reasonable concern that fibers will dissociate upon ionization giving an
artificially elevated apparent concentration of precursor. It is therefore important to establish that the ionization conditions do not result in dissociation of
the fibrils. One method for doing this is to measure the concentration of precursor at the end of the reaction. The reaction endpoint, in this case, must be
assessed by an orthogonal measurement, e.g., ThT binding. The reaction should
then be doped with internal standard and then assessed for precursor concentration before and after the fibers have been separated (by centrifugation or
filtration). Any difference between the measurements may be attributed to gas
phase dissociation of the fibers.

3.2.1. Protocol for IAPP

1. Prepare reaction buffer: Buffer E doped to 2 μM with rodent IAPP.
2. Minimize dead time by tuning instrument on reaction buffer.
3. Dilute 1 mM stock solution of IAPP into reaction buffer.
4. Load nanospray needle and begin data acquisition.

It is reasonable to expect a given reaction mixture to give a sustained signal
for an hour or more. Provided the entire sample is not loaded into the electrospray needle, the reaction may be further sampled at discrete time points on
longer timescales as well. In such cases, the temperature of incubation should be
closely matched to the temperature at the needle source. Care should be taken
here as ambient instrument heat can elevate the needle temperature by several
degrees. All acquisitions are then expressed on a time axis by integrating the

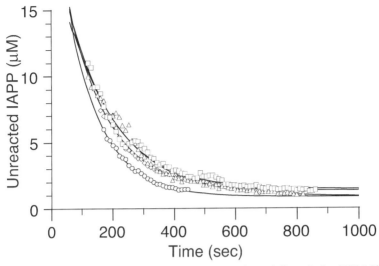

Fig. 3. Kinetic assay of IAPP fiber formation monitored directly by ESI-MS (from **ref. *3*** with permission). Fiber formation reaction was initiated by dilution of stock solution in HFIP to 25 μ*M* in buffer E and 2 μ*M* (monomer units) of preformed IAPP fibers. Sample was immediately loaded into nanoelectrospray needles and analyzed by comparing the relative areas of the +4 charge states of human and rodent IAPP.

time dependent area of human IAPP and renormalizing to the time dependant are of rat IAPP (**Fig. 3**).

4. Notes

1. A variety of alternatives to C18 ZipTips exist. C4 ZipTips are available which are likely superior for larger protein systems. PrepTips™ from Harvard Apparatus are also available. This tip is not a column, but rather a coating on the interior surface of the yellow tip. In practice, it has reduced binding capacity, however, it is much more readily loaded, washed and is less prone to contaminating eluent with salt if it is run dry.

2. The dynamic range of mass spectral intensity does not necessarily vary linearly with ion flux. In practice, the internal standard ion counts should be within a factor of 10 of the target analyte ion counts. For the measurement of residual concentrations of unreacted precursor, we typically mix ratios by trial and error until our internal standard intensity is within a factor of 2 of the target analyte intensity. Triplicate measurements are then made at this ratio.

3. Clogging of nanospray needles is common in real-time fibrillogenesis reactions. Typically, the operator must observe the spectra as they are collected. Sudden loss of signal reflects a clog which can be expelled if caught early by transiently raising the backing pressure.

Acknowledgments

We thank C. Eakin, J. Hall (formerly Larson), S. Jaswal, B. Koo, and J. Knight for assistance in developing and preparing these protocols. This work was supported by the NIH (DK54899) and the Pew charitable trusts (PO219SC).

References

1. Kelly, J. W. (1998) The alternative conformations of amyloidogenic proteins and their multi- step assembly pathways. *Curr. Opin. Struct. Biol.* **8,** 101–106.
2. Rochet, J. C. and Lansbury, P. T. Jr. (2000) Amyloid fibrillogenesis: themes and variations. *Curr. Opin. Struct. Biol.* **10,** 60–68.
3. Larson, J. L., Ko, E., and Miranker, A. D. (2000) Direct measurement of islet amyloid polypeptide fibrillogenesis by mass spectrometry. *Protein Sci.* **9,** 427–431.
4. Hoppener, J. W. M., Ahren, B., and Lips, C. J. M. (2000) Islet amyloid and type 2 diabetes mellitus. *N. Engl. J. Med.* **343,** 411–419.
5. Jaikaran, E. T. and Clark, A. (2001) Islet amyloid and type 2 diabetes: from molecular misfolding to islet pathophysiology. *Biochim. Biophys. Acta* **1537,** 179–203.
6. Hooke, S. D., Eyles, S. J., Miranker, A., Radford, S. E., Robinson, C. V., and Dobson, C. M. (1995) Cooperative elements in protein-folding monitored by electrospray- ionization mass-spectrometry. *J. Am. Chem. Soc.* **117,** 7548–7549.
7. Canet, D., Last, A. M., Tito, P., et al. (2002) Local cooperativity in the unfolding of an amyloidogenic variant of human lysozyme. *Nat. Struct. Biol.* **9,** 308–315.
8. Eakin, C. M., Knight, J. D., Morgan, C. J., Gelfand, M. A., and Miranker, A. D. (2002) Formation of a copper specific binding site in non-native states of beta-2-microglobulin. *Biochemistry* **41,** 10646–10656.
9. Oda, Y., Huang, K., Cross, F. R., Cowburn, D., and Chait, B. T. (1999) Accurate quantitation of protein expression and site-specific phosphorylation. *Proc. Natl. Acad. Sci. USA* **96,** 6591–6596.
10. Gygi, S. P., Rist, B., Gerber, S. A., Turecek, F., Gelb, M. H., and Aebersold, R. (1999) Quantitative analysis of complex protein mixtures using isotope-coded affinity tags. *Nat. Biotechnol.* **17,** 994–999.
11. LeVine, H. D. (1993) Thioflavine T interaction with synthetic Alzheimer's disease beta- amyloid peptides: detection of amyloid aggregation in solution. *Protein Sci.* **2,** 404–410.
12. Padrick, S. B. and Miranker, A. D. (2002) Islet amyloid: phase partitioning and secondary nucleation are central to the mechanism of fibrillogenesis. *Biochemistry* **41,** 4694–4703.
13. Kebarle, P. (2000) A brief overview of the present status of the mechanisms involved in electrospray mass spectrometry. *J. Mass Spectrom* **35,** 804–817.
14. Cole, R. B. (2000) Some tenets pertaining to electrospray ionization mass spectrometry. *J. Mass Spectrom* **35,** 763–772.
15. Westermark, P., Engstrom, U., Johnson, K. H., Westermark, G. T., and Betsholtz, C. (1990) Islet amyloid polypeptide: pinpointing amino acid residues linked to amyloid fibril formation. *Proc. Natl. Acad. Sci. USA* **87,** 5036-5040.
16. Larson, J. L. and Miranker, A. D. (2004) The mechanism of insulin action on islet amyloid polypeptide fiber formation. *J. Mol. Biol.* **35,** 221–231.

II

Cell Culture Assays

13

Isolation and Culturing of Human Vascular Smooth Muscle Cells

Finnbogi R. Thormodsson and Ingvar H. Olafsson

Summary

Cerebral amyloid angiopathy (CAA) results from amyloid accumulation within arteries of the cerebral cortex and leptomeninges. This condition is age-related, especially prevalent in Alzheimer's disease (AD), and the main feature of certain hereditary disorders (i.e., HCHWA-I). The vascular smooth muscle cells (VSMCs) appear to play a vital role in the development of CAA, which makes them well-suited as an experimental model to study the disease and screen for possible remedies. We describe two different methods for isolating and culturing human VSMCs. First, using the human umbilical cord as an easy source of robust cells, and secondly, using brain tissue that provides the proper cerebral VSMCs, but is more problematic to work with. Finally, the maintenance, preservation, and characterization of the isolated VSMCs are described.

Key Words: Amyloid; cell culture; human; cystatin C; vascular smooth muscle; vascular smooth muscle cells (VSMC); umbilical cord; cerebral amyloid angiopathy (CAA); HCHWA-I; immunostaining; leptomeninges; cerebral blood vessels.

1. Introduction

Cerebral amyloid angiopathy (CAA) is characterized by the presence of amyloid within the vessel walls, most commonly composed of amyloid β protein (Aβ), and is found in over 90% of brains from patients with Alzheimer's disease. Similarly, CAA is associated with Down's syndrome, few other cerebrovascular maladies and otherwise unaffected aging brains (1). Apart from those principally sporadic CAAs, an increasing number of rare familial conditions have been identified (2), two of which are Hereditary Cerebral Hemorrhage with Amylodosis-Dutch type (HCHWA-D) (3) and Icelandic type (HCHWA-I), also known as Hereditary Cystatin C Amyloid Angiopathy (HCCAA) (4,5). The amy-

From: *Methods in Molecular Biology, vol. 299: Amyloid Proteins: Methods and Protocols*
Edited by: E. M. Sigurdsson © Humana Press Inc., Totowa, NJ

loid fibrils accumulate in the media of the vessel walls eventually replacing the vascular smooth muscle cells (VSMC) suggesting a causal relationship between the muscle cells and the amyloid *(6,7)*. The VSMC might be the source or a partial source of the amyloid material and subsequently the victims of its accumulation. In an effort to study different aspects of this interaction several research groups have established cellular models, utilizing VSMC from various sources. Cerebral VSMC from aged dogs have served as a model for Alzheimer's CAA to study their ability to produce β-protein *(8)*, human vascular smooth muscle cells have been used to establish the toxic effects of the amyloid β-protein *(9)* and we are using cerebral VSMC for studying HCHWA-I. Here we present methods to obtain human VSMC from two distinct sources. The umbilical cord, which is a good source of robust VSMC and human brain leptomeningal tissue containing the small arteries that play central role in the pathology of CAA.

2. Materials

2.1. Special Equipment and Supplies

1. Autoclave.
2. Inverted microscope.
3. Laminar flow cabinet (clean hood).
4. CO_2 Incubator.
5. Water bath at 37°C.
6. Centrifuge for 12–15 mL tubes.
7. Dissecting microscope (optional).
8. Fluorescence microscope.
9. Hemocytometer.
10. Sterile scalpels.
11. For the umbilical cord: Two sets of sterile blunt needles with a valve (*see* **Note 1**) and straight 5-in to 7-in hemostatic forceps.
12. For the brain: Sterile micro dissecting forceps and scissors.
13. Disposable sterile syringes, 25 and 50 mL.
14. Syringe filter 0.2 μm.
15. Sterilized 250-mL glass beakers.
16. Disposable sterile centrifuge tubes, 12–15 mL and 50 mL.
17. Glass cover slips, 22 × 22 mm, sterilized by autoclaving.
18. Disposable plastic Petri dishes about 55 cm².
19. Disposable 9 cm² culture dishes or 25 cm² flasks.

2.2. Reagents

1. Hank's Balanced Salt Solution (HBSS) (Gibco 24020-083).
2. HBSS Ca^{++} and Mg^{++} (Gibco 14170-070).
3. Sterile Saline Solution (150 m*M* NaCl).

4. Chymotrypsin solution (0.2 mg/mL): Dissolve 8 mg α-chymotrypsine (type II at 50 U/mg, Sigma C-4129) in 40 mL of Ca^{++} and Mg^{++}-free HBSS and sterilize by using a syringe and an 0.2 µm ultra syringe filter. Make fresh each time.

5. Collagenase solution (0.4 mg/mL): Dissolve 16 mg collagenase (type IA at ≥125 U/mg, Sigma C-2674) in 40 mL Ca^{++} and Mg^{++}-free HBSS and filter sterilize as previously. Make fresh each time.

6. Dulbecco's Modified Eagle Medium (DMEM) with GLUTAMAXI™ (Gibco 31966-021).

7. Fetal Bovine Serum (FBS) (Gibco 10108-165).

8. Penicillin/Streptomycin solution (5000 U/5000 mg Gibco 15070-022).

9. Culture medium for umbilical VSMC: Place 4 mL of FBS into 50 mL disposable centrifuge tube add 400 µL of Penicillin/Streptomycin and fill to the 40 mL mark with DMEM. Make fresh each time.

10. Enzyme solution for cerebral VSMC: Place 10 mL of DMEM in a sterile centrifuge tube and add 0.1 mL of penicillin/streptomycin solution along with 150 µL of 1 *M* HEPES buffer (Gibco 15630-049). Weigh 10 mg of Protease (Dispase) (at 0.45 U/mg, Sigma P-3417) and 20 mg of Collagenase (type IA, Sigma C-2674) and add to the tube. Mix the enzyme solution thoroughly and filter sterilize.

11. Insulin-like growth factor-I (IGF-I) (Sigma I-3769). Prepared a stock solution by reconstituting the 50 µg of IGF-I in 100 µL of 0.1 *M* sterile filtrated acetic acid (228 µL acetic acid 100% in 40 mL water). In this form it will keep for maximum of 3 mo in the refrigerator but for an extended period in the freezer. Since repeated freezing and thawing is not recommended it should be frozen in working aliquots.

12. Hydrocortisone (Sigma H-0135). Hydrocortisone stock solution is prepared by dissolving 1 mg hydrocortisone in 1.0 mL absolute ethanol and adding 19 mL DMEM. Keep frozen in working aliquots.

13. Culture medium for cerebral VSMC: Place 4 mL of FBS into a 50 mL sterile centrifuge tube and filling to 40 mL with DMEM. Then add 0.4 mL of the Penicillin/Streptomycin solution, 40 µL of the hydrocortisone solution and 4 µL of the Insulin-like growth factor solution. Make fresh each time.

14. Phosphate buffered saline (PBS):

    3.1 g $NaH_2PO_4 \cdot H_2O$;
    10.9 g Na_2HPO_4 (anhydrous);
    9.0 g   NaCl;
    fill to 1 L with distilled water.
    The pH should be 7.4 (adjust if needed), autoclave before use.
    Will keep for 1 mo at 4°C.

15. PBSA: Add 0.5 g bovine serum albumin (BSA) to 50 mL PBS. Make fresh and sterilize by filtration before use.

16. Trypsin solution (0.25%): Add 1 mL of 2.5% Trypsin (Gibco 25090-028) to 9 mL PBSA. Should be made fresh each time.

17. Triton X-100.
    Since Triton X-100 is particularly viscous make up 20% stock solution by placing 1 mL into a graduated cylinder and fill to 5 mL with PBS.

18. Blocking solution: For each culture dish you need:
    1 mL   PBSA;
    50 µL   Normal goat serum (NGS);
    5 µL   20% Triton X-100.
19. Trypan blue stain, 0.4% (Gibco 15250-061).
20. Dimethyl sulfoxide (DMSO) (Sigma D-5879).
21. Triton X-100.
22. Methanol at –20°C.
23. Monoclonal anti-α-smooth muscle actin (Sigma A-2547).
24. Alexa Fluor 546 goat anti-mouse IgG (Molecular Probes A-1018).
25. Vectrashield mounting medium with DAPI, hard set (Vector H-1500).
26. Clear nail polish.

3. Methods

This subheading describes methods for isolating human vascular smooth muscle cell from two different sources. Cells from umbilical cord are relatively easy to work with and more readily available. These cells can be used as experimental paradigm to tackle various questions related to CAA. We also describe a method for extracting VSMC from human brain tissue, since in many instances it may be necessary to go right to the heart of the matter and work with cerebral VSMC, which are more difficult to isolate and adequate tissue samples more difficult to secure.

3.1. Vascular Smooth Muscle Cells Extracted From Umbilical Cord

Culture of pure smooth muscle cells is obtained by stripping away the endothelial lining from the blood vessel lumen prior to the smooth muscle cell extraction. Chymotrypsin detaches the endothelial cells without penetrating the collagen rich basal lamina leaving the smooth muscle cells intact *(10)*. Digestion with collagenase will subsequently yield smooth muscle cell culture with degree of purity that rests solely on the success of the endothelial removal. The following method is restricted to one umbilical cord, but in practice we usually harvest cells from few cords in each session.

3.1.1. Preparing the Umbilical Cord

1. Start the procedure by taking all the necessary reagents out of the freezer or refrigerator and place in the clean hood. Then fill a 250-mL glass beaker with sterilized saline and secure in the 37°C water bath. While things are warming up the umbilical cord is fetched.
2. Cut the umbilical cord from the placenta, placed in a plastic bag and transported on ice to the laboratory.

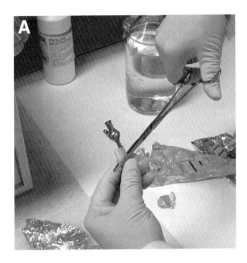

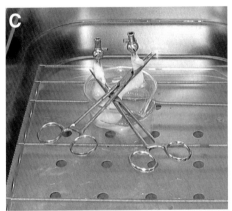

Fig. 1. (**A**) Thread a blunt-ended needle, fitted with a valve, into one of the arteries and seal with hemostatic forceps. (**B**) Fill a blood vessel of the cord to its full capacity with the enzyme solution. (**C**) The hemostatic forceps are used to secure the cord by crossing them on top of the saline-filled beaker. The saline solution in the beaker is maintained at 37°C in the water bath.

3. Rinse the cord thoroughly in running tap water, making sure that no water enters the blood vessels and do one final wash with sterile saline.
4. Wipe the cord with a clean paper towel. Now, all subsequent work is performed on a clean surface using only sterile material and instruments, starting by cutting few centimeters from each end of the cord with a sterile scalpel.
5. Thread a blunt needle, attached to a valve, into one of the two arteries (*see* **Note 2**), and sealed with haemostatic forceps (**Fig. 1A**).
6. Attach a saline filled 50-mL syringe to the valve and rinse the blood vessel clean of all blood. The other end of the same blood vessel is fitted in the same way with

a blunt needle plus a valve and now the umbilical cord is ready for the enzymatic treatment.

3.1.2. Digestion

1. Close the valve at one end of the umbilical cord and attach a syringe with the chymotrypsin solution to the other end. Subsequently, the blood vessel is filled with the enzyme solution under enough pressure to balloon the cord (**Fig. 1B**), but without breaking the seal. Remove the syringe after closing the valve.
2. Place the cord in the saline filled beaker in the 37°C water bath, arranging the hemostatic forceps across the beaker to secure the cord (**Fig. 1C**).
3. Leave to incubate for 20 min and then decant the enzyme solution through one of the valves and discard (the endothelial cells can be harvested and cultured).
4. Fill the blood vessel twice with HBSS and squeeze out the liquid to wash away any residual endothelial cells.
5. Now, the blood vessel is filled with the collagenase solution and placed in the 37°C saline, as described earlier, and incubate for 10 min.
6. The enzyme solution containing the smooth muscle cells is decanted into sterile 12–15 mL centrifuge tubes and the cells sedimented at 140g for 5 min in the centrifuge.

3.1.3. Culturing the Cells

1. Discard the supernatant and resuspend the cells in 3 mL of the tissue culture medium prepared as described before (*see* **Subheading 2.2., step 9.**).
2. Transfer the cell suspension into one 25-cm^2 culture flask and add 3 mL of prepared culture medium to make total of 6 mL (*see* **Note 3**). On the other hand, if the cells are being prepared for microscopic evaluation, place autoclaved coverslip (22 × 22 mm) into each of three disposable 9 cm^2 culture dishes, and put 1 mL of the cell suspension, along with 2 mL of the prepared culture medium, in each dish (*see* **Notes 3** and **4**).
3. The cells are placed in the CO_2 incubator, set at 37°C and 5% CO_2.

3.1.4. Maintaining the Cells

On the second day in culture, the medium is changed to remove any cell debris. Wash the cells once with HBSS before adding the fresh culture medium. In few days the cells will lose the typical smooth muscle phenotype and enter a fibroblast-like synthetic state and commence exponential growth reaching confluence in the second week. If the cells will be passed on to produce more cells in subcultures they are not allowed to become fully confluent and proceeded as described in **Subheading 3.3.1.** However, if the cells are being prepared for staining, or other experimental procedure, grow them to full confluence, and, for the last 2 d exchange the culture media with serum-free media (prepare the media as before, but omitting the FBC). This should return them to the spindle-shape contractile phenotype (**Fig. 2**).

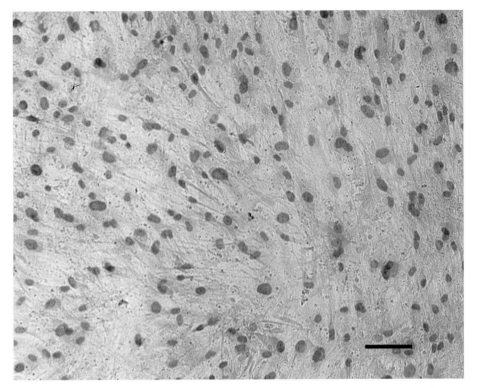

Fig. 2. Confluent culture of H & E stained vascular smooth muscle cells. Bar = 100 μm.

3.2. Vascular Smooth Muscle Cells Isolated From Human Brain Tissue

For isolation and culturing of human cerebral VSMC, we use a method outlined by Van Nostrand and coworkers *(11)* and present it here as we have come to apply it.

3.2.1. Setting Up

Prior to fetching the tissue sample, make sure that you have autoclaved fine instruments (forceps and scissors), arranged a tray for crushed ice under a dissecting microscope (with practice we find it possible to forgo the microscope), and prepared the protease/collagenase solution (*see* **Subheading 2.2., step 10**) aseptically in the laminar hood and place it on ice.

3.2.2. Preparation and Digestion of the Sample

Brain sample should be as fresh as possible, either from autopsy few hours postmortem, or from surgical brain operation. Make sure that the sample is from the cortical surface containing mostly leptomeningal tissue.

1. Place the sample aseptically into a bottle of cold sterile HBSS and transport on ice to the tissue culture facility.
2. Transfer the tissue sample into a sterile Petri dish containing ice cold HBSS and peel the leptomeninges from the residual cortical tissue. Move 1–2 cm^2 meningal tissue to another HBSS-filled Petri dish and place it on crushed ice under the microscope.
3. Take fine forceps in each hand and tease the meningal tissue apart to free the blood vessels from the supporting tissue. This is the most critical part of this procedure requiring both patience and great care.
4. Transfer the fragments of blood vessels to the cold enzyme solution and cut the larger ones into smaller pieces with fine scissors.
5. Incubate the tube on ice for 90 min to ensure that the enzymes penetrate the tissue fragments before they are activated at elevated temperature.
6. Transfer the tube to the 37°C water bath and incubate for additional 2 h to digest the tissue.
7. While the sample is incubating prepare the culture medium as described in **Subheading 2.2., step 13**.

3.2.3. Culturing the Cells

1. Following the incubation at 37°C, the tissue fragments are triturated by vigorously pipetting up and down to break up the tissue and liberate the cells.
2. Transfer the tube to a centrifuge and pull down the tissue debris at 140g for 5 min.
3. Discard the supernatant and suspend the tissue pellet in 1 mL of the prepared culture medium. Triturate again by pulling the mixture few times into the pipet tip.
4. Place the tissue suspension in 25 cm^2 culture flask (*see* **Note 5**) making the final volume 6 mL by additonal prepared culture medium. Alternatively, the cells can be seeded into a dish with a cover slip as described in **Subheading 3.1.3.** (*see* **Note 4**).
5. Finally, the flasks are placed in the incubator at 37°C and 5% CO_2.

3.2.4. Maintaining the Cells

All the same things apply as in **Subheading 3.1.4.**, except that these cells are usually not as vigorous and grow a little slower.

3.3. Miscellaneous Techniques

3.3.1. Subcultures of the Cells

When the cells are close to confluence they can be split into subcultures, which is also referred to as passing the cells. It is our experience that a 1:3 splitting ratio suits the vascular smooth muscle cells well, that is, one culture is divided into three, but if needed they can easily be stretched into four cultures.

1. Transfer the culture flasks from the incubator to the laminar hood and discard the medium.

2. Wash the cells once in PBSA and place 3 mL of the diluted trypsin solution into each flask.
3. After 30 s the trypsin is decanted and the flasks incubated for 5–15 min or till the cells start to round up. Monitor the process using the inverted microscope and help the cells to detach by taping the flask gently on the tabletop. Trypsin causes cellular damage, so keep the time of exposure to a minimum (*see also* **Note 6**).
4. When most of the cells have rounded up add 1 mL culture medium to the flask. Squirt the medium a few times down the culture surface of the flask, held at an angle, to pry the cells loose.
5. Dilute the cell suspension further with 2 mL of medium and divide them by transferring 1 mL to each of three flasks. Finally, add 5 mL of medium to each flask.
6. Place the flasks into the CO_2 incubator at 37°C.

3.3.2. Cell Counting

To determine the cell seeding density, the cells need to be counted (*see* **Note 5**). Moreover, quantification is frequently a part of the experimental procedure.

1. The process starts by putting 0.5 mL of 0.4% trypan blue solution into a small tube along with 0.3 mL of HBSS and then adding 0.2 mL of the cell suspension (dilution factor = 5) and mixing carefully (*see* **Note 6**). Keep in mind that trypan blue is harmful and protective gloves should be worn.
2. The hemocytometer is thoroughly cleaned with alcohol and the coverslip put in place on top of it to cover the two chambers.
3. Transfer a small amount of the cell mixture by a pipet tip to the edge of the coverslip and the chambers are allowed to fill by capillary action making sure they do not over- or under fill.
4. Count all the cells in the center 1×1 mm square and the four corner squares (**Fig. 3**) in both chambers, ten squares in all. Cells that stain blue are either dead or not viable and are counted separately as such. Cells that touch the middle line of the triple lines at the top and on the left side of each square are not counted.

Since each square represent 0.1 mm^3 (approx 10^{-4} mL) the number of cells per mL is calculated by:

$$\text{Cells per mL} = \text{the average count per square} \times \text{dilution factor} \times 10^4$$

Best accuracy is acquired if each square contains between 20 and 50 cells. Adjust by changing the dilution factor if needed.

3.3.3. Freezing and Thawing the Cells

Instead of passing the cells right away it is possible to freeze them and use them later.

1. Detach the cells by trypsin, as described in **Subheading 3.3.1.**, and suspend them in 3 mL of freezing solution, made up of 95% FCS and 5% DMSO, instead of culture medium. Caution: do not let DMSO come in contact with any part of you (it will penetrate many types of gloves).

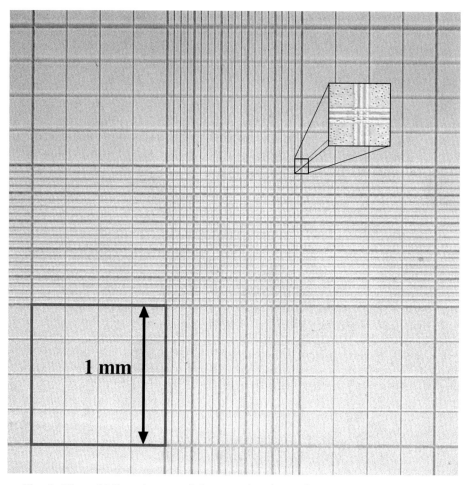

Fig. 3. The grid lines in one of the two chambers of a standard hemocytometer. Count the cells in the center and the four corner 1 × 1 mm squares. Do not count any cells touching the middle of the triple line (*see* insert) at bottom and right of each square.

2. Split the 3 mL of cells suspension between three 1.5-mL freezing tubes.
3. Freeze the cells in a −70°C freezer. To slow down the freezing rate, to the optimal −1°C/minute, the tubes can be fitted into a close Styrofoam cube with a 2-cm wall thickness on each side.
4. Next day, the tubes are transported to liquid nitrogen, where they will keep for a long time.

When the cells are needed thaw them rapidly.

1. Drop the tube into a bucket containing 37°C water and cover with a lid. Protective goggles and gloves must be worn since the tubes can explode violently if liquid nitrogen is trapped inside.
2. Transfer the cell suspension to a culture flask and add 5 mL of medium slowly, drop by drop over 1 min.
3. Place the flask into the CO_2 incubator at 37°C.

3.3.4. Immunostaining the Cells

The cells can be immunostained with a host of different antibodies, but as an example we describe staining for a smooth muscle actin that serves to confirm the smooth muscle phenotype of the cells.

1. Discard the culture medium and wash the cells once in few mL of cold PBS.
2. After pouring of the washing solution the cells are fixed in few mL of −20°C methanol for 10 min.
3. Remove the methanol and wash the cells 3 times in PBS for 5 minutes each time and then do one final wash in PBSA.
4. To block unspecific staining add 1 mL of blocking solution to each 9 cm^2 culture dish and incubate for 30 min at room temperature.
5. Dilute the primary antibody 1:400 by adding 2.5-μL mouse anti-α-smooth muscle actin to 1 mL PBSA.
6. Remove the blocking solution and add the diluted antibody, without washing the cells, and incubate at room temperature for 30 min or overnight at 4°C (*see* **Note 7**).
7. Then the cells are washed three times with PBSA for 5 min each time and the secondary antibody at appropriate dilution added. We use Alexa Fluor 546 goat antimouse at 1:500 dilution and incubate for 30 min in the dark.
8. The secondary antibody is rinsed out by two 5 min washes in PBSA and one with PBS.
9. Take the coverslip with the stained cells out of the culture dish with forceps and shake excess wetness from the cells. The cell-free side and edges of the coverslip are dried with a paper towel.
10. Placed the coverslip on a table edge, put one drop of the mounting medium on the center and lower a reversed microscope slide on top of it. Try to let the coverslip attach without introducing any air bubbles.
11. Allow the mounting medium to harden for 15 min in the dark and then the cells are ready for the microscope (**Fig. 4**).

The mounted slides can be kept for later use at 4°C and protected from the light. It is advisable to seal the edges of the coverslip with clear nail polish to protect the sample from drying. The mounting medium will slow the fading of the fluorescence intensity, making the stained slides usable for several weeks for most fluorescent reagents.

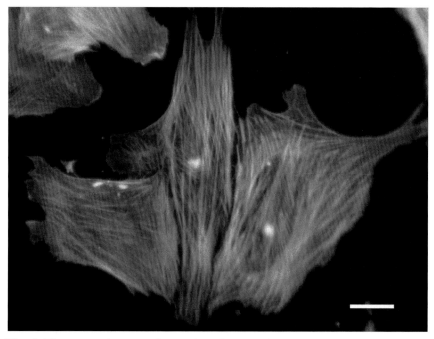

Fig. 4. Three vascular smooth muscle cells grouped together in a culture, immuno-stained with anti-α-smooth muscle actin. Bar = 10 μm.

4. Notes

1. In our laboratory we have old reusable 16-gage 1.5-in needles with a 2-mm ball, to make the end blunt, and a valve with a Luer lock attached at the inlet end. We have not been able to find this item in any catalog, but as an alternative there are separate blunt needles and valves that can be locked together with Luer fittings. George Tiemann & Co. carries a valve with female Luer to male Luer needle lock (160-7021) and an appropriate straight ball pointed 18-gage, 1.5-in long needle, with a 2.25-mm ball (160-8904) or 16-gage, 3-in long needle, with a 3.0-mm ball (160-8908). Roboz has 16G, 2-in needles with 3-mm ball, reusable (FN-7923) or disposable (FN-9918). Similar needles can also be obtained from Fine Science Tools.

2. We occasionally use the large central vein, since it is more accessible and easier to work with. However, the arteries are richer in VSMC and there may be additional reasons for preferring cells of that origin. As an extra bonus, there are two of them and we have succeeded in harvesting both in the same cord.

3. Seeding densities of smooth muscle cells are extremely important and should be between 7000 to 70,000 cells per cm^2. With too few cells they will go through too many population doublings and exceed their ability to return to the contractile phenotype before reaching confluence. However, with too many cells, confluence

will be reach with very limited proliferation. To determine the proper dilution the cells are counted in a hemocytometer (*see* **Subheading 3.3.2.**).

4. If the cells in the primary culture will in due course be prepared for light microscopic evaluation, then we grow them on glass coverslips and proceed as explained in **Subheading 3.3.4.** However, since the cells are usually passed a few times before they are utilized for experiments, a better choice is the 25-cm^2 culture flask. In the final passing, the cells are seeded on coverslips as described.

5. As explained in **Note 3**, minimum seeding density is crucial for a successful culture. However, the digested leptomeningal sample contains tissue fragments along with the liberated cells, making counting difficult. We bypass this problem by seeding them in a limited area, usually only one culture flask per digested sample. As soon as the cells have attached, the debris is washed away by changing the medium.

6. The smooth muscle cells tend to clump together, especially cells from confluent cultures. This makes them more difficult to count, and for that reason it is important to trypsinize completely and then try to disperse the cells as possible in the trypan blue cell suspension by thorough mixing. But keep in mind that trypsin is harmful to the cells.

7. Double immunostaining can be performed by adding a second polyclonal antibody in appropriate dilution and incubating along with the anti-α-actin monoclonal antibody. A suitable secondary antibody is then used along with the anti-mouse Alexa Fluor 546, such as Alexa Fluor 488. The mounting medium that we employ contains DAPI nuclear stain so in a microscope with the right filter set the cells will be triple stained.

Acknowledgments

We would like to thank The Icelandic Research Council and Heilavernd (The Icelandic HCHWA-I Foundation) for their support. Also, Professor H. Blöndal and the rest of the staff at the Department of Anatomy for their assistance.

References

1. Vinters, H. V. (1987) Cerebral amyloid angiopathy. A critical review. *Stroke* **18,** 311–324.

2. Revesz, T., Holton, J. L., Lashley, T., et al. (2002) Sporadic and familial cerebral amyloid angiopathies. *Brain Pathol.* **12,** 343–357.

3. Levy, E., Carman, M. D., Fernandez-Madrid, I. J., et al. (1990) Mutation of the Alzheimer's disease amyloid gene in hereditary cerebral hemorrhage, Dutch type. *Science* **248,** 1124–1126.

4. Gudmundsson, G., Hallgrimsson, J., Jonasson, T. A., and Bjarnason, O. (1972) Hereditary cerebral haemorrhage with amyloidosis. *Brain* **95,** 387–404.

5. Bjarnadottir, M., Nilsson, C., Lindstrom, V., et al. (2001) The cerebral hemorrhage-producing cystatin C variant (L68Q) in extracellular fluids. *Amyloid: Int. J. Exp. Clin. Invest.* **8,** 1–10.

6. Wisniewski, H. M., Fraçkowiak, J., Zòltowska, A., and Kim, K. S. (1994) Vascular β-amyloid in Alzheimer's disease angiopathy is produced by proliferating and deghenerating smooth muscle cells. *Amyloid: Int. J. Exp. Clin. Invest.* **1,** 8–16.

7. Wang, Z. Z., Jensson, O., Thorsteinsson, L., and Vinters, H. V. (1997) Microvascular degeneration in hereditary cystatin C amyloid angiopathy of the brain. *Apmis* **105,** 41–47.

8. Wisniewski, H. M., Frackowiak, J., and Mazur-Kolecka, B. (1995) In vitro production of β-amyloid in smooth muscle cells isolated from amyloid angiopathy-affected vessels. *Neuroscience Letters* **183,** 120–123.

9. Van Nostrand, W. E., Davis-Salinas, J., and Saporito-Irwin, S. M. (1996) Amyloid β-protein induces the cerebrovascular cellular pathology of Alzheimer's disease and related disorders. *Ann. NY Acad. Sci.* **777,** 297–302.

10. Mulder, A. B., Blom, N. R., Smit, J. W., et al. (1995) Basal tissue factor expression in endothelial cell cultures is caused by contaminating smooth muscle cells. Reduction by using chymotrypsin instead of collagenase. *Thromb. Res.* **80,** 399–411.

11. Van Nostrand, W. E., Rozemuller, A. J. M., Chung, R., Cotman, C. W., and Saporito-Irwin, S. M. (1994) Amyloid β-protein precursor in cultured leptomeningeal smooth muscle cells. *Amyloid: Int. J. Exp. Clin. Invest.* **1,** 1–7.

14

Murine Cerebrovascular Cells as a Cell Culture Model for Cerebral Amyloid Angiopathy

Isolation of Smooth Muscle and Endothelial Cells From Mouse Brain

Sonia S. Jung and Efrat Levy

Summary

The use of murine cerebrovascular cells, that is, endothelial and smooth muscle cells, has not been widely employed as a cell culture model for the investigation of cellular mechanisms involved in cerebral amyloid angiopathy (CAA). Difficulties in isolation and propagation of murine cerebrovascular cells and insufficient yields for molecular and cell culture studies have deterred investigators from using mice as a source for cerebrovascular cells in culture. To date, most of the literature has described isolation of smooth muscle cells or endothelial cells from human, canine, rat, guinea pig, or other large animals.

In recent years, several transgenic mice have been established that show CAA pathology; therefore, it is necessary to re-examine the use of mouse cerebrovascular cells as an important model for cell culture studies. We have optimized the isolation procedure of (1) murine microvessels, (2) smooth muscle cells, and (3) endothelial cells to yield a sufficient population of cells for experimentation purposes. Comparisons with rat and human isolation procedures are also noted. Murine smooth muscle cells isolated using the methodology described herein exhibit the classic "hill and valley" morphology and are immunoreactive for smooth muscle cell-specific α-actin, whereas endothelial cells demonstrate a more "cobblestone" appearance and stain for von Willebrand factor or factor VIII-related antigen.

Key Words: Cerebral blood vessel; microvessel; cerebrovascular cell; smooth muscle cell; endothelial cell; primary cells; cystatin C; amyloid; α-actin; von Willebrand factor; factor VIII-related antigen.

1. Introduction

The deposition of amyloid β (Aβ) in the cerebrovasculature is one of the hallmarks of Alzheimer's disease (AD) and related disorders, termed cerebral amyloid angiopathy (CAA). In hereditary cerebral hemorrhage with amyloidosis of

From: *Methods in Molecular Biology, vol. 299: Amyloid Proteins: Methods and Protocols*
Edited by: E. M. Sigurdsson © Humana Press Inc., Totowa, NJ

the Icelandic type (HCHWA-I) *(1)*, also known as hereditary cystatin C amyloid angiopathy (HCCAA) *(2)*, the deposited protein is a variant form of cystatin C *(3,4)*. Amyloid deposition in cerebral vasculature results in the degeneration of smooth muscle cells and endothelial cells, which leads to a decrease in the integrity of the blood vessel wall, ultimately resulting in hemorrhagic stroke and death *(5,6)*.

Studies on cerebrovascular smooth muscle cells from human brain (obtained at autopsy) *(7,8)*, canine *(9–11)*, rat *(12–14)* and other large animals *(15,16)* have been reported. However, brain tissue from these sources can be difficult to obtain and the methods for the isolation of smooth muscle cells from these sources are widely varied in the literature. Isolation and propagation difficulties of murine cerebrovascular cells along with insufficient yields for practical molecular and cell-culture studies have deterred investigators from using mice as a source for starting material in the past *(17)*. Several of the transgenic mice that have been created as models for AD in recent years, have demonstrated significant CAA pathology *(18–20)*. CAA is a common pathological hallmark of AD, and a recent population-based study demonstrated that having both CAA and AD contributes to substantially worse cognitive performance than AD alone *(21,22)*, re-emphasizing the need to further understand the mechanisms that underlie CAA pathogenesis. We have generated transgenic mice carrying the human cystatin C gene. Neuropathological examination of aged transgenic mice revealed mice with cerebral or subarachnoid hemorrhages. None of the mice had amyloid deposition either in the vessel walls or in the neuropil, demonstrating that elevated brain and/or blood levels of cystatin C can cause hemorrhagic strokes in the absence of vascular amyloid deposits *(23)*. It is evident that these mice are an important resource for the investigation of mechanisms involved in the degeneration of cerebrovascular cells.

Here, we describe the isolation and conditions for propagation of murine microvessels, smooth muscle cells, or endothelial cells. Gentle enzymatic digestion and filtration through a glass-bead column result in the isolation of microvessels, which can be further treated to yield vessel fragments that will produce either smooth muscle cells or endothelial cells depending on the culture media. Murine cerebrovascular cells provide a valuable model for the investigation of CAAs.

2. Materials

1. Murine brain tissue.
2. 70% Ethanol.
3. Sterile scalpels, forceps, tweezers, scissors.
4. 1X Sterile Dulbecco's Phosphate Buffered Saline (PBS), without calcium and magnesium (Invitrogen Corporation, Piscataway, NJ, no. 14190-144).

5. 1X Sterile Hank's Balanced Salt Solution (HBSS, Invitrogen Corporation, no. 14175-095).
6. Fetal Bovine Serum (FBS), heat-inactivated (Gemini Bio-Products, Woodland, CA, no. 100-106).
7. Dulbecco's Modified Eagle's Medium (DMEM, Invitrogen Corporation, no. 11960-044).
8. Penicillin/Streptomycin, 10,000 U/mL penicillin base and 10,000 µg/mL streptomycin base (Invitrogen Corporation, no. 15140-122).
9. L-glutamine, 200 mM (Invitrogen Corporation, no. 25030-081).
10. Non-essential amino acids, 10 mM (Invitrogen Corporation, no. 11140-050).
11. 1 M N-2-Hydroxyethylpiperazine-N'-2-ethane sulfonic acid (HEPES), pH 7.5 (Invitrogen Corporation, no. 15630-080). Prepare a solution of 20 mM HEPES in DMEM and store at 4°C.
12. DNase, Type I, from bovine pancreas, grade II (Roche Diagnostics Corporation, no. 104159). Prepare a 10 mg/mL stock solution in PBS, filter sterilize through a 0.22-µm membrane filter, aliquot and store at −20°C.
13. Endothelial cell growth supplement (ECGS) from Bovine Neural Tissue (Sigma-Aldrich Co., E-2759). Prepare stock solution of 3 mg/mL in sterile PBS, filter sterilize through a 0.22-µm membrane filter.
14. Prepare media for culturing microvessels and smooth muscle cells: DMEM containing 10% FBS, 2 mM L-glutamine, 100 U/mL penicillin, 100 µg/mL streptomycin, 0.1 mM non-essential amino acids, and 50 µg/mL DNaseI. Store at 4°C.
15. Prepare media for culturing endothelial cells: DMEM containing 10% FBS, 2 mM L-glutamine, 100 U/mL penicillin, 100 µg/mL streptomycin, 1 mM non-essential amino acids, 50 µg/mL DNaseI, and 20 µg/mL endothelial cell growth factor (ECGS). Recommended ECGS concentration is 75–300 µg/mL, but 20 µg/mL is sufficient. Store at 4°C.
16. Collagenase/dispase (Roche Diagnostics Corporation, Indianapolis, IN, no. 1097113). Prepare a 100 mg/mL stock in ddH$_2$O and dilute with PBS to a concentration of 10 mg/mL. Sterilize through a 0.22-µm membrane filter, aliquot and store at −20°C. Use 0.05% as the working concentration.
17. 0.5 M EDTA, pH 8.0, sterile (Invitrogen Corporation, no. 15575-020).
18. Dextran (Sigma-Aldrich Co., St. Louis, MO, D-3759). Prepare a 17% dextran solution on the day of use in DMEM containing 20 mM HEPES. Sterilize through a 0.22-µm membrane filter unit and store at 4°C.
19. Trypsin, 2.5% (Invitrogen, no. 15090-046).
20. Poly-D-lysine (Sigma-Aldrich Co., P-7886). Prepare a stock solution of 1 mg/mL poly-D-lysine in ddH$_2$O. Sterilize through a 0.22-µm membrane filter, aliquot and store at 4°C. Use at a working concentration of 20 µg/mL to coat glass slides before plating cells.
21. Albumin, bovine, Fraction V (Sigma-Aldrich Co., A-7906). Prepare a 5% BSA solution in PBS to use as blocking solution for immunocytochemistry.
22. Tissue culture dishes, 100 mm (Falcon, no. 353003).
23. Tissue culture dishes, 35 mm (Falcon, no. 353001).

24. 50 mL Oak Ridge polycarbonate tubes (Nalgene, no. 3118-0050). Sterilize by auto-claving.
25. 70-μm nylon cell strainer (Falcon, no. 352350).
26. 50-mL conical tubes (Falcon, no. 352098).
27. Glass beads, 425–600 μm, acid-washed (Sigma-Aldrich Co., G-8772). Autoclave glass beads in a small beaker. When ready to use, transfer some into a 70-μm nylon cell strainer and wash thoroughly with either sterile PBS or HBSS.
28. 8-Well glass chamber slides (Lab-Tek II, no. 154534).
29. Gel/Mount (Biomeda Corp., Foster City, CA, no. M01).
30. Cover slips (Fisher Scientific, Suwanee, GA).
31. Methanol, 100%, chilled at –20°C.
32. Anti-α smooth muscle actin antibody (Sigma-Aldrich Co., A-2547).
33. Anti-human von Willebrand factor (Factor VIII-related antigen) antibody (Sigma-Aldrich Co., F-3520).
34. Sheep anti-mouse Ig, fluorescein-linked whole antibody (Amersham Biosciences, Piscataway, NJ, N1031).
35. Texas Red Anti-rabbit IgG (H+L) antibody (Vector Laboratories, Burlingame, CA, TI-1000).

3. Methods

The methods described in the following outline the isolation of microvessels from mouse brain, the isolation of smooth muscle cells, the propagation of endothelial cells, and the identification of the cells once in culture. This protocol describes the procedure for isolation of vessels and cells from mouse brain. Although vessels and cells can be isolated from one mouse brain, a better yield is obtained when at least two mouse brains are used for the isolation. All steps must be performed in a sterile environment, including the removal of the mouse brain. All solutions should be filtered through a 0.22-μm membrane filter.

3.1. Isolation of Mouse Brain and Microvessels

1. Sacrifice a mouse by a method approved by your animal care committee.
2. In a laminar flow hood, wet the fur of the mouse with 70% ethanol.
3. Remove mouse brain into ice-cold DMEM containing 20 mM HEPES in a 50-mL conical tube (*see* **Note 1**).
4. Mince the brain into small pieces using a sterile scalpel.
5. Centrifuge the tissue for 5 min at 500g, at room temperature.
6. Resuspend the tissue in ample 0.05% collagenase/dispase in PBS; use a minimum of 3 mL per mouse brain (*see* **Notes 2** and **3**).
7. Digest for 15 min at 37°C, pipetting every few minutes to dissociate the tissue.
8. Stop the digestion with cold DMEM containing 20 mM HEPES and 1 mM EDTA (*see* **Note 4**).
9. Centrifuge for 10 min at 1000g at room temperature.
10. Discard the supernatant.

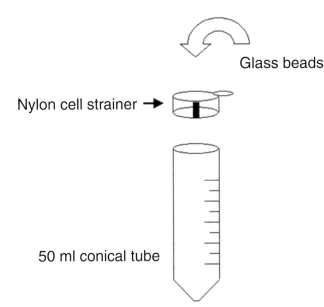

Fig. 1. A 70-μm nylon cell strainer fits perfectly into the opening of a sterile 50-mL conical tube.

11. Resuspend the tissue in 17% dextran in DMEM containing 20 mM HEPES and transfer the suspension to a sterile polycarbonate Oak Ridge tube (*see* **Note 5**).
12. Ultracentrifuge at 10,000g for 30 min at 4°C using a fixed-angle rotor (e.g., Sorvall centrifuge RC2-B, Rotor no. SS34).
13. Remove the supernatant and fatty top layer by vacuum aspiration. The red pellet contains vessels.
14. Wash sterile glass beads several times with sterile PBS or HBSS to remove debris and acid.
15. Transfer autoclaved glass beads into a 70-μm nylon-cell strainer and fit the strainer into a 50-mL conical tube (*see* **Fig. 1** and **Note 6**).
16. Resuspend the pellet with vessels (from **step 13**) in 10 mL culture medium (DMEM containing 10% FBS, 2 mM L-glutamine, 100 U/mL penicillin, 100 μg/mL streptomycin, 0.1 mM nonessential amino acids, and 50 μg/mL DNaseI).
17. Pass the mixture over the sterile glass beads.
18. Transfer the strainer to a new 50 mL tube.
19. Pass the flow-through over the beads a few more times.
20. Transfer the glass beads to a 100-mm tissue culture dish.
21. Rinse out the strainer with culture medium to dislodge any remaining glass beads.
22. Gently shake the dish to disturb the beads—vessels will dissociate from the glass beads.
23. Collect vessels with a sterile pipet and place in a new 50-mL tube.
24. Centrifuge for 10 min at 1000g at room temperature.

25. Resuspend the vessels in culture medium and plate into a 35-mm tissue culture dish.
26. Incubate vessels at 37°C, 5% CO_2 in a humidified incubator.

3.2. Culturing Smooth Muscle Cells

1. Resuspend the pellet with vessels from **step 23** (*see* **Subheading 3.1.**) in 0.05% collagenases/dispase and incubate for 2.5 to 3 h in a 37°C water bath, gently pipetting every 15 min, taking care to not introduce bubbles into the mixture.
2. Add EDTA, pH 8.0, to a final concentration of 1 mM to stop the reaction.
3. Centrifuge for 10 min at 1000g, at room temperature.
4. Discard the supernatant.
5. Resuspend the pellet in 5 mL culture medium.
6. Plate in a 35-mm tissue culture dish.
7. Incubate at 37°C, 5% CO_2 humidified incubator.
8. 48 h after plating, change the medium to remove cellular debris.
9. Examine daily.
10. Smooth muscle cells appear after 2 to 3 d in culture.
11. Change the medium every 3 to 4 d.
12. Passage every 7 to 10 d using 0.25% trypsin (*see* **Note 7**).
13. Cells can be frozen at early passages and stored in liquid nitrogen.

3.3. Propagation of Endothelial Cells

1. Follow the protocol for culturing smooth muscle cells (*see* **Subheading 3.2.**), except use the media for culturing endothelial cells that contain a minimum of 20 µg/mL ECGS.
2. Changing medium for endothelial cells: Endothelial cells secrete factors necessary for their growth. Feed every 3 to 4 d by replenishing 50% of medium and adding 50% fresh medium.
3. Endothelial cells appear after 2 to 3 d in culture.

3.4. Immunocytochemistry for the Identification of Cell Type

1. When cells are near confluence, split into 10 µg/mL poly-D-lysine coated glass chamber slides.
2. When cells are near confluence, gently wash cells three times with filter-sterilized PBS to remove medium.
3. Fix the cells in 100% methanol that was pre-chilled at −20°C for 15 min at room temperature.
4. Block nonspecific binding sites using 5% BSA in PBS for 1 h at room temperature.
5. Wash cells three times with PBS.
6. Add primary antibody: either anti-α smooth muscle actin or anti-von Willebrand factor antibodies, 1:200 dilution in PBS containing 5% BSA.
7. Incubate for 1 h at room temperature.
8. Wash cells three times with PBS.
9. Add fluorescently labeled secondary antibodies, 1:100 dilution in PBS containing 5% BSA.

10. Incubate for 1 h at room temperature.
11. Wash cells three times with PBS.
12. Mount cover slip with Gel/Mount and view using a fluorescence microscope.

4. Notes

4.1. General Notes

1. Vessels adhere to glass, therefore, plastic disposables should be used at all steps. Glass beakers, Pasteur pipets, etc. should be avoided.
2. Higher concentrations of collagenases/dispase are used in the isolation of smooth muscle cells from rat and human brains. For example, it has been reported that for the isolation of rat cerebral vessels 0.1% collagenases/dispase is used for 1 h at 37°C *(12–14)* and for human brains 0.5% collagenases/dispase at 4°C for 2 h followed by 37°C for 1.5 h *(8)*. These higher concentrations have been shown to be too harsh for mouse brains resulting in lower yields.
3. Elastase and hyaluronidase are often used in isolation methods in addition to collagenases/dispase *(12–14,25)*.
4. Collagenase/dispase is not inhibited by serum. Use 1 mM EDTA or EGTA to stop the reaction.
5. The percentage of dextran varies in the literature. While for rats 13 to 15% dextran has been cited *(12–14)* and 15% dextran has been used for humans *(24)*, a higher percentage (minimum 17%) is used to isolate vessel material from murine brain.
6. If separation of the vessel population on the basis of size is desired, separation can be done using a nylon-cell strainer without using glass beads. While leptomeninges and large cortical vessels do not pass through the 70-μm nylon-cell strainer, smaller cortical vessels pass through it *(25)*.
7. Murine smooth muscle cells require several weeks (>3 wk) before a suitable doubling time is reached for experimental purposes. The doubling time is very long in the first passages and the split ratio should be 1:2. In later passages, the doubling time shortens and a split ratio of 1:4 can be used.

4.2. Experimental Notes

1. The choice of the source of smooth muscle cells should be made with great consideration. Human cerebrovascular smooth muscle cells behave very differently from those obtained from human umbilical artery, and this could have a great effect on the outcome of the experiments. For example, human umbilical artery smooth muscle cells produce free radicals in response to Aβ treatment, whereas human adult smooth muscle cells do not *(26*, Jung and Van Nostrand, unpublished observations).
2. Experiments performed on smooth muscle cells often require incubation of the cells in serum-free medium. For example, treatment of human smooth muscle cells with Aβ is done in the absence of serum *(8)*. Furthermore, serum contains high levels of cystatin C, and therefore, for the purpose of investigating its effects on murine smooth muscle cells, cystatin C is reconstituted in medium without serum. Murine smooth muscle cells, however, do not survive in the absence of serum (Jung and

Levy, unpublished observations). Therefore, we have successfully used Opti-MEMI Reduced Serum Media (Invitrogen, no. 51985-034) containing 2 mM L-glutamine, 100 U/mL penicillin, 100 μg/mL streptomycin and 0.1 mM non-essential amino acids. Cells are viable up to 3 to 4 d in this medium in the presence or absence of cystatin C. After 4 d, smooth muscle cell morphology changes dramatically as cells start to degenerate.

Acknowledgments

This work was supported by grants from the National Institute of Neurological Disorders and Stroke (NS42029) and the American Heart Association (0040102N).

References

1. Gudmundsson, G., Hallgrimsson, J., Jonasson, T. A., and Bjarnason, O. (1972) Hereditary cerebral hemorrhage with amyloidosis. *Brain* **95,** 387–404.
2. Olafsson, I., Thorsteinsson, L., and Jensson, O. (1996) The molecular pathology of hereditary cystatin C amyloid angiopathy causing brain hemorrhage. *Brain Pathol.* **6,** 121–126.
3. Cohen, D. H., Feiner, H., Jensson, O., and Frangione, B. (1983) Amyloid fibril in hereditary cerebral hemorrhage with amyloidosis (HCHWA) is related to the gastro-entero-pancreatic neuroendocrine protein, γ trace. *J. Exp. Med.* **158,** 623–628.
4. Ghiso, J., Jensson, O., and Frangione, B. (1986) Amyloid fibrils in hereditary cerebral hemorrhage with amyloidosis of Icelandic type is a variant of γ trace basic protein (cystatin C). *Proc. Natl. Acad. Sci. USA* **83,** 2974–2978.
5. Vinters, H. V. (1987) Cerebral amyloid angiopathy. A critical review. *Stroke* **18,** 311–324.
6. Kalaria, R. N. (2001) Advances in molecular genetics and pathology of cerebrovascular disorder. *Trends Neurosci.* **24,** 392–400.
7. Vinters, H. V., Reave, S., Costello, P., Girvin, J. P., and Moore, S. A. (1987) Isolation and culture of cells derived from human cerebral microvessels. *Cell Tissue Res.* **249,** 657–667.
8. Van Nostrand, W. E., Rozemuller, A. J. M., Chung, R., Cotman, C. W., and Saporito-Irwin, S. M. (1994) Amyloid β-protein precursor in cultured leptomeningeal smooth muscle cells. *Amyloid: Int. J. Exp. Clin. Invest.* **1,** 1–7.
9. Frackowiak, J., Mazur-Kolecka, B., Wisniewski, H. M., et al. (1995) Secretion and accumulation of Alzheimer's β-protein by cultured vascular smooth muscle cells from old and young dogs. *Brain Res.* **676,** 225–230.
10. Mazur-Kolecka, B., Frackowiak, J., and Wisniewski, H. M. (1995) Apolipoproteins E3 and E4 induce, and transthyretin prevents accumulation of the Alzheimer's β-amyloid peptide in cultured vascular smooth muscle cells. *Brain Res.* **698,** 217–222.
11. Mazur-Kolecka, B., Frackowiak, J., Krzeslowska, J., et al. (1999) Apolipoprotein E alters metabolism of AβPP in cells engaged in β-amyloidosis. *J. Neuropathol. Exp. Neurol.* **58,** 288–295.

12. Diglio, C. A., Grammas, P., Giacomelli, F., and Wiener, J. (1982) Primary culture of rat cerebral microvascular endothelial cells. Isolation, growth, and characterization. *Lab. Invest.* **46,** 554–563.

13. Diglio, C. A., Grammas, P., Giacomelli, F., and Wiener, J. (1986) Rat cerebral microvascular smooth muscle cells in culture. *J. Cell. Physiol.* **129,**131–41.

14. Diglio, C. A., Liu, W., Grammas, P., Giacomelli, F., and Wiener, J. (1993) Isolation and characterization of cerebral resistance vessel endothelium in culture. *Tissue Cell* **25,** 833–846.

15. Seidel, M. F., Simard, J. M., Hunter, S. F., and Campbell, G. A. (1991) Isolation of arteriolar microvessels and culture of smooth muscle cells from cerebral cortex of guinea pig. *Cell Tissue Res.* **265,** 579–587.

16. Reisner, A., Olson, J. J., Yang, J., Assietti, R., Klemm, J. M., and Girard, P. R. (1995) Isolation and culture of bovine intracranial arterial endothelial cells. *Neurosurgery* **36,** 806–813.

17. Tontsch, U. and Bauer, H.-C. (1989) Isolation, characterization and long-term cultivation of porcine and murine cerebral capillary endothelial cells. *Microvascular Res.* **37,** 148–161.

18. Sturchler-Pierrat, C., Abramowski, D., Duke, M., et al. (1997) Two amyloid precursor protein transgenic mouse models with Alzheimer disease-like pathology. *Proc. Natl. Acad. Sci. USA* **94,** 13287–13292.

19. Calhoun, M. E., Burgermeister, P., Phinney, A. L., et al. (1999) Neuronal overexpression of mutant amyloid precursor protein results in prominent deposition of cerebrovascular amyloid. *Proc. Natl. Acad. Sci. USA* **96,** 14088–14093.

20. Winkler, D. T., Bondolfi, L., Herzig, M. C., et al. (2001) Spontaneous hemorrhagic stroke in a mouse model of cerebral amyloid angiopathy. *J. Neurosci.* **21,** 1619–1627.

21. Greenberg. S. M. (2002) Cerebral amyloid angiopathy and dementia: Two amyloids are worse than one. *Neurology* **58,** 1587–1588.

22. Pfeifer, L. A., White, L. R., Ross, G. W., Petrovitch, H., and Launer, L. J. (2002) Cerebral amyloid angiopathy and cognitive function: The HAASS autopsy study. *Neurology* **58,** 1629–1634.

23. Pawlik, M., Sastre, M., Calero, M., et al. (2004) Overexpression of human cystatin C in transgenic mice does not affect levels of endogenous brain amyloid β peptide. *J. Mol. Neurosci.* **22,** 13–18.

24. Grammas, P., Roher, A. E., and Ball, M. J. (1991) Decreased α-adrenergic receptors at the blood-brain barrier in Alzheimer's disease, in *Alzheimer's Disease: Basic Mechanisms, Diagnosis and Therapeutic Strategies* (Iqbal, K., McLachlan, D. R. C., Winblad, B., and Wisniewski, H. M. eds.), John Wiley & Sons Ltd., New York, NY.

25. Frackowiak, J., Miller, D. L., Potempska, A., Sukontasup, T., and Mazur-Kolecka, B. (2003) Secretion and accumulation of Aβ by brain vascular smooth muscle cells from AβPP-Swedish transgenic mice. *J. Neuropathol. Exp. Neurol.* **62,** 685–696.

26. Jung, S. S. and Van Nostrand, W. E. (2002) Aβ does not induce oxidative stress in human cerebrovascular smooth muscle cells. *NeuroReport* **13,** 1309–1312.

15

Purification of Human Wild-Type or Variant Cystatin C From Conditioned Media of Transfected Cells

Frances Prelli, Monika Pawlik, Blas Frangione, and Efrat Levy

Summary

The characterization of proteins in their native state is essential for the understanding of pathogenic isoforms. A variant of the cysteine protease inhibitor cystatin C is the major constituent of the amyloid deposited in the cerebral vasculature of patients with the Icelandic form of hereditary cerebral hemorrhage with amyloidosis (HCHWA-I) *(1,2)*. In order to study the nature of the biophysical changes owing to the Leu68Gln substitution in cystatin C, we have developed a purification procedure of human cystatin C in its native state. The protein is isolated from media of stably transfected tissue culture cells using physiological conditions that preclude protein denaturation. The importance of mild purification conditions is underscored by the finding that denaturation of the wild-type and variant proteins facilitates a similar folding of both molecules, diminishing their differences in structure and biophysical properties. Following native purification conditions, variant cystatin C has a distinct structure compared to the wild-type protein.

Key Words: Cystatin C; cysteine protease inhibitor; cerebral amyloid angiopathy; amyloid fibrils; protein conformation.

1. Introduction

Human cell lines stably overexpressing the cystatin C genes were used to compare wild-type and variant cystatin C production, processing, secretion, and clearance. While production and secretion of both proteins are similar, the proteins differ in their stability and tendency to dimerize and to form aggregates *(3,4)*. Utilizing full-length cystatin C purified in mild conditions from media of cells stably transfected with either the wild-type or variant cystatin C genes, we demonstrated that: (1) while cystatin C formed concentration dependent dimers, the variant dimerized at lower concentrations than the wild-type protein *(3)*; (2) we demonstrated by circular dichroism, steady-state fluorescence and Fourier-transformed infrared spectroscopy that the amino acid substitution

From: *Methods in Molecular Biology, vol. 299: Amyloid Proteins: Methods and Protocols*
Edited by: E. M. Sigurdsson © Humana Press Inc., Totowa, NJ

modifies cystatin C structure by destabilizing α-helical structures and exposing the tryptophan residue to a more polar environment, yielding a more unfolded molecule. These spectral changes demonstrated that variant cystatin C has a three-dimensional structure different from that of the wild-type protein *(5)*; (3) the variant cystatin C forms fibrils in vitro detectable by electron microscopy in conditions in which the wild-type protein forms amorphous aggregates *(5)*. The structural differences between variant and wild-type cystatin C account for the susceptibility of the variant protein to unfolding, proteolysis, and fibrillogenesis.

In order to test whether denaturation affects the structure of cystatin C, wild-type and variant cystatin C isolated from conditioned media were denatured in 6 *M* guanidinium hydrochloride and refolded by passing over a gel filtration column at pH 7.8 *(5)*. We found that this treatment increased the content of un-ordered structures of both the variant and wild-type proteins, completely exposing the tryptophan residue of both proteins to the solvent, and thus minimizing the difference in secondary structure between the two molecules *(5)*. These data indicate the importance of avoiding steps of denaturation and refolding in the process of cystatin C purification.

2. Materials

2.1. Establishment of Stably Transfected Cell Lines

1. pcDNA3.1 vector (Invitrogen, Carlsbad, CA).
2. Human kidney HEK293 cells.
3. Lipofectamine transfection reagent (GIBCO/BRL, Frederick, MD).
4. Geneticin (G418, GIBCO, Frederick, MD).

2.2. Purification of Cystatin C From Tissue Culture Media

1. Dulbecco's Modified Eagle's Medium (DMEM).
2. Complete medium: DMEM supplemented with 10% Fetal Bovine Serum, 100 U/mL of penicillin and 100 μg/mL streptomycin sulfate.
3. Phosphate buffered saline (PBS), pH 7.3.
4. Spectra/Por membrane (MWCO 6000–8000) (Spectrum Medical Industries, Houston, TX).
5. 20 m*M* Ammonium bicarbonate, pH 9.4.
6. DEAE-Sephacel column (Pharmacia, Piscataway, NJ).
7. 0.1% Coomassie blue R-250 (BioRad, Richmond, CA) in 40% methanol, 1% acetic acid.
8. Colloidal gold assay (Quantigold, Diversified Biotech, Boston, MA).
9. Human urinary cystatin C (Calbiochem, San Diego, CA).

2.3. Western Blot Analysis

1. Sample buffer: 80 m*M* Tris-HCl, pH 6.8, 2% sodium dodecyl sulfate (SDS), 10% glycerol, 2% β-mercaptoethanol and 0.001% Bromophenol blue.

2. 10% Tricine-Tris-polyacrylamide gel electrophoresis (PAGE).
3. Nitrocellulose membrane (BioRad, Richmond, CA).
4. Transfer buffer: 10 mM CAPS (3-cyclohexylamino-l-propanesulfonic acid) buffer, pH 11.0 containing 10% methanol.
5. TBST: 10 mM Tris-HCl, pH 7.5, 150 mM sodium chloride, with 0.1% Tween-20.
6. Blocking solution: 5% nonfat dry milk in TBST.
7. Rabbit anti-cystatin C antibody (Accurate, Westbury, NY).
8. Horseradish peroxidase-linked anti-rabbit IgG (Amersham. Piscataway, NJ).
9. SuperSignal® West Pico Chemiluminescent Substrate (Pierce, Rockford, IL).
10. Kodak Blue XB-1 Autoradiography film.

3. Methods

The methods described in the following outline the construction of the expression plasmid and its transfection into tissue culture cells, the isolation and purification of the protein, and the characterization of the protein by Western blot, and amino acid sequence analyses.

3.1. Establishment of Stably Transfected Cell Lines

Subclone either the full-length cystatin C genes or cystatin C cDNA into a pcDNA3.1 vector designed for expression in mammalian hosts. The vector contains a neomycin (G418) resistance sequence affording selection of stable transfectants in mammalian cells. Transfect the plasmids into mammalian cell using Lipofectamine transfection reagent. Test overexpression and secretion of cystatin C by Western blot analysis of cell lysates and cell culture media with anti-cystatin C polyclonal antibody (**Fig. 1**). Select lines with the highest levels of secretion of either wild-type (**Fig. 1, lane 2**) or variant (**Fig. 1, lane 4**) cystatin C for purification of the proteins.

3.2. Purification of Cystatin C From Tissue Culture Media

1. Grow stably transfected cells to near confluency in 150-mm plates in complete medium at 37°C in 5% CO_2 atmosphere.
2. Wash cells with PBS and incubate in 13 mL medium without serum for 24 h.
3. Harvest conditioned media, spin at 4500g for 10 min at 4°C (*see* **Note 1**) to remove cellular debris and store cleared media at −80°C.
4. Defrost the media when the desired volume is accumulated (*see* **Note 2**) and dialyze in Spectra/Por membrane (MWCO 6000–8000) against deionized distilled (dd) H_2O at 4°C for 2–3 d with at least two changes of ddH_2O per day.
5. Lyophilize the media without drying it to completion. Pool the lyophilized samples and lyophilize again to reach a final volume of 100 mL.
6. Following lyophilization, adjust the sample to 20 mM ammonium bicarbonate, pH 9.4, and dialyze overnight against the same buffer at 4°C (*see* **Note 3**).
7. Adjust the pH of the sample to pH 9.4, spin at 4500g for 10 min and apply the supernatant to a DEAE-Sephacel column equilibrated in 20 mM ammonium bicarbonate,

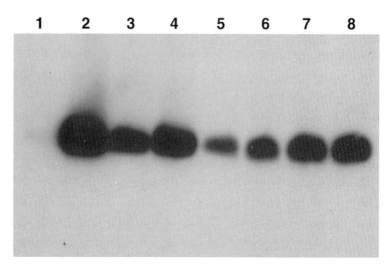

Fig. 1. Western blot analysis of cystatin C secretion by human kidney HEK293 cells stably transfected with either wild-type (lanes 2, 3, 6) or variant (lanes 4, 5, 7, 8) cystatin C genes and untransfected cells (lane 1). The samples were resolved in a 10% Tricine-Tris-PAGE and transferred to nitrocellulose membrane for staining with anti-cystatin C antibody.

pH 9.4 buffer at 4°C. Collect the flow-through and monitor fractions containing cystatin C by UV spectrometry at 280 nm and Western blot analysis.

8. Estimate protein concentration in solution by a colloidal gold assay according to the manufacturer's protocol, using commercial urinary cystatin C as standard.
9. Determine the purity of cystatin C in the samples by silver staining of 10% Tricine-Tris-PAGE or by staining transfer membranes with 0.1% Coomassie blue R-250 in 40% methanol, 1% acetic acid (**Fig. 2**).
10. Confirm the isolation of full-length cystatin C by microsequence analysis.

3.3. Western Blot Analysis

1. Prepare equal amounts of total proteins to be applied in each lane, in order to enable controlled comparison of cystatin C levels.
2. Boil the samples in sample buffer and separate by 10% Tricine-Tris-PAGE.
3. Electrophoretically transfer the proteins (2 h at 200 mA at 4°C) to nitrocellulose membrane using transfer buffer.
4. Block the membranes for 1–2 h at room temperature.
5. Incubate the membranes with rabbit anti-cystatin C antibodies (1:600) for 2 h at room temperature.
6. Wash the membranes with TBST.
7. Incubate the membranes with Horseradish peroxidase-linked anti-rabbit IgG (1: 10,000).

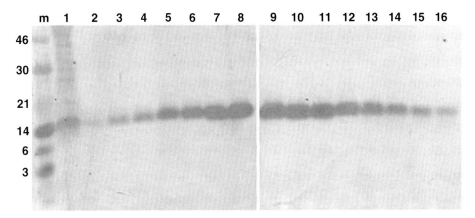

Fig. 2. Determination of the purity of wild-type cystatin C by staining transfer membranes with Coomassie blue. A sample before loading the column (lane 1) and every other sample starting at sample 120 collected from the column (lanes 2–16) were loaded on 10% Tricine-Tris-PAGE. Molecular mass standards (m) in kDa are shown on the left.

8. Wash the membranes with TBST.
9. Incubate the membranes with enhanced chemiluminescent substrate as specified by the manufacturer.
10. Visualize blots on Autoradiography film.

3.4. Amino Acid Sequence Analysis

To determine the purity and full length of the cystatin C protein, apply approx 20 pmol of purified cystatin C solution to a micro filter pretreated with Biobrene and perform N-terminal sequence analysis on a Procise 494 microsequencer (PE Applied Biosystems, Foster City, CA). The resulting phenylthiohydantin amino acid derivatives are identified using an online model 140C Microgradient Delivery System analyzer, 785A Programmable Absorbance Detector, and the standard program.

4. Notes

1. The samples should be kept at 4°C at all times in order to protect the integrity of the full-length protein. Amino terminal sequence analysis revealed that the variant and wild-type proteins purified from media conditioned by transfected cells are full length cystatin C *(3)*. The cystatin C isolated from the leptomeninges of HCHWA-I patients *(1,2)* and from urine *(6)* were amino-terminally truncated. It was proposed that a serine protease with the specificity of elastase could be the protease responsible for the truncation of the amino-terminal ten amino acids *(7)*.

2. The volume of cultured media used depends on the level of expression and secretion of the transgene and should be monitored in order to optimize the purification of the protein. We used 2 L media conditioned by cells transfected with wild-type cystatin C (**Fig. 1, lane 2**) and 4 L media conditioned by cells transfected with variant cystatin C (**Fig. 1, lane 4**)

3. Cystatin C is a basic protein *(8)* and therefore the DEAE-Sephacel column and the samples applied to it should be calibrated to pH 9.4. Under these conditions cystatin C does not bind to the column and elutes with the flow-through, while most of the other proteins present in the medium absorb onto the column.

Acknowledgments

This work was supported by grants from the National Institute of Neurological Disorders and Stroke (NS42029) and the American Heart Association (0040102N).

References

1. Cohen, D. H., Feiner, H., Jensson, O., and Frangione, B. (1983) Amyloid fibril in hereditary cerebral hemorrhage with amyloidosis (HCHWA) is related to the gastroentero-pancreatic neuroendocrine protein, γ trace. *J. Exp. Med.* **158,** 623–628.

2. Ghiso, J., Jensson, O., and Frangione, B. (1986) Amyloid fibrils in hereditary cerebral hemorrhage with amyloidosis of Icelandic type is a variant of γ trace basic protein (cystatin C). *Proc. Natl. Acad. Sci. USA* **83,** 2974–2978.

3. Wei, L., Berman, Y., Castano, E. M., et al. (1998) Instability of the amyloidogenic cystatin C variant of hereditary cerebral hemorrhage with amyloidosis, Icelandic type. *J. Biol. Chem.* **273,** 11806–11814.

4. Abrahamson, M. and Grubb, A. (1994) Increased body temperature accelerates aggregation of the Leu-68→Gln mutant cystatin C, the amyloid-forming protein in hereditary cystatin C amyloid angiopathy. *Proc. Natl. Acad. Sci. USA* **91,** 1416–1420.

5. Calero, M., Pawlik, M., Soto, C., et al. (2001) Distinct properties of wild-type and the amyloidogenic human cystatin C variant of hereditary cerebral hemorrhage with amyloidosis, Icelandic type. *J. Neurochem.* **77,** 628–637.

6. Hall, A., Dalboge, H., Grubb, A. O., and Abrahamson, M. (1993) Importance of the evolutionarily conserved glycine residue in the N-terminal region of human cystatin C (Gly-11) for cysteine endopeptidase inhibition. *Biochem. J.* **291,** 123–129.

7. Abrahamson, M., Mason, R. W., Hansson, H., Buttle, D. J., Grubb, A., and Ohlsson, K. (1991) Human cystatin C. Role of the N-terminal segment in the inhibition of human cysteine proteinases and its inactivation by leukocyte elastase. *Biochem. J.* **273,** 621–626.

8. Grubb, A. and Lofberg, H. (1982) Human γ-trace, a basic microprotein: amino acid sequence and presence in the adenohypophysis. *Proc. Natl. Acad. Sci. USA* **79,** 3024–3027.

16

Prion Propagation in Cell Culture

Sylvain Lehmann

Summary

During the past two decades, considerable efforts have been made to set up cellular cultures supporting the replication of prions, the infectious agents responsible of transmissible spongiform encephalopathies. As a matter of fact, prion-infected cell lines are very valuable to investigate the cell biology of both the normal and the pathological isoform of the prion protein or to develop and screen new therapeutics. In this chapter, we present a detailed protocol for the generation of prion-infected cells. We also give step-by-step procedures to test the biochemical properties (mainly protease resistance and insolubility) of abnormal PrP molecules, which detection represents a biochemical marker of prion propagation.

Key Words: Prion; scrapie; PrP; transmissible spongiform encephalopathies; Creutzfeldt-Jakob disease; protease resistance; cell culture; murine neuroblastoma cell line N2a.

1. Introduction

In this chapter, we describe the generation of cellular cultures supporting the replication of prions, the infectious agents responsible for transmissible spongiform encephalopathies (**Fig. 1**). We also give step-by-step procedures to test the acquisition of partial protease resistance and insolubility by the cellular prion protein (PrP^C). These two biochemical properties are characteristic of the pathologic isoform, PrP^{Sc}, which detection serves as a biochemical marker of prion propagation. Prion-infected cell lines have been widely used to investigate the cell biology of both the normal and the pathological isoforms of the prion protein *(1)*. They have also contributed to the comprehension of the pathogenic processes occurring in Transmissible Spongiform Encephalopathies and in the development of new therapeutic approaches of these diseases.

2. Materials

1. BCA protein assay (Pierce/Perbio Science, Brebières, France).
2. Dulbecco Modified Eagle's Medium (DMEM) (Invitrogen, Pontoise, France).

From: *Methods in Molecular Biology, vol. 299: Amyloid Proteins: Methods and Protocols*
Edited by: E. M. Sigurdsson © Humana Press Inc., Totowa, NJ

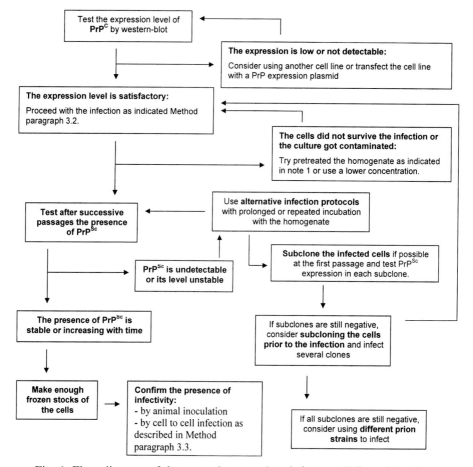

Fig. 1. Flow diagram of the general approach to infect a cell line with prions.

3. Penicillin/streptomycin (Invitrogen).
4. Phosphate-buffered saline (PBS) (Invitrogen).
5. Fetal Bovine Serum (Invitrogen).
6. OptiMEM (Invitrogen).
7. Pefabloc (Roche Diagnostic, Meyan, France). Stock at 100 mM in ddH$_2$O, make small aliquots conserved at −80°C.
8. Proteinase K (Roche Diagnostic; Meyan, France). Stock at 2 mg/mL in ddH$_2$O, make small aliquots conserved at −20°C.
9. Prion Protein Monoclonal Antibodies SAF-32 and SAF-70 (Spi-Bio, Massy, France).
10. Lysis buffer (LB), composition: 150 mM NaCl, 0.5% Triton X-100, 0.5% sodium deoxycholate, and 50 mM Tris-HCl (pH 7.5). To prepare mix: 0.5 g of sodium deoxycholate (deoxycholic acid), 0.5 mL of Triton X-100, 3 mL of 5 M NaCl,

5 mL of 1 M Tris-HCl, pH 7.5, then bring volume to 100 mL with ddH$_2$O, store at 4°C.

11. 2X Laemmli sample buffer with DTT (2X LMSB). To prepare mix: 2 mL of 1 M dithiothreitol (DTT) (1 M = 1.542 g/10 mL), 1.2 g of sodium dodecyl sulphate (SDS), 10 mL of ddH$_2$O, 4 mL of glycerol, 2.5 mL of Tris-HCl (1 M, pH 7,4), 0.5 mL of bromphenol blue at 10 mg/mL, then bring volume to 20 mL with ddH$_2$O. Make 1 mL aliquots, store at −20°C. Dilute v/v with water to make 1X LMSB.

12. Murine neuroblastoma cell line N(euro-)2a: N2a cells can be obtained from the ATCC (reference CCL-131) and were grown in DMEM containing 10% fetal calf serum and penicillin/streptomycin in an atmosphere of 10% CO$_2$. It is wise to split the cells before they reach complete confluence, to record the number of passages and to make frozen stocks of the cells.

3. Methods

3.1. Preparation of Brain Homogenates for the Infection

Brain homogenates are prepared from control and from mice (CD1 or C57/Bl6) in the terminal phase of prion infection with the Chandler/RML or the 22-L prion strains *(2)*.

1. Prepare either by Dounce homogenization or by serial passages through syringes, a 10 % (10 g/100 mL) brain homogenate in a sterile solution of NaCl 0.9%, glucose 5%.
2. Store this homogenate frozen at −80°C in small (0.5-mL) aliquots to avoid repeated freeze/thaw.
3. Dilutions of brain homogenates to a final concentration of 2% or below for infection are made in OptiMEM (*see* **Note 1**).

3.2. Infection of N2a Cells

We describe here the generation of prion infected N2a cells. A similar protocol can be used for other cell lines like the GT-1 cells *(4)*. N2a cells are grown in 6-well plates and plated at 2×10^5 cells/well 2 d before infection. The day of the infection, the cells are expected to be 70 to 80% confluent. To infect the cells, different dilutions of a 10% brain homogenate can be used (*see* **Notes 1** and **2**). We suggest starting with two dilutions: 2% and 0.2% of infectious and control brain homogenate (*see* **Note 3**). In our hand the 22 L strain is more efficient at infecting cells than the Chandler/RML one.

1. Rinse each well with warm sterile OptiMEM for 3 min, remove the OptiMEM and add 1 mL per well of the dilution of the brain homogenate in OptiMEM. Incubate for 5 h in the incubator, then add 1 mL of complete medium and incubate the plate overnight (14–16 h) at 37°C.
2. Rinse each well separately twice with PBS, put 2 mL of fresh medium, incubate for an additional 24 h.

3. Split each well in a new well of a 6-well plate, as well as in one or two additional 25 cm^2 flasks. At confluence, lyse one 25 cm^2 flask to perform a proteinase K (PK) resistance assay. The other flask could serve as a backup or its cells could be frozen for further subcloning to isolate infected clones.
4. Passage the cells up to 10 times and monitor the presence of PrP^{Sc}. In the first or the second passage after infection, the PrP^{Sc} from the inoculum could still be detected.

3.3. Confirmation of the Infection by Cell-to-Cell Transmission

To confirm the generation of infectivity by the newly infected cells after multiple passages, the best way is to inoculate cell lysates to mice. However, an alternative way is represented by the possibility of using cell cultures to detect the presence of infectivity as follows.

1. Grow control and infected cells to obtain confluent 75 cm^2 flasks.
2. Rinse the cultures twice in PBS, scrape the cells in 5 mL of cold PBS under sterile conditions, and collect them in a conical 15-mL falcon tube.
3. Pellet the cells by spinning the tubes 6 min at 800g. Remove the supernatant and resuspend the pellet in 100 µL of cold sterile PBS with 5% glucose.
4. To prepare the cell extract, submit this suspension to four cycles of freezing-thawing in liquid nitrogen.
5. The inoculum is then passed through a 27-gauge needle several times. Fifty microliters of the preparation is diluted in 1 mL of Opti-MEM and used to infect cells following the method in **Subheading 3.2.**

3.4. Detection of Proteinase K-Resistant PrP in Cell Culture

PK digestion of PrP is the most common method to detect PrP^{Sc} and the presence of prions. This protocol describes the detection of PrP^{Sc} in scrapie infected N2a or GT1 cells as in **ref. 4** (*see* **Note 4**).

1. Starts from a sub-confluent tissue culture flask of 25 cm^2 or equivalent. Gently rinse the flask twice for 2 min with cold PBS.
2. Aspirate the PBS, put 0.6 mL of lysis buffer (LB) in the flask eventually completed with protease inhibitors (pepstatin and leupeptin, 1 µg/mL and EDTA, 2 mM). Rock the flask 10 s and lay flat at 4°C for 15 min. The protease inhibitor phenylmethylsulfonylfluoride (PMSF) is a PK inhibitor and must be avoided.
3. Collect the cell lysate in a 1.5-mL eppendorf tube, spin 3 min at 10,000g in a microcentrifuge, a procedure that removes debris/DNA but does not pellet significant amounts of PrP. Put the supernatant in a new eppendorf tube, keep on ice.
4. Measure the total protein concentration with the BCA assay (Pierce). Adjust the protein concentrations of the samples to 0.6 mg/mL by adding LB to the tubes.
5. Take 50 µL of each lysate and mix it in a new eppendorf with an equal volume of Laemmli sample buffer with DTT (LMSB) 2X and boil 5 min at 90°C. These tubes will be used to evaluate PrP^C levels in the samples.
6. To detect PrP^{Sc}, put 500 µL of the samples (the equivalent of 300 µg of protein) in eppendorf tubes containing 2.4 µL of the stock solution of PK at 2 mg/mL (ratio:

PK/protein 1/62, equivalent to 16 µg of PK/mg protein). Gently vortex and put at 37°C for 30 min. Put the tubes on ice, add 5 µL of Pefabloc in each (stock at 100 mM: 1 mM final), mix, and wait 5 min.

7. Centrifuge the eppendorfs at 14,000 rpm (20,000*g*) for 45 min at 4°C, discard the supernatant very gently by returning the tubes and resuspend the pellet first in 30 µL of LB, then by adding 30 µL of LMSB 2X, vortex, and boil 5 min at 90°C.

8. Perform the detection of PrP by Western blot using a 12% SDS-PAGE as indicated by the company. Always load a positive control (PK-digested prion-infected brain homogenate from **Subheading 3.2.,** for example). To detect PrP^{Sc}, an anti-PrP antibody raised against the carboxy-terminus of the protein like SAF-70 must be used. In fact, in prion infected cells, most of PrP^{Sc} is cleaved around codon 90 and is not recognized by antibodies raised against the amino-terminus, including those raised against the octapeptide repeats. To detect total PrP before PK digestion, use an antibody raised against the amino-terminus of the protein like SAF-32.

3.5. Rapid Protocol for PrP^{Sc} Detection

This protocol is an alternative to the previous protocol. It permits a rapid detection of PrP^{Sc} in small culture wells and can be useful to screen therapeutic agents. However, this protocol, which does not include a measure of the total protein concentration or a centrifugation step, does not permit a precise quantitation of PrP^{Sc}.

1. Starts from confluent wells of a 12-well plate. Gently rinse the wells once with cold PBS. Put 1 mL of cold PBS in the wells, scrap the cells, and collect them in 1.5-mL eppendorf tubes.

2. Pellet the cells by spinning the tubes 3 min at 5000*g* in a micro-centrifuge. Remove the supernatant; add 50 µL of LB no. 2, vortex to put the pellet in suspension in the buffer, put on ice for 10 min.

3. Vortex the tube again and pellet the debris by spinning the tubes 3 min at 10,000*g* in a micro-centrifuge.

4. Put 30 µL of the supernatant in a new eppendorf containing 5 µL of a solution of PK at 0.2 mg/mL. Gently vortex and put at 37°C for 30 min.

5. Add to each tube 30 µL of LMSB 2X, vortex, boil 5 min at 90°C.

6. Perform the detection of PrP by Western blot as described in **Subheading 3.4.**

3.6. Detection of Insoluble PrP in Cell Culture

Insolubility in non-ionic detergent is a common property of PrP^{Sc} that can be used to detect this protein in infected cells without using PK digestion (*see* **Note 5**).

1. Perfom **steps 1** to **4** as in **Subheading 3.4.**

2. To detect insoluble PrP, put 400 µL of the samples in Beckmann eppendorf tubes and centrifuge at 70,000 rpm (200,000*g*) for 30 min in the TLA 100.4 rotor of a Beckman Optima TL ultracentrifuge to separate detergent-soluble and detergent-

insoluble protein. An alternate method valid for PrP^{Sc} in N2a cells consists in spinning the samples in a regular refrigerated centrifuge at 14,000 rpm (20,000g) for 45 min at 4°C.

3. At the end of the centrifugation, take 200 µL of the supernatant, put it new eppendorf tubes and methanol precipitate the samples as follows: fill the eppendorf with methanol stored at −20°C, mix by inversion several times, put at −20°C for at least 2 h. Centrifuge the eppendorf tubes at 14,000 rpm (20,000g) for 10 min at 4°C, remove the supernatant by inversion, let the pellet dry (but not too much!), resuspend the pellet first in 30 µL of LB, then by adding 30 µL of LMSB 2X, vortex, and finally, boil 5 min at 90°C.

4. Remove carefully the supernatant left in the eppendorfs after the spinning of **step 2**. The pellet often looks like a small oil drop on the bottom side of the tube. Resuspend this pellet in 40 µL of LMSB 1X, vortex, boil 5 min at 90°C.

5. Perform a Western blot detection of soluble and insoluble PrP on 12% SDS PAGE: load equal volume of each samples and Western blot with an anti-PrP antibody (either against the amino-terminus (SAF-32) to detect full length PrP or against the carboxy-terminus (SAF-70) to also detect cleaved PrP). Load on the gels a positive control (brain from scrapie infected animals).

3.7. Other Methods Relevant to Prion Infected Cells

- Detection of PrP^{Sc} by immunofluorescence after guanidine treatment *(6)*.
- Subcloning of cells to isolate prion susceptible cell lines *(5)*.
- Filter retention assay *(7)*.
- Scrapie cell assay *(8)*.
- Steel wire assay *(9)*.
- Screening of drugs *(10–16)*.

4. Notes

1. To avoid contaminations by conventional agents, we advise to "decontaminate" the homogenates with one of these two alternative methods:
   - Warm the 10% homogenate 20 min at 80°C in a heat-block and sonicate the sample for 3 min to break the clumps resulting from the heating.
   - After dilution of the homogenate in the OptiMeM, filter the homogenate through a 0.22-µm sterile filter. This filtration can be done also following the warming procedure.

   In our hands, pre-treatment of the homogenate reduced the efficiency of the cell culture infection by one log *(3)*. However, we also noticed that this procedure significantly reduced the toxicity of the homogenate for the cultured cells.

2. It is necessary to confirm the presence of abnormal prion protein (PrP^{Sc}) in the homogenate aliquots at the time of the infection of the cells. Briefly, mix 100 µL of brain homogenate with an equal volume of 2X Lysis buffer on ice. Incubate for 20 min, spin 3 min at 10,000g, and digest the supernatant with 100 µg of PK per mL at 37°C for 30 min. The reaction is stopped with 1 mM Pefabloc for 5 min on ice.

The samples are then mixed with an equal volume of 2X LMSB (Laemmli sample buffer with DTT), boiled for 5 min, and then loaded onto a 12% polyacrylamide gel. Proteins are electroblotted onto Immobilon membranes and PrP^{Sc} detected by Western blot using the monoclonal antibodies SAF 70 as indicated by the company.

3. Toxicity and contamination: depending on the origin of the homogenate, the cultured cells may not survive to the infection or be contaminated by conventional agents. To minimize this problem, we suggest using lower homogenate concentration or pre-treating the homogenate by heating and/or filtration as indicated earlier. Subcloning of the infected cells is sometimes needed and may be performed preferably at the first passage after the inoculation to maximize the chance of getting infected clones.
   - As described in **ref. 5**, it may be useful to subclone the cells prior to the infection to generate highly prion susceptible clones.
   - The length of incubation of the cells with the homogenate can be extended if the cells can handle it. Repeated incubation with the homogenate can also increase the chance of infection.

4. In some protocols, the following alternate lysis buffer is used: 100 mM NaCl, 0.5% NP-40, 0.5% sodium deoxycholate and 10 mM Tris-HCl (pH 8.0). As a shorter alternative to steps 2 and 3 of **Subheading 3.4.**, avoiding the centrifugation step, add the lysis buffer in the flask, rock and leave a few minutes, at room temperature. Collect the cell lysate in a 1.5-mL eppendorf tube, and remove the white material (nuclei+DNA) with a pipet. The most critical point is to optimize PK digestion in order to completely digest PrP^{C} from control, non-infected cells. To fine tune the digestion, one can vary the volume of the lysis buffer from 0.4 to 0.8 mL/flask, the ratio of PK:protein from 1:300 to 1:30 and the length of the digestion from 15 to 60 min.

5. Always perform an insolubility assay on non-infected cells expressing detectable amount of PrP^{C} to test the efficacy of the separation. One can expect to have more than 90% of PrP^{C} in the soluble fraction.

Acknowledgments

Our group is supported by the CNRS ("Centre National de la Recherche Scientifique") and by grants from the GIS prion ("Groupement d'intérêt scientifique" sur les Prions) and the European Community (Biotech, PL976064). I also thank Véronique Perrier and Jérôme Solassol and Noriyuki Nishida in contributing to the validation of these protocols.

References

1. Béranger, F., Mangé, A., Solassol, J., and Lehmann, S. (2001) Cell culture models of Prion Diseases. *Biochem. Biophys. Res. Communications* **289**, 311–316.
2. Somerville, R. A., Chong, A., Mulqueen, O. U., Birkett, C. R., Wood, S. C., and Hope, J. (1997) Biochemical typing of scrapie strains. *Nature* **386**, 564.

3. Lehmann, S., Laude, H., Harris, D. A., et al. (2001) *Ex vivo transmission of mouse adapted prion strains to N2a and GT1-7 cell lines,* in Alzheimer's Disease: Advances in Etiology, Pathogenesis and Therapeutics (K. Iqbal, S. S. Sisodia, and B. Winblad, eds.), John Wiley & Sons, Ltd., UK.

4. Nishida, N., Harris, D. A., Vilette, D., et al. (2000) Successful transmission of three mouse-adapted scrapie strains to murine neuroblastoma cell lines overexpressing wild-type mouse prion protein. *J. Virol.* **74,** 320–325.

5. Bosque, P. J. and Prusiner, S. B. (2000) Cultured cell sublines highly susceptible to prion infection. *J. Virol.* **74,** 4377–4386.

6. Taraboulos, A., Serban, D., and Prusiner, S. B. (1990) Scrapie prion proteins accumulate in the cytoplasm of persistently infected cultured cells. *J. Cell Biol.* **110,** 2117–2132.

7. Winklhofer, K. F., Hartl, F. U., and Tatzelt, J. (2001) A sensitive filter retention assay for the detection of PrP(Sc) and the screening of anti-prion compounds. *FEBS Lett.* **503,** 41–45.

8. Klohn, P. C., Stoltze, L., Flechsig, E., Enari, M., and Weissmann, C. (2003) A quantitative, highly sensitive cell-based infectivity assay for mouse scrapie prions. *Proc. Natl. Acad. Sci. USA* **100,** 11666–11671.

9. Flechsig, E., Hegyi, I., Enari, M., Schwarz, P., Collinge, J., and Weissmann, C. (2001) Transmission of scrapie by steel-surface-bound prions. *Mol. Med.* **7,** 679–684.

10. Caughey, B. and Raymond, G. J. (1993) Sulfated polyanion inhibition of scrapie-associated PrP accumulation in cultured cells. *J. Virol.* **67,** 643–650.

11. Mange, A., Nishida, N., Milhavet, O., McMahon, H. E. M., Casanova, D., and Lehmann, S. (2000) Amphotericin B inhibits the generation of the scrapie isoform of the prion protein in infected cultures. *J. Virol.* **74,** 3135–3140.

12. Kocisko, D. A., Baron, G. S., Rubenstein, R., Chen, J., Kuizon, S., and Caughey, B. (2003). New inhibitors of scrapie-associated prion protein formation in a library of 2000 drugs and natural products. *J. Virol.* **77,** 10288–10294.

13. Korth, C., May, B. C., Cohen, F. E., and Prusiner, S. B. (2001). Acridine and phenothiazine derivatives as pharmacotherapeutics for prion disease. *Proc. Natl. Acad. Sci. USA* **98,** 9836–9841.

14. Rudyk, H., Vasiljevic, S., Hennion, R. M., Birkett, C. R., Hope, J., and Gilbert, I. H. (2000). Screening Congo Red and its analogues for their ability to prevent the formation of PrP-res in scrapie-infected cells. *J. Gen. Virol.* **81,** 1155–1164.

15. Supattapone, S., Nguyen, H. O., Cohen, F. E., Prusiner, S. B., and Scott, M. R. (1999). Elimination of prions by branched polyamines and implications for therapeutics. *Proc. Natl. Acad. Sci. USA* **96,** 14529–14534.

16. Perrier, V., Wallace, A. C., Kaneko, K., Safar, J., Prusiner, S. B., and Cohen, F. E. (2000). Mimicking dominant negative inhibition of prion replication through structure-based drug design. *Proc. Natl. Acad. Sci. USA* **97,** 6073–6078.

III

In Vivo-Related Assays

17

Preparation and Propagation of Amyloid-Enhancing Factor

Robert Kisilevsky

Summary

Amyloid-enhancing factor (AEF) is a biological "activity" that is defined in the context of inflammation-associated amyloidogenesis (AA). When administered intravenously to mice followed by an inflammatory stimulus, such primed mice deposit substantial AA amyloid in spleen within 36–48 h. Since experimental induction of AEF is dependent on amyloidogenic protocols, and rapid AA amyloid induction is dependent on AEF, a strategy for AEF isolation is required to break into this circular process. AEF activity may be prepared from a variety of human forms of amyloid that include tissue containing any of Aβ, AA, ATTR, and AL amyloids. The preparation of an AEF extract from such human tissue is described using 4 *M* glycerol, which then may be used to induce splenic AA amyloid fibrils in mice as a source for the propagation of additional AEF and/or for the study of amyloidogenesis. The glycerol and AA fibril preparations are stable frozen for many years.

Key Words: Amyloid-enhancing factor (AEF); AA amyloid; amyloid; inflammation; mice; preparation; propagation; prions.

1. Introduction

Amyloid-enhancing factor (AEF) is defined as a biological "activity" which when delivered to mice by an intravenous or intra-peritoneal route followed by an inflammatory stimulus reduces the induction time of inflammation-associated amyloidogenesis (AA amyloid) to 36–48 h *(1)*. In this respect, it is defined only in the context of AA amyloid formation. AEF has been described by Cathcart as "an enigma wrapped in a mystery" *(2)* because its precise nature has never been determined. Nevertheless, sufficient information has been gathered to conclude that AEF is protein in nature and its biological activity is conformation dependent *(3)*. AEF appears to function as a nidus, or scaffold, for the generation and growth of AA amyloid fibrils in a fashion analogous to

From: *Methods in Molecular Biology, vol. 299: Amyloid Proteins: Methods and Protocols*
Edited by: E. M. Sigurdsson © Humana Press Inc., Totowa, NJ

the action and propagation of prions (4). In contrast to prions AEF activity in relation to the induction of AA amyloid sets the stage for very rapid initiation of amyloid deposition (36–48 h as opposed to months for prions) (cf., **ref. 5**). Furthermore, AEF activity can be exhibited by extracts from tissues containing various forms of amyloid, in addition to the AA type (6,7). This activity has been extracted not only from tissue containing AA amyloid (the form of amyloid which AEF provokes), but also from brain tissue containing Aβ amyloid in Alzheimer's disease, transthyretin amyloid and light-chain amyloid, among others.

AEF activity provides the means to accelerate AA amyloid induction and therefore allows for comparisons to be made between equivalently treated animals that possess but an inflammatory reaction and those actually depositing AA amyloid. It allows one to distinguish epi-phenomena related to inflammation from those factors that are directly involved in AA amyloid deposition. It has therefore proven to be exceedingly valuable in the study of amyloidogenesis in vivo.

Though AEF activity has been described as being present in normal mouse and rat tissues (8,9) and in strains of mice that are resistant to AA amyloid formation (9), the experimental generation of AEF and its preparation from mouse tissues is usually dependent on protocols that induce AA amyloid in mice (1). Furthermore, AA amyloid fibrils themselves possess potent AEF activity (10,11). Therefore, obtaining an active AEF preparation becomes a circular process. One requires a rapid reproducible technique for inducing AA fibrils, and one requires AEF to induce this process rapidly. The following strategy is suggested as a way around this dilemma. One should begin with human tissue that is known to contain any one of a variety of forms of amyloid (e.g., Aβ, AL, ATTR, or AA), and apply the classical AEF extraction protocol described 20–25 yr ago (1). As will be described later, this preparation will allow for the rapid induction of AA amyloid in mouse spleen, which in turn can be a source of AA amyloid fibrils for use as AEF in the study of amyloidogenesis and/or for the propagation of additional fibrils as AEF.

2. Materials

2.1. Classical AEF Extraction From Human or Mouse Tissue

1. 4 M Glycerol, 10 mM Tris-HCl, pH 7.5.
2. 10 mM Phosphate-buffered saline, pH 7.5.
3. Potter-Elvejhem tissue homogenizer (hand held or motor driven).
4. Shaking platform.
5. Refrigerated ultracentrifuge.
6. Dialysis tubing, 3500 mw cut-off.

2.2. Mouse AA Amyloid Induction

1. AEF 1 mg/mL, prepared as described later in **Subheading 3.1.**
2. 2% AgNO₃.
3. 1-mL syringes and 27 gage needles.

2.3. Crude AA Amyloid Fibril Extraction From Mouse Spleen Which Can Be Used as AEF

1. 0.9% NaCl.
2. Distilled de-ionized water (DDW).
3. Tissue homogenizer.
4. Low-speed refrigerated centrifuge.
5. Spectrophotometer.

3. Methods

3.1. Glycerol Extraction of AEF Activity

1. Amyloid containing tissues may be used fresh or frozen.
2. Eight milliliters of ice-cold extractant (4 M glycerol) are used per gram of tissue.
3. The tissue is thoroughly homogenized in a hand-held or motor driven homogenizer after which it is placed in an Erlenmeyer flask and shaken vigorously on a platform shaker for 1 h at 4°C.
4. The homogenate is then centrifuged at 250,000g for 1 h at 4°C.
5. The supernatant is harvested and may be stored at −20°C for up to 5 yr without loss of AEF activity.
6. An aliquot (several milliliters) of the glycerol extract is dialyzed extensively against phosphate buffered saline at 4°C and sonicated before determining the protein concentration and its use for amyloid induction (*see* **Note 1**).
7. Dilute the AEF preparation to 1 mg/mL.
8. Two hundred and fifty (250)–500 µg protein of the preparation injected intravenously per mouse is sufficient to prime the animal for rapid, splenic AA amyloidogenesis in the presence of an acute inflammatory reaction *(1)* (*see* **Subheading 3.2.**).

3.2. AA Amyloid Induction in Mice as a Test for AEF Activity

1. Use female mice 8–10 wk old and from a mouse strain that is susceptible to AA amyloid deposition, such as CBA/J (very susceptible but expensive) or CD1 (*see* **Note 2**).
2. Mice are warmed under an infra-red lamp for a few minutes to dilate the lateral tail veins before administering the intravenous AEF (250–500 µg if using the glycerol extract; 100 µg if using the amyloid fibril preparation).
3. Use a 1-mL syringe and 27 gauge needle and ensure that the needle is in the lateral tail vein.
4. Immediately after administering the AEF, inject 0.5 mL 2% AgNO₃, the inflammatory stimulus, subcutaneously in the loose skin of the back between the two shoulder blades to provoke a sterile abscess.

5. After 2–3 d animals will develop easily demonstrable perifollicular splenic AA amyloid when the tissue is fixed and prepared for histological sections and stained with the alkaline Congo red technique described by Puchtler et al. *(12)*.

3.3. AEF as AA Amyloid Fibrils Extracted From Mouse Spleen

1. Induce AA amyloid in 20–30 mice as described in **Subheading 3.2.**
2. Do not sacrifice the mice after 2–3 d, but allow them to survive for about 10 d (or longer if possible) to accumulate substantial quantities of splenic amyloid fibrils.
3. Sacrifice mice by cervical dislocation or CO_2 narcosis.
4. Collect spleens and determine their weight.
5. Homogenize spleens in 10 volumes of ice-cold 0.9% NaCl per weight of spleen using a motor-driven homogenizer (e.g., a Polytron).
6. Centrifuge at 24,000*g* for 30 min at 4°C using a Beckman J2-21 centrifuge and JA-17 rotor (or their equivalent).
7. Repeat **steps 5** and **6** until the supernatant reads less than 0.1 OD_{280}. This may require 10–12 repeats.
8. Homogenize pellet in 10 volumes DDW.
9. Centrifuge at 30,000*g* for 1 h at 4°C and discard supernatant.
10. Homogenize pellet in 10 volumes DDW.
11. Centrifuge at 30,000*g* for 1 h at 4°C using a 50.2 TiRotor and a Beckman ultracentrifuge.
12. Save supernatant and upper gelatinous layer on pellet = fraction 2.
13. Repeat **steps 10** and **11**.
14. Save supernatant and upper gelatinous layer on pellet = fraction 3.
15. Pool fractions 2 and 3 and mix thoroughly.
16. Determine protein concentration of pooled fractions and dilute to 1 mg/mL.
17. Divide into 1–2-mL aliquots and freeze at −70°C.
18. To test for the AEF activity of the AA fibril preparation, follow the procedure described in **Subheading 3.2.** (*see* **Note 3**).

4. Notes

1. A very fine, white, fluffy precipitate will accumulate at the bottom of the dialysis bag. Make sure to include this material in your preparation before sonicating and determining the protein concentration. Use any one of several standard techniques to determine the protein concentration.
2. Mice of the CE/J strain fail to deposit AA amyloid because they only produce a non-amyloidogenic form of the AA precursor, SAA2.2. A/J mice, though resistant to AA amyloidogenesis when using standard protocols, are quite susceptible to AA amyloid deposition after being treated with AEF. CD1 mice, being out-bred, are by far the cheapest mice for this purpose.
3. After thawing a frozen aliquot of AEF fibrils re-sonicate the aliquot with short, 5 s, bursts for about 15–30 s while it is immersed in ice to ensure that there is a uniform dispersal of the fibrils in solution.

Acknowledgments

This work was supported financially by the Medical Research Council of Canada (MRC), now the Canadian Institutes for Health Research (CIHR), grants numbers MT-3153 and MOP-3153, respectively. The author is indebted to Mrs. Ruth Tan and Mr. Lee Boudreau for their able technical assistance.

References

1. Axelrad, M. A., Kisilevsky, R., Willmer, J., Chen, S. J., and Skinner, M. (1982) Further characterization of amyloid enhancing factor. *Lab. Invest.* **47,** 139–146.
2. Cathcart, E. S. (1995) AEF: an enigma wrapped in a mystery. *Amyloid* **2,** 126–127.
3. Kisilevsky, R., Gruys, E., and Shirahama, T. (1995) Does amyloid enhancing factor (AEF) exist? Is AEF a single biological entity? *Amyloid* **2,** 128–133.
4. Caughey, B. (2000) Transmissible spongiform encephalopathies, amyloidoses and yeast prions: Common threads? *Nat. Med.* **6,** 751–754.
5. Kisilevsky, R. (2000) Review: Amyloidogenesis-unquestioned answers and unanswered questions. *J. Struct. Biol.* **130,** 99–108.
6. Varga, J., Flinn, M. S., Shirahama, T., Rodgers, O. G., and Cohen, A. S. (1986) The induction of accelerated murine amyloid with human splenic extract. Probable role of amyloid enhancing factor. *Virchows Arch. B Cell Pathol.* **51,** 177–185.
7. Ali-Khan, Z., Quriion, R., Robitaille, Y., Alizadeh-Khiavi, K., and Du, T. (1988) Evidence for increased amyloid enhancing factor activity in Alzheimer brain extract. *Acta Neuropathol.* **77,** 82–90.
8. Axelrad, M. A., Kisilevsky, R., and Beswetherick, S. (1975) Acceleration of amyloidosis by syngeneic spleen cells from normal donors. *Am. J. Pathol.* **78,** 277–284.
9. Gonnerman, W. A., Kandel, R., and Cathcart, E. S. (1996) Amyloid enhancing factor is produced by rats and amyloid-resistant CE/J mice. *Lab. Invest.* **74,** 259–264.
10. Baltz, M. L., Caspi, D., Hind, C. R. K., Feinstein, A., and Pepys, M. B. (1986) Isolation and characterization of amyloid enhancing factor (AEF), in *Amyloidosis* (Glenner, G. G., Osserman, E. F., Benditt, E. P., Calkins, E., Cohen, A. S., and Zucker-Franklin, D., eds.), Plenum Press, New York, pp. 115–121.
11. Niewold, Th. A., Hol, P. R., van Andel, A. C. J., Lutz, E. T. G., and Gruys, E. (1987) Enhancement of amyloid induction by amyloid fibril fragments in hamster. *Lab. Invest.* **56,** 544–549.
12. Puchtler, H., Sweat, F., and Levine, M. (1962) On the binding of Congo red by amyloid. *J. Histochem. Cytochem.* **10,** 355–364.

18

Purification of Amyloid Protein AA Subspecies From Amyloid-Rich Human Tissues

Gunilla T. Westermark and Per Westermark

Summary

Protein AA, the major amyloid fibril protein in reactive (secondary) systemic amyloidosis is derived from the acute phase reactant liver-produced apolipoprotein serum AA (SAA) by proteolytic cleavage, usually in the C-terminal half of the 104 amino acid residues long precursor. The cleavage points in SAA vary between patients and the deposited protein AA is often quite heterogeneous. In this chapter, we describe methods to extract amyloid fibrils and to purify protein AA by sequential gel filtration. Further purification of subspecies of protein AA is best achieved by the use of differences in charge and chromatofocusing is described as the method of choice. Analytic methods include sodium dodecylsulfate polyacrylamide gel electrophoresis and analytic isoelectric focusing.

Key Words: Secondary amyloidosis; apolipoprotein; fibril; gel filtration; isoelectric point; protein fragment; isoelectric focusing; chromatofocusing; polybuffer; AA-subtypes; amyloid extraction.

1. Introduction

Amyloid Protein (Protein AA) is the major amyloid fibril protein in AA-(secondary, systemic, or reactive) amyloidosis *(1)*. Protein AA is derived from the precursor apolipoprotein serum AA (SAA) by proteolytic cleavage towards the C-terminus. It is not yet understood whether cleavage in vivo comes before or after fibril formation and whether cleavage is of any importance for the formation of amyloid deposits. There are, however, clear associations between patterns of amyloid depositions and variations in protein cleavage which indicate that cleavage is depending not only on postfibril formation trimming *(2–4)*. For this reason, it is of interest to purify the different protein cleavage products for further studies of their nature and properties.

From: *Methods in Molecular Biology, vol. 299: Amyloid Proteins: Methods and Protocols*
Edited by: E. M. Sigurdsson © Humana Press Inc., Totowa, NJ

1.1. Amyloid Protein AA

Human protein AA is an N-terminal cleavage product of SAA, which is an apolipoprotein of high density lipoproteins (HDL). SAA is an acute phase reactant, mainly produced by the liver under regulation of cytokines such as interleukin (IL)-6 and tumor necrosis factor (TNF)-α, and is under normal healthy conditions found in plasma at a very low concentration *(5)*. In association with inflammation, the production in the liver may increase dramatically yielding very high plasma content of SAA bound to HDL. Persistently high plasma concentration of SAA is one, but not the sole, factor in the pathogenesis of AA-amyloidosis. There are two different genes for major human plasma SAA, *SAA1* and *SAA2 (6)*. Both *SAA1* and *SAA2* exist as different allelic forms with small differences in amino acid sequences. All these protein forms have been found as amyloid deposits.

SAA is a phylogenetically conserved protein; AA-amyloidosis is the most common systemic amyloidosis found in other mammals and also in birds. Mouse AA-amyloidosis has long been used as a model of the human disease and has many similarities with human AA-amyloidosis.

The many amino acid sequence analyses performed on protein AA purified from humans have revealed a considerable heterogeneity in the length of the purified protein *(4,7,8)*. Almost always, a piece of the C-terminus of SAA is missing and protein AA often consists of a mixture of SAA-derived peptides of different sizes. In addition, the N-terminal Arg is also often removed. The cleavage pattern varies between patients and while a 76 amino acid residue long AA-variant may be most common *(9,10)*, protein AA species from 44 to 104 amino acid residues long have been described.

Interestingly, there is an association between the compositional mixture of protein AA species and the deposition pattern of amyloid. Thus, a mixture of particularly long and short protein AA variants is characteristic of a form of AA-amyloidosis with severe vascular involvement, but little or no amyloid deposition in renal glomeruli *(2,4)*. Other associations may exist but are less well-studied.

2. Materials

1. Amyloid-rich tissue, e.g. spleen, kidney.
2. Motor-driven and dounce homogenizers.
3. 6 M Guanidine HCl in 0.1 M Tris-HCl buffer.
4. 5 M Guanidine HCl in distilled water.
5. Sepharose 6B CL and Sephacryl S-300 (Amersham Biosciences, Uppsala, Sweden).
6. Chromatography columns.
7. UV-recorder.
8. Fraction collector.
9. pH meter.

10. Pump for running gel separations including chromatofocusing.
11. Dialysis equipment and lyophilizer.
12. Dithiothreitol.
13. Equipment for sodium dodecylsulfate gel electrophoresis (SDS-PAGE).
14. Basic chemicals and buffers.
15. Multiphor II system with gel-casting mold (Amersham Biosciences).
16. Acrylamide.
17. *N,N*-metylene-bis-acrylamide.
18. Ammonium persulfate.
19. 1 *M* NaOH and 1 *M* H_3PO_4.
20. Ampholytes (Ampholine; Amersham Biosciences) for the pH gradient of choice: broadest range 3.5–10.
21. Trichloroacetic acid, sulfosalicylic acid.
22. Coomassie Brilliant Blue R250.
23. Büchner flask.
24. PBE gel 96 or 118 (Amersham Biosciences).
25. Polybuffer (Amersham Biosciences). For separation of proteins with high IP, use Pharmalyte (Amersham Biosciences).
26. Urea.

3. Methods

3.1. Extraction of Fibrillar Material

First, amyloid fibrils have to be extracted. This is principally performed according to methods described in the 1960s *(11,12)*. The source may be any tissue containing substantial amounts of amyloid. It is difficult or impossible to extract reasonably pure amyloid fibril preparations from tissue with low contents of amyloid. The best tissue to start with is the spleen from a case with severe, diffuse splenic amyloid infiltration. Alternatively, kidneys, thyroid, and adrenals contain considerable amounts of amyloid and are suitable for fibril extraction (*see* **Note 1**). Although often good for AL-amyloid, the liver rarely is heavily involved in AA-amyloidosis and is not a good source for AA-amyloid fibrils.

1. In order to remove water-soluble proteins, tissue pieces are homogenized repeatedly in 0.15 *M* NaCl, preferably containing 0.02% NaN_3 as bacteriostatic agent. Amyloid P-component is calcium dependently bound to amyloid fibrils *(13),* and can easily be removed by the addition of 0.05 *M* sodium citrate to the homogenization solution. We use a tissue:solvent proportion of 1:15 (w:v).
2. The first homogenization can be performed in a motor-driven homogenizer while the subsequent ones are done by hand in dounce homogenizers (*see* **Note 2**). This allows a good control over the results and the possibility to remove connective tissue pieces. After each homogenization, the material is centrifuged (Beckman centrifuge equipped with a JA20 rotor) at 27,000*g* at 4°C for 30 min. About ten subsequent homogenization steps are usually necessary to remove most soluble proteins.

3. Finally, the pellet material is homogenized three times in distilled water. Supernatants from these steps contain some fibrillar material and should be lyophilized and saved. In many forms of amyloidosis, however, most of the amyloid is still in the pellet after the last homogenization in water. The pellet material should therefore be divided into a top and a bottom layer that are lyophilized separately.

3.2. Purification of Protein AA

According to earlier protocols, only water-extracted fibrils should be used for purification. However, most of the amyloid remains in the pellet, which is a good source for amyloid protein AA.

1. Delipidize material in two changes of acetone (*see* **Note 3**). Let dry.
2. Dissolve (15 mg protein/mL solvent) overnight at room temperature in 6 *M* guanidine HCl, 0.1 *M* Tris-HCl, pH 8.0, containing 0.1 *M* dithiothreitol. Centrifuge 30 min at 27,000*g*.
3. Gel filtration on a Sepharose 6B-CL column, equilibrated with 5 *M* guanidine HCl in distilled water. Another gel filtration matrix may also be used. Choose the column dimension depending on the amount of material. A slow flowrate (generally 1–8 mL/h depending on column size) is preferable for good resolution. Absorbance at 280 nm is monitored and protein AA elutes as a single or double distinct retarded peak (**Fig. 1**).
4. Pool fractions, dialyze (3.5 kDa cut-off) against distilled water, several changes (twice a day for 3 days), and lyophilize.
5. For further purification, redissolve protein (10–15 mg/mL) in 6 *M* guanidine HCl and apply to a Sephacryl S300 HR column with other conditions as in **step 3**. Repeat as in **step 4**.

3.3. Analysis of Results

This method yields reasonably pure protein AA, containing a mixture of protein AA species of varying length. Analysis of purity may be performed by SDS-PAGE. Edman degradation may also be used to assess purity and possible N-terminal heterogeneity. Mass spectrometry can also be used.

3.3.1. Analysis of Protein AA Subspecies by Isoelectric Focusing

Protein AA is not a homogenous protein *(14)*. The composition of the protein AA material can be analyzed by thin-layer isoelectric focusing (**Fig. 2**). This procedure works because amino acids are Zwitterions and proteins can be positively or negatively charged or uncharged depending on the pH of the direct environment. The mobility of a protein in a pH gradient depends on the protein's isoelectric point (IP). The IP for a specific protein is where the net charge is zero and the protein will not migrate in a pH gradient in an electrical field. In isoelectric focusing, a pH gradient is formed by carrier ampholytes using poly-

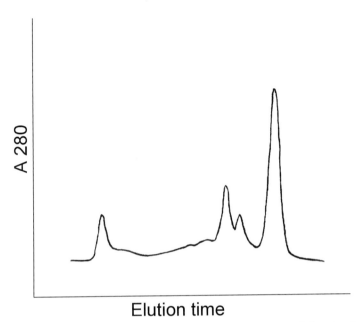

Fig. 1. Gel filtration through a column with Sepharose 6B-CL of dissolved AA-amyloid fibrils. Protein AA elutes as a retarded major peak. In this case, protein AA consists of species with two different molecular masses, explaining the double peak. The last peak contains oxidized dithiothreitol.

acrylamide or agarose as support matrix. Different pH gradients can be established using various carrier ampholytes. Isoelectric focusing can be performed in 6 *M* urea, an advantage, since amyloid proteins often are difficult to dissolve. The proteins are separated on top of the gel and can easily be passively transferred onto a nitro cellulose sheet and used for Western blot analysis with specific antibodies *(2)*.

Precast isoelectrical focussing gels with different pH intervals can be purchased from various suppliers. Amersham Biosciences has two different systems: the Multiphor II system that can handle up to 26 samples in each separate run and the PhastSystem that can handle 8 samples in each run. Both systems have two different kinds of precast gels with different pH gradients and dry gels that can be equilibrated with the appropriate buffer selected by the user. This allows equilibration with 6 *M* urea, important owing to the low solubility of amyloid proteins in other solutions. However, we prefer to make our own urea-containing gels.

pH 10

pH 4

Fig. 2. Analytic isoelectric focusing of protein AA purified from four different individuals. The variability and heterogeneity of protein AA is evident. This heterogeneity depends mainly on variation in cleavage points in the precursor protein, SAA.

3.3.2. Mounting the Mold

Wash the glass plates with detergent and warm tap water. Rinse in distilled water and wipe dry with a clean cloth. Apply some water on the glass plate and place the polyacrylamide gel (PAG) molding sheet on top of it. Place the second glass plate on top of the PAG molding sheet. The supporting glass plate and the PAG molding sheet are 2 cm longer than the top plate, leaving a 1-cm edge on each side. The parts of the mold are held together by two clamps.

3.3.3. Solutions

Solutions are to be prepared fresh prior to use and 20-mL gel mixture is sufficient for one gel. If the isoelectric focusing is performed in the presence of 6 M urea, all stock solutions are made up in 6 M urea (*see* **Note 4**).

Solutions are prepared as follows:

1. 29.1% Acrylamide in 6 *M* urea.
2. 0.9% *N,N*-methylene-bis-acrylamide in 6 *M* urea.
   The acrylamide solutions are filtered prior to use.
3. 1% Ammonium persulfate in 6 *M* urea.

Mix in a Büchner flask:

1. 3.5 mL of acrylamide solution.
2. 3.5 mL of % *N,N*-methylene-*bis*-acrylamide.
3. 1.5 mL of ampholine carrier ampholytes.
4. 12 mL 6 *M* urea.
5. Degas for 5 min.
6. Add 0.5-mL ammonium persulfate.

When the ammonium persulfate has been added, the gel has to be poured immediately.

3.3.4. Casting the Gel

The degassed gel is poured on one edge and the capillary forces will fill the space between the glass plates. Take care not to introduce any air bubbles into the gel. Let the gel polymerize for 1 h (*see* **Note 5**). A non-urea containing gel can be saved for 2–3 d at 4°C.

3.3.5. Running the IEF

Separate the gel from the top plate. This can be achieved with the help of a scalpel. Place the gel on the Multiphor and connect the cooling water. The cathode electrode strip should be immersed with 1 *M* NaOH and the anode electrode strip with 1 *M* H_3PO_4. Prior to sample application, the gel has to be run for 20 min to establish a pH gradient. The running effect should be 25 W. The samples are dissolved in 20–40 µL 6 *M* urea and absorbed into the sample application pieces. These pieces are placed directly on the gel. Samples can be applied anywhere on the flatbed gel, but the gel becomes slightly disarranged at the application site and therefore, the samples should be applied to an area different from the IP of the proteins of interest. The focusing continues at the same conditions for 1 h. Hemolyzed red blood cells can be used as a visible marker. The red colored sample is applied at the cathode and the focusing is ready when the sample has reached the anode.

The pH gradient can be directly measured on the gel with the use of a flat bottom electrode and a pH meter.

3.3.6. Staining and Destaining of the Gel

Fixation solution is prepared by dissolving 34 g trichloroacetic acid and 10.4 g sulfosalicylic acid in 250 mL distilled water.

The gels are fixed in the fixation solution for 30 min. Note that samples soluble in acids can be dissolved and disappear from the gel.

The proteins are stained by 0.1% Coomassie Brilliant Blue R250 at 60°C for 30 min. Thereafter, the gel is destained with 25% ethanol with 8% acetic acid in distilled water. A complete destaining can take 24 h to reach.

3.4. Purification of Protein AA Species by Chromatofocusing

The protein AA species may be purified according to their isoelectric points. This may be achieved with several methods, including ion exchange chromatography, preparative isoelectric focusing, or chromatofocusing. Chromatofocusing usually gives excellent results and is described here.

3.4.1. Chromatofocusing

Chromatofocusing is a method that allows a rapid and high resolution separation of proteins based on differences in IP. The separation takes place in a linear pH gradient and the pH interval is selected so the IP of sample protein of interest falls in the middle of the gradient for optimal separation. The self-generated pH gradient is established with a cation exchange matrix (PBE 97 and PBE 118) and Polybuffer. Polybuffer contains charged groups with high buffering capacity and it is supplied for two different pH regions, Polybuffer 9-6 and Polybuffer 7-4. Separation at a higher pH range is performed with Pharmalyte 8-10.5. The gel exchange matrix is calibrated with a buffer with high pH and sample is applied to the column and eluted with elution buffer with low pH. Proteins will be eluted according to their IPs and proteins with high IPs will be eluted first (**Figs. 3** and **4**). A narrower pH gradient will give higher resolution (*see* **Note 6**). If the IP is unknown, it can be determined by isoelectric focusing (*see* **Subheading 3.3.**) or the sample can be separated in a gradient with a longer pH interval that can be obtained by combining Polybuffer 9-6 with Polybuffer 7-4. In this way, a pH gradient that allows separation of proteins with IPs between pH 9.4 can be produced. Chromatofocusing can be performed in 8 *M* urea, which allows separation of the amyloid proteins. The system can be used for separation of large quantities of proteins. We use a 60-cm long column with a 0.9-cm diameter and separate up to 1 g of protein at one run. This column is adequate for 30–35 mL of gel.

3.4.2. Procedure

1. Equilibrate PBE 97 gel with the degassed start buffer (pH 9.4) and pour onto the column.
2. Allow the gel to pack at high pressure obtained by a fast flow (100 mL/h) for 20 min (*see* **Note 7**).

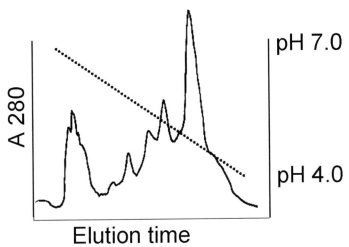

Fig. 3. Elution pattern of chromatofocusing of purified protein AA in which acidic (long) protein AA subspecies predominate.

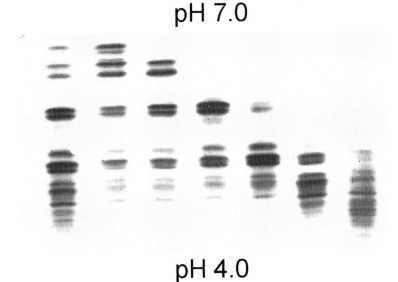

Fig. 4. Analytic isoelectric focusing of eluted protein AA from a chromatofocing run. Subspecies can be further purified by a second cycle of chromatofocusing with a more narrow pH span.

3. On top of the PBE gel, pour about 1 cm of preswollen coarse Sephadex G 25.
4. Dissolve the sample in the elution buffer with low pH. Centrifuge at 10,000g with Eppendorf (or similar) bench top centrifuge for 5 min.
5. Apply the dissolved sample to the column and pump an additional 10 mL of the start buffer onto the column.

6. Thereafter, the elution buffer is pumped in with pump speed set to 20 mL/h (*see* **Note 8**).
7. Monitor the elution at 280 nm and plot the profile.
8. Collect fractions.
9. The pH gradient can be determined during elution if the eluent is lead through a pH meter.

The pH gradient ends after 10–15 column volumes of elution buffer. This separation procedure can easily be set up and operated at any laboratory (*see* **Note 9**).

4. Notes

1. In order to obtain a good amyloid fibrillar material, it is crucial to use tissue with heavy amyloid deposits, constituting at least 25 to 50% of the tissue, estimated on histological sections. The fibril extraction method by Pras et al. *(12)* is not useful for concentration of amyloid fibrils.
2. Extraction of fibrils can be performed entirely with motor-driven homogenizers and is preferred by some researchers.
3. Extracted fibrillar material contains some lipids (the amount depends on the tissue origin) and we usually delipidize the material before dissolving in guanidine hydrochloride. For material from pancreas and other lipid-rich tissues, this is best performed with chloroform-methanol (2:1) mixture followed by acetone; for less lipid-containing tissues, e.g., the spleen, delipidization in acetone is usually enough.
4. When urea is included, this solution must be made fresh. Boiling the redistilled water facilitates dissolving urea.
5. It will take about 1 h for a 0.5-mm acrylamide gel to polymerize if the solution is properly degassed. The degassing step is important and, if inappropriately performed, the polymerization time will be extended or the polymerization incomplete. Remember: acrylamide in contact with air will not polymerize. Therefore, acrylamide at the ends of the mold will never polymerize since it is in contact with air. To check if the gel is ready; carefully tilt the mold. Also remember that unpolymerized acrylamide is neurotoxic. Always use gloves in all steps when acrylamide is handled.
6. The choice and composition of start and elution buffers for chromatofocusing depend on the pH interval selected for the separation. Example: If the protein as an IP of 6 the following procedure can be used. The selected pH gradient can be 7.0–5.0. A complete list with recommended buffers and mixing proportions is supplied with the Polybuffer. A booklet is also available on the net: http://www5.amershambiosciences.com. Separation is performed on a PBE 97 exchange gel matrix equilibrated with the start buffer. In this example the start buffer can be 0.0025 *M* bis-Tris (bis(2-hydroxyethyl)aminotris(hydroxymethyl)methane) with the pH adjusted to 7.1 with CH_3COOH. The elution buffer consists of 10% Polybuffer 74 diluted in water and the pH adjusted to 4.0 with iminodiacetic acid. If 8 *M* urea is to be used, it has to be added to all solutions. Remember that urea will

add a significant volume to the solution and final volume adjustment should not be done before the substance is completely dissolved.

7. When packing the chromatofocusing gel, it is of great importance to do so at a very high buffer flow. If the packing is performed at a slower flow, the column will not give the high resolution, which otherwise is one of the great advantages with the method. Note that guanidine hydrochloride cannot be used instead of urea.

8. The gel possesses a high buffering capacity and it will take 10–15 column volumes before the pH of the eluted solution reaches the low pH of the elution buffer. A good hallmark for the elution buffer volume is 15 times the gel volume. Do not be too restricted with the volume of elution buffer. It is better to be slightly generous and make some extra elution buffer instead of letting the column run dry. The gel cracks easily if it gets dry, and the packing procedure is rather time-consuming. A well-packed column can be reused multiple times. Rinse the gel with two volumes of 1 *M* NaCl between runs.

9. The given methods include material from Amersham Biosciences. Other ampholytes and equipments than those from this company may be used with equally good results.

Acknowledgments

This work was supported by the Swedish Research Council (projects Nos. 5941 and 14040) and the Swedish Heart Lung Association.

References

1. Benditt, E. P., Eriksen, N., Hermodson, M. A., and Ericsson, L. H. (1971) The major proteins of human and monkey amyloid substance: common properties including unusual N-terminal amino acid sequences. *FEBS Lett.* **19,** 169–173.

2. Westermark, G. T., Westermark, P., and Sletten, K. (1987) Amyloid fibril protein AA. Characterization of uncommon subspecies from a patient with rheumatoid arthritis. *Lab. Invest.* **57,** 57–64.

3. Westermark, G. T., Sletten, K., and Westermark, P. (1989) Massive vascular AA-amyloidosis: A histologically and biochemically distinctive subtype of reactive systemic amyloidosis. *Scand. J. Immunol.* **30,** 605–613.

4. Westermark, G. T., Sletten, K., Grubb, A., and Westermark, P. (1990) AA-amyloidosis. Tissue component-specific association of various protein AA subspecies and evidence of a fourth SAA gene product. *Am. J. Path.* **137,** 377–383.

5. Uhlar, C. M. and Whitehead, A. S. (1999) Serum amyloid A, the major vertebrate acute-phase reactant. *Eur. J. Biochem.* **265,** 501–523.

6. Husby, G., Marhaug, G., Dowton, B., Sletten, K., and Sipe, J. D. (1994) Serum amyloid A (SAA): biochemistry, genetics and the pathogenesis of AA amyloidosis. *Amyloid: Int. J. Exp. Clin. Invest.* **1,** 119–137.

7. Møyner, K., Sletten, K., Husby, G., and Natvig, J. B. (1980) An unusually large (83 amino acid residues) amyloid fibril protein AA from a patient with Waldenström's macroglobulinaemia and amyloidosis. *Scand. J. Immunol.* **11,** 549–554.

8. Liepnieks, J. J., Kluve-Beckerman, B., and Benson, M. D. (1995) Characterization of amyloid A protein in human secondary amyloidosis: the predominant deposition of serum amyloid A1. *Biochim. Biophys. Acta* **1270,** 81–86.

9. Levin, M., Franklin, E. C., Frangione, B., and Pras, M. (1972) The amino acid sequence of a major nonimmunoglobulin component of some amyloid fibrils. *J. Clin. Invest.* **51,** 2773–2776.

10. Sletten, K. and Husby, G. (1974) The complete amino-acid sequence of non-immunoglobulin amyloid fibril protein AS in rheumatoid arthritis. *Eur. J. Biochem.* **41,** 117–125.

11. Shirahama, T. and Cohen, A. S. (1967) High-resolution electron microscopic analysis of the amyloid fibril. *J. Cell. Biol.* **33,** 679–708.

12. Pras, M., Schubert, M., Zucker-Franklin, D., Rimon, A., and Franklin, E. C. (1968) The characterization of soluble amyloid prepared in water. *J. Clin. Invest.* **47,** 924–933.

13. Skinner, M., Shirahama, T., Cohen, A. S., and Deal, C. L. (1983) The association of amyloid P-component (AP) with the amyloid fibril: an updated method for amyloid fibril protein isolation. *Prep. Biochem.* **12,** 461–476.

14. Westermark, P. (1982) The heterogeneity of protein AA in secondary (reactive) systemic amyloidosis. *Biochim. Biophys. Acta* **701,** 19–23.

19

Purification of Transthyretin and Transthyretin Fragments From Amyloid-Rich Human Tissues

Per Westermark and Gunilla T. Westermark

Summary

Transthyretin is the major amyloid fibril protein in many forms of familial systemic amyloidosis where a missense mutation creates an amyloidogenic protein, and in senile systemic amyloidosis in which wild-type transthyretin aggregates into amyloid fibrils. The amyloid deposits may consist of full-length transthyretin but is very often, in senile systemic amyloidosis always, a mixture of full-length transthyretin and C-terminal transthyretin fragments. The amyloid fibril protein mixture can be purified by extraction of fibrils followed by sequential gel filtration after solubilization in a solution of guanidine hydrochloride. Since the C-terminal transthyretin fragments lack cysteine residues, a method to separate full-length transthyretin from fragments by covalent chromatography has been developed.

Key Words: Familial amyloidosis; senile systemic amyloidosis; fibril; gel filtration; heart; amyloid extraction; covalent chromatography; thiopropyl sepharose.

1. Introduction

Transthyretin (TTR) constitutes the fibrils in most forms of familial amyloidosis *(1)* and in senile systemic amyloidosis (SSA) *(2)*. In familial TTR-amyloidosis, there is a missense mutation in the TTR gene *(3)* while no such abnormality exists in senile systemic amyloidosis. TTR can form fibrils as a full-length protein or after proteolytic removal of an N-terminal piece of the protein.

The amyloid fibrils in SSA and in most forms of familial amyloidosis contain a mixture of TTR and proteins derived by proteolytic processing of TTR *(4,5)*. In many cases, always in SSA, fragments predominate and full-length TTR is a minor component. The predominant TTR fragments are truncated at the N-terminal part and start at positions 46, 49, and 52 *(2)*. Other fragments may occur as well.

From: *Methods in Molecular Biology, vol. 299: Amyloid Proteins: Methods and Protocols*
Edited by: E. M. Sigurdsson © Humana Press Inc., Totowa, NJ

Normal TTR as well as almost all mutant TTR variants have a single cysteine residue residing at position 10. Since the fibrillar TTR-fragments start long after this position, the Cys residue can be used for the separation of full-length TTR from TTR fragments *(2)*. In normal TTR, the Cys residue may have a free SH-group or be covalently linked by a disulfide group to several different compounds, including glutathione. In fibrils, at least in SSA and in Swedish amyloidotic polyneuropathy with a V30M mutation, the Cys residue is more or less always covalently linked to other, yet not completely identified, components. Some of the disulfide bridges may be owing to homodimer formation but most probably, TTR is also bound to other proteins. In any case, at gel filtration in non-reducing conditions, most of fibril TTR is eluted corresponding to a much higher than normal molecular mass. The presence of a cysteine residue in full-length TTR, but not in the C-terminal fragments, can be used for a simple method to separate these two components.

2. Materials

1. Amyloid-rich tissue, e.g., myocardium.
2. Motor-driven and Dounce homogenizers.
3. 6 *M* guanidine HCl in 0.1 *M* Tris-HCl buffer (*see* **Note 1**).
4. 5 *M* guanidine HCl in distilled water.
5. Sepharose 6B CL, Sephacryl S-200 and Sephadex G25 (Amersham Biosciences, Uppsala, Sweden).
6. Chromatography equipment.
7. Chromatography columns for gel filtration, desalting and covalent chromatography.
8. Dialysis equipment and lyophilizer.
9. Dithiothreitol (DTT).
10. Thiopropyl-Sepharose 6B (Amersham Biosciences).
11. Equipment for sodium dodecylsulfate gel electrophoresis (SDS-PAGE).
12. L-Cysteine.

3. Methods

3.1. Extraction of Amyloid Fibril-Rich Material

Amyloid fibril-rich material, usually from heart tissue, is extracted using the same procedure as described for AA-amyloid in chapter 18. Most of the amyloid is to be found in the pellet material (*see* **Notes 2** and **3**). The material should be lyophilized.

3.2. Initial Purification of Fibril Protein

1. Delipidize material in two changes of acetone. Let dry.
2. Dissolve (15 mg protein/mL solvent, *see* **Note 4**) for 2 d with constant stirring at room temperature in 6 *M* guanidine HCl, 0.1 *M* Tris-HCl, pH 8.0, without addition of any reducing agent. Centrifuge 30 min at 27,000*g*.

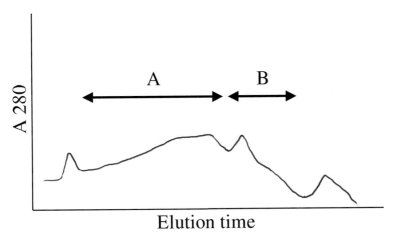

Fig. 1. Gelfiltration of amyloid-material from heart tissue of an individual with senile systemic (wild-type TTR-derived) amyloidosis. Amyloid was dissolved in 6 *M* guanidine HCl without reduction and proteins separated through a Sepharose 6B-CL column, equilibrated with 5 *M* guanidine HCl in distilled water. Full-length TTR can be purified from material in fraction A while fraction B contains TTR-fragments. For further purification, material in fraction A can be separated by covalent chromatography.

3. Gel filtration through a Sepharose 6B-CL (Amersham Biosciences) column, equilibrated with 5 *M* guanidine HCl in distilled water. Other gel filtration media may also be used. Choose the dimension of column depending on the amount of material, but generally a 1.6 × 100 cm column should be sufficient. A low flowrate is preferable to get good resolution. Absorbance at 280 nm is monitored. TTR fragments elute as a late retarded distinct peak (Fraction B, **Fig. 1**).

4. Pool all fractions between the V_0-material and the TTR-fragment peak (Fraction A, **Fig. 1**). Dialyze (cut-off 3.5 kDa) against distilled water, several changes, and lyophilize.

5. Redissolve in 6 *M* guanidine HCl containing 0.1 *M* Tris-HCl buffer, pH 8.0 and add DTT to 100 m*M*. Incubate overnight at room temperature.

6. Gel filtrate through a short Sephadex G25 column, equilibrated with well-degassed 5 *M* guanidine HCl in 0.1 *M* Tris-HCl buffer, pH 8.0. Monitor at 280 nm and collect the protein peak. This protein material with free-SH groups will be used immediately for covalent chromatography.

3.3. Covalent Chromatography

In order to separate cysteine-containing proteins from those without, covalent chromatography can be performed. In this procedure, proteins are immobilized by a disulfide bridge to a matrix from which it can be eluted by the use of a reducing agent. The treatment of the TTR-containing material with DTT in **steps 5** and **6** above have created free SH-groups and the proteins with these can be directly bound covalently (*see* **Note 5**).

1. Apply the reduced protein material directly to a column of Thiopropyl Sepharose CB-CL (Amersham Biosciences), equilibrated with degassed 5 *M* guanidine HCl, 0.1 *M* Tris-HCl, pH 8.0, solution. Thiopropyl-Sepharose 6B comes as a freeze-dried powder and has to be reswollen according to the manufacturer's description before it is equilibrated with guanidine HCl. Wash with 2–3 column volumes 5 *M* guanidine HCl, 0.1 *M* Tris-HCl, pH 8.0 (*see* **Note 6**).
2. Elute bound protein with 25 m*M* cysteine in 5 *M* guanidine HCl, 0.1 *M* Tris-HCl, pH 8.0. Dialyze and lyophilize.
3. Redissolve in degassed 6 *M* guanidine HCl in water, containing 0.1 *M* DTT and gel filter through a 0.9 × 60 cm Sephacryl S-300 column. Alternatively, other gel filtration columns may be used. Dialyze against distilled water and lyophilize.

With this method, it is possible to separate full length TTR from a mixture of TTR-derived peptides which commonly are found in both familial forms of TTR-amyloidosis and in SSA (**Fig. 2**).

3.4. Further Purification of TTR Fragments

The material, eluted in the last protein peak at the non-reduced gel filtration, contains mainly C-terminal TTR fragments, starting at positions 46, 49, and 52 and ending at positions 126 or 127. Additional fragments may also be present. These fragments have fairly similar isoelectric points and are difficult to separate from each other, although reversed-phase HPLC on a C4 or C8 column may give partial resolution.

4. Notes

1. An alternative to 5 *M* guanidine hydrochloride is 8 *M* urea, which is cheaper. Urea gives equally good results but must be of high quality and freshly dissolved.
2. Extraction of ATTR amyloid fibrils from heart is not easy. The heart tissue is often very rich in collagen and homogenization by hand is sometimes difficult and should be avoided until a reasonably homogenous slurry has been obtained. Otherwise, there is a great risk for accidents with broken Dounce homogenizers. Several cycles with an electric homogenizer are often necessary.
3. Like some other amyloid forms, TTR-amyloid fibrils tend to remain bound to each other and are difficult to extract with the water extraction method of Pras et al. *(4)*. Most amyloid material stays in the pellet, which therefore usually is the best source for protein purification. Since the pellet will contain many other components, including insoluble connective tissue and nuclei, the start material should be rich in amyloid.
4. The lyophilized pellet material, containing most of the amyloid but also a lot of other components, is only partially soluble in 6 *M* guanidine hydrochloride, even after an extended period of time. We therefore often start with a comparably large

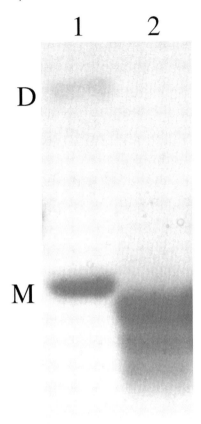

Fig. 2. SDS-PAGE of full length TTR (lane 1) and TTR fragments (lane 2) purified from cardiac amyloid fibrils of an individual with senile systemic amylodosis. M corresponds to monomeric and D to dimeric TTR.

amount of lyophilized material (e.g., 300 mg) and stir this in a large volume (e.g., 40 mL) of 6 M guanidine hydrochloride in 0.1 M Tris-HCl, pH 8.0, for 2–3 d. After centrifugation, the supernatant is dialyzed against saturated ammonium sulfate (to precipitate all proteins in dialysis bags), followed by water, and then lyophilized. This material is thereafter redissolved in 6 M guanidine hydrochloride, 0.1 M Tris-HCl, pH 8.0, and applied to a column for gel filtration. In this way, we get a more concentrated material, which is easier to work with.

5. Without reduction, all full-length TTR-molecules remain bound to other molecules. They are eluted in virtually all fractions before the TTR fragment peak. When gel filtration is run in the presence of DTT, the free TTR molecules are eluted very close to the fragment peak. It is possible to perform covalent chromatography with this material as an alternative.

6. The Thiopropyl-Sepharose 6B column cannot be used directly again but can be reactivated by reduction and treatment with 2,2'-dipyridyl disulfide according to the manufacturer's instructions. However, it is our experience that recycling does not give an optimal gel, and is therefore not recommended.

Acknowledgments

This work was supported by the Swedish Research Council (projects Nos. 5941 and 14040) and the Swedish Heart Lung Association

References

1. Ikeda, S., Nakazato, M., Ando, Y., and Sobue, G. (2002) Familial transthyretin-type amyloid polyneuropathy in Japan: clinical and genetic heterogeneity. *Neurology* **58**, 1001–1007.
2. Westermark, P., Sletten, K., Johansson, B., and Cornwell, G. G. III. (1990) Fibril in senile systemic amyloidosis is derived from normal transthyretin. *Proc. Natl. Acad. Sci. USA* **87**, 2843–2845.
3. Connors, L. H., Lim, A., Prokaeva, T., Roskens, V. A., and Costello, C. E. (2003) Tabulation of human transthyretin (TTR) variants. *Amyloid* **10**, 160–184.
4. Pras, M., Franklin, E. C., Prelli, F., and Frangione, B. (1981) A variant of pre-albumin from amyloid fibrils in familial polyneuropathy of Jewish origin. *J. Exp. Med.* **154**, 989–993.
5. Westermark, P., Sletten, K., and Olofsson, B.-O. (1987) Prealbumin variants in the amyloid fibrils of Swedish familial amyloidotic polyneuropathy. *Clin. Exp. Immunol.* **69**, 695–701.

20

Extraction and Chemical Characterization of Tissue-Deposited Proteins From Minute Diagnostic Biopsy Specimens

Fernando Goñi and Gloria Gallo

Summary

The compelling need for early detection and chemical characterization of protein deposits in conformational disorders has always relied on immunohistochemical techniques performed on diagnostic biopsy specimens. Although the identity of the culprit can be assessed, the molecular nature of the defect still requires the analysis of the extracted material. The purpose of this chapter is to provide investigators with a method to extract and analyze—from minute residual diagnostic tissue biopsies—the deposited proteins. If done successfully, the technique will allow further investigation by molecular studies, while the patient is still being evaluated.

Key Words: Human; micro biopsy specimens; amorphous deposition; amyloid; protein extraction; protein sequencing; immunohistochemistry; immunoblot.

1. Introduction

Conformational disorders are characterized by the deposition of normal/ abnormal proteins in tissue. The most common form of conformational disease is amyloidosis in which proteins with extensive β-sheet structure deposit as insoluble fibrils *(1)*. However, in the case of immunoglobulins, the proteins can precipitate and deposit in tissues as fibrillar or nonfibrillar material (**Fig. 1**) *(2)*. The nonfibrillar forms are referred to as nonamyloid monoclonal immunoglobulin deposition disorders (NAMIDD) *(3)*. In any event, the nature and the chemical identity of the protein are necessary for accurate diagnosis.

Most often, the chemical characterization of the protein deposits has been performed on large amounts of tissue obtained from post-mortem specimens. Such methodologies have been described *(2,3)* and some are reviewed in chapters 18 and 19 of this book. Most of them require at least several milligrams of

From: *Methods in Molecular Biology, vol. 299: Amyloid Proteins: Methods and Protocols*
Edited by: E. M. Sigurdsson © Humana Press Inc., Totowa, NJ

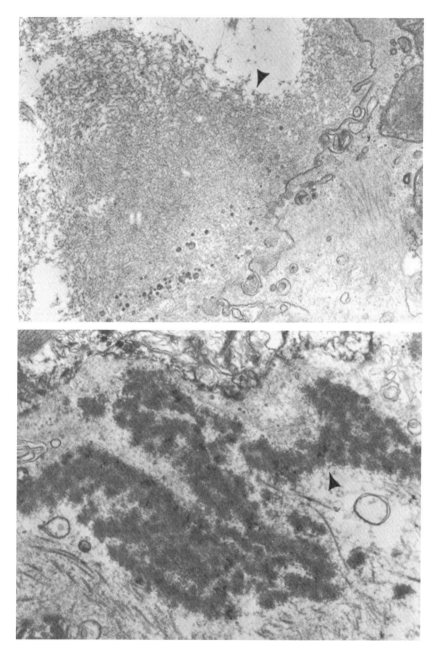

Fig. 1. Top panel: Electron micrograph of myocardial tissue showing fibrillar deposits (arrowhead) of amyloid overlying collagen and myocardial cells. Bottom panel: Electron micrograph of myocardial tissue demonstrating punctuate granular deposits of non-amyloid κ light chain deposits (arrowhead); a case of LCDD.

tissue to perform the technique *(2–4)*. However, it is becoming increasingly necessary to develop methods that can be used on small diagnostic biopsy specimens from living patients to determine the nature of the tissue deposits that are relevant for specific treatment and prognosis. We briefly examine the methodologies available to extract precipitated proteins, amorphous or fibrillar, from tiny biopsies. Although the methodology was developed to characterize the deposits in renal biopsy tissue with nonamyloid light chain deposition disease (LCDD) *(5)*, the same technique with minor appropriate modifications can be applied to other tissues with amyloid deposits.

2. Materials

1. Warm bath for 37°C.
2. 1.5 or 2.0 mL siliconized/low retention plastic tubes.
3. OCT compound (Optimal Cutting Temperature, Embedding medium for frozen tissue specimens, Sakura Finetek USA Inc., Torrance, CA). Contents: 10.24% w:w polyvinyl alcohol; 4.26% w:w polyethylene glycol; 85.50% w:w nonreactive ingredients.
4. Microcentrifuge for 1.5–2.0 mL tubes.
5. Micropipets for 1–20 µL, 20–200 µL and 100–1000 µL.
6. Long end pipet tips for 1–200 µL (*see* **Note 1**).
7. Thin plastic or glass rods, 1 mm in diameter or less (*see* **Note 2**).
8. Saline solution: 0.9% NaCl w:v.
9. Small thin tweezers with plastified ends (*see* **Note 3**).
10. Vortex.
11. Dialysis tubing of low molecular weight cut off (1000 to 2000), (Spectrapor, Gardena, CA).
12. Distilled deionized water (DDW).
13. Freeze-drying equipment, lyophilizer (*see* **Note 4**).
14. Sample buffer for Tris-Tricine polyacrylamide electrophoresis: 2% SDS, 50 m*M* Tris-HCl, pH 6.8, 5% glycerol, 0.1% bromophenol blue (can be purchased ready made from various companies).
15. Thermomixer (Eppendorf) up to at least 14,000 rpm and 90°C (*see* **Note 5**).
16. Dithiothreitol (DTT).
17. Polyacrylamide gels 10–20%, 12.5%, 16.5%.
18. Polyvinyledine difluoride membranes (PVDF) (can be purchased from various sources).
19. 10 m*M* CAPS (3-cyclo-hexyl-amino-1-propanesulphonic acid) pH 11.0.
20. Methanol HPLC grade.
21. Tween-20.
22. 20 m*M* Phosphate buffer saline (PBS) pH 7.2 (PBS tablets [Sigma]: dissolve 1 tablet in 200 mL of DDW).
23. Coomassie Brilliant Blue R-250 (CBBR-250).
24. Glacial acetic acid, ultra pure.
25. Amino acid sequencer.

3. Methods

The methodology described is intended primarily for the extraction of amorphous material from residual diagnostic frozen tissue biopsies (~ 0.5 to 1 mm) embedded in OCT, following immunohistochemistry.

1. The frozen specimen embedded in OCT is transferred to a 1.5-mL tube.
2. The sample is brought to 37°C until the OCT is melted. The tube is centrifuged for 1 min at 900g.
3. The supernatant is gently pipetted off and the tissue at the bottom of the tube is washed once—gently stirring—with 0.5 mL of saline solution (*see* **Note 6**).
4. After removing the supernatant, the tissue is gently loosened, preferably with a thin plastic rod avoiding its rupture or piercing. Otherwise, a glass rod can be used (*see* **Note 7**).
5. One milliliter of saline solution is added to the tube and the mixture is slowly vortexed (no more than power 3 in a 12 scale) for 10 s before centrifuging it for 1 min at 900g. The tissue is washed twice in the same manner.
6. All saline supernatants are pooled, dialyzed against DDW, lyophilized and stored at −20°C until used as controls.
7. 100 μL of Tris-Tricine polyacrylamide sample buffer are added to the washed tissue as described in **Subheading 2.** (*see* **Note 8**).
8. The mixture is brought to 80°C and incubated for at least 2 h with vigorous agitation in a thermomixer block.
9. The sample is spun for 5 min at 900g, and the supernatant transferred to another 1.5-mL tube.
10. DTT is added to a final concentration of 100 mM (*see* **Note 9**), and the mixture boiled for at least 5 min before being applied onto a 16.5% SDS-PAGE Tris-Tricine (*see* **Note 10**).
11. The samples are then electrophoresed for 3 h at 30 mA/gel and electrotransferred for 4 h at 150 mA onto PVDF membranes, using 10 mM CAPS buffer pH 11.0 containing 20% methanol (*see* **Note 11**).
12. A strip representing not more than 20% of the total material in the membrane is subsequently cut, blocked for one hour in 5% nonfat milk in PBS containing 0.1% Tween 20 (PBS-T) and incubated with the specific antibody as per the identity of the protein determined by immunohistochemistry (**Fig. 2**).
13. The remainder of the material and the accompanying molecular weight markers are stained for 1 min in 0.3% Coomassie BBR-250 in 40% methanol-1% acetic acid. The membrane is then quickly destained using a few changes of 40% methanol in DDW and finally washed twice with DDW and air dried.
14. The bands matching the positive material on the immunoblot are excised and subjected to N-terminal sequencing (**Fig. 2**) on a 477 A microsequencer (Applied Biosystems). The resulting phenylthiohydantoin derivatives are then identified using an online PTH derivative analyzer (Applied Biosystems).

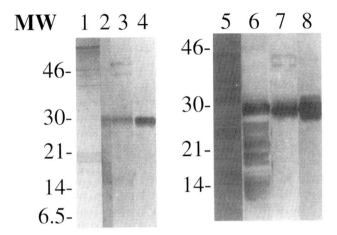

Fig. 2. Fifteen percent SDS-PAGE and Western blot analysis of GLA and CHO κ light chains. Extracted light chains from GLA kidney biopsy tissue (lane 1) and CHO cardiac tissue (lane 5) were stained with Coomassie blue. Immunoreactivity was demonstrated for the extracted proteins GLAA (lane 2) and CHO (lane 6) with anti-κ. Urine eluates are for GLA (lane 3) and CHO (lane 7) after kappaLock purification, stained with Coomassie blue. Western blot analysis was made of GLA (lane 4) and CHO (lane 8) urine eluates immunostained with anti-κ. The 30Kd band seen for GLA (lane 1) was sequenced with results: DIQMTQSPSTLSA**F**VGD including a mutation from gene L12a shown in bold. (Reproduced from **ref. 5** with permission from the American Journal of Pathology.)

4. Notes

1. Use the same tips that are used for loading the polyacrylamide gels.
2. The rods can be replaced by either a Pasteur pipet with the narrow end closed by heating with a lab burner, or by plastic disposable loops for bacteriology (Fisher).
3. The tweezers can be replace by a small plastic cell scraper.
4. Lyophilizer can be replaced by other equipment like a SpeedVac or filtration/concentration devices; however, the use of these methodologies is discouraged because they might result in the loss of protein that already is scarce.
5. The thermomixer can be replaced by a water bath with shaker that can be brought to 80°C.
6. It has to be emphasized that the sample should be treated very gently to wash off only material that could be loosely attached, but is not part of the pathological deposit.
7. It is very important not to mechanically disrupt, dislodge, or homogenize the tissue, which may result in the solubilization of proteins that might interfere with the

proper characterization of the pathological deposits. Owing to the tiny size of the sample, sometimes it is better to perform this part of the procedure under a magnifier with a good light source.

8. Before **step 7** there could be introduced alternative extractions for different amyloids as follows: (1) Three 1 h repetitive extractions using 500 μL of 0.1% trifluoroacetic acid in 20% acetonitrile, and centrifugation at 900*g*. The supernatants should be pooled and lyophilized before proceeding to the separation on SDS-PAGE; (2) Extraction with 200 μL of formic acid for 10 min, centrifugation at 900*g* and removal of the supernatant. In both cases, after performing the extractions the methodology has to continue from **step 7** to ensure there are not any other non specific adherent proteins.

9. Prepare a solution of 1 *M* DTT in DDW (154 mg/mL) and add 1/10 of the sample volume. The DTT solution has to be kept frozen but can be thawed repeatedly.

10. The choice of the exclusion pore of the polyacrylamide gel to be prepared depends on the molecular weight of the protein being extracted, and, more importantly, on the fragments that may have been generated in the deposited material.

11. When samples will be used for amino acid sequencing it is better not to use a transfer buffer containing glycine. The use of CAPS gives the best option and with pH 11.0 most of the proteins in a biological sample will be transferred to the membrane. The transfer can also be performed at 350 mA for 1.5 h with a cooling system.

References

1. Gallo, G., Picken, M., Buxbaum, J., and Frangione, B. (1989) The spectrum of monoclonal immunoglobulin deposition disease associated with immunocytic dyscrasias. *Seminars Hematol.* **26,** 234–245.
2. Kaplan, B., Vidal, R., Kumar, A., Ghiso, J., Frangione, B., and Gallo, G. (1997) Amino-terminal identity of co-existent amyloid and non-amyloid immunoglobulin κ light chain deposits. A human disease to study alterations of protein conformation. *Clin. Exp. Immunol.* **110,** 472–478.
3. Gallo, G., Goñi, F., Boctor, F., et al. (1996) Light chain cardiomyopathy. Structural analysis of the light chain tissue deposits. *Am. J. Pathol.* **148,** 1397–1405.
4. Kaplan, B., Hrncic, R., Murphy, C., Gallo, G., Weiss, D., and Solomon, A. (1999) Microextraction and purification techniques applicable to chemical characterization of amyloid proteins in minute amounts of tissue. *Methods Enzymol.* **309,** 67–81.
5. Vidal, R., Goñi, F., Stevens, F., et al. (1999) Somatic mutations of the L12a gene in V-κ1 light chain deposition disease. Potential effects on aberrant conformation and deposition. *Am. J. Pathol.* **155,** 2009–2017.

21

Tissue Processing Prior to Protein Analysis and Amyloid-β Quantitation

Stephen D. Schmidt, Ying Jiang, Ralph A. Nixon, and Paul M. Mathews

Summary

Amyloid-containing tissue, whether from human patients or an animal model of a disease, is typically characterized by various biochemical and immunohistochemical techniques, many of which are described in detail in this volume. In this chapter, we describe a straightforward technique for the homogenization of tissue prior to these analyses. The technique is particularly well-suited for performing a large number of different biochemical analyses on a single mouse brain hemisphere. Starting with this homogenate, multiple characterizations can be done, including Western blot analysis and isolation of membrane-associated proteins, both of which are described here. Additional analyses can readily be performed on the tissue homogenate, including the ELISA quantitation of Aβ in the brain of a transgenic mouse model of β-amyloid deposition. The ELISA technique is described in detail in the following chapter.

Key Words: Aβ; Alzheimer's disease; amyloid; amyloid precursor protein (APP); brain; dissection; fractionation; homogenization; membrane protein; SDS-PAGE; Western blot analysis.

1. Introduction

It is often valuable to characterize and quantitate multiple proteins from a tissue sample in addition to the amyloid species of interest. Unfortunately, the tissue extraction techniques used to solubilize amyloid, such as formic acid extraction of β-amyloid to fully recover Aβ in human Alzheimer's disease brain or in β-amyloid-depositing transgenic mice (*1*), are frequently incompatible with other methods of protein analysis because of the extensive denaturation to which the tissue is subjected. The extraction protocol described here, however, is compatible with numerous biochemical analyses while allowing for quantitation of Aβ levels both prior to and following β-amyloid deposition (*2–6*). In this chapter, we describe the preparation of a sucrose homogenate from a mouse brain hemisphere and the subsequent Western blot analysis of various proteins.

From: *Methods in Molecular Biology, vol. 299: Amyloid Proteins: Methods and Protocols*
Edited by: E. M. Sigurdsson © Humana Press Inc., Totowa, NJ

Additionally, fractionation of membrane-associated proteins from soluble proteins, which may be used to separate amyloid precursor protein (APP) from the soluble, secreted N-terminal fragment of APP (sAPP) *(7–10)*, is described.

As illustrated in Chapter 22, a portion of this tissue homogenate can also be subjected to standard amyloid extraction techniques, which allows for the quantitation of Aβ in β-amyloid plaques. This reduces the amount of tissue required to obtain multiple readouts and allows for direct comparison of the levels of an amyloid precursor to the aggregated amyloid within a single sample. This approach is particularly valuable in the analysis of β-amyloid depositing transgenic mice, where a single hemibrain can be used as the starting material for multiple biochemical analyses, preserving the other hemibrain for additional purposes, such as fixation and immunohistochemistry. This greatly increases the numbers and types of analyses one can obtain from a single mouse, and thus represents significant savings in the number of animals necessary for an informative data set.

2. Materials

1. Tissue of interest or, as described here, a relevant β-amyloid mouse model for dissection of the brain. A generally useful quantity of tissue is approx 200 mg.
2. Homogenization equipment, such as a 5-mL Wheaton Potter-Elvehjem tissue grinder with a piston-type polytetrafluoroethylene (Teflon) pestle and glass vessel (available from Fisher Scientific, Pittsburgh, PA, cat. no. 08-414-16C), motorized by an electric stirrer. Other homogenization techniques can also be employed (*see* **Note 1**).
3. Neutral pH, phosphate buffered 10% formalin (Fisher Diagnostics, Middletown, VA, cat. no. 245-685).
4. Ultracentrifuge and rotor (e.g., Beckman Coulter Optima TLX or MAX-E centrifuge and TLA-100.1 rotor).
5. 0.5-mL Thick-walled polycarbonate ultracentrifuge tubes (Beckman Instruments, Palo Alto, CA, cat. no. 343776).
6. Tissue Homogenization Buffer (THB): 250 mM sucrose, 20 mM Tris-HCl (pH 7.4), 1 mM EDTA, 1 mM EGTA at 4°C. Stable for months when sterile filtered, handled aseptically, and stored at 4°C.
7. 100 mM PMSF (phenylmethylsulfonyl fluoride; Sigma, St. Louis, MO) in 100% ethanol. Stable for months when stored at −20°C.
8. 1000 x LAP: 5 mg each of leupeptin hemisulfate salt, antipain HCl, and pepstatin A dissolved in 1 mL *N-N*-dimethylformamide (all available from Sigma). Stable for months when stored at −20°C.
9. DC Protein Assay Reagents Package (Bio-Rad, Hercules, CA, cat. no. 500-0116).
10. 96-Well EIA/RIA plates, medium binding (Corning, Corning, NY, cat. no. 9017).
11. 10 mg/mL Bovine serum albumin (BSA). Store at −20°C in multiple aliquots.
12. Microplate spectrophotometer capable of reading the optical density at 750 nm of samples in a 96-well plate.

13. 2 x Sample Buffer containing urea (2 × SB): 9.6% sodium dodecyl sulfate (SDS), 4 M urea, 16% sucrose, 0.5 mg/mL bromphenol blue. Add 46 μL of 2-mercapto-ethanol per 1 mL of sample buffer to the amount of 2 × SB to be used daily (2-mer-captoethanol final concentration: 700 mM).
14. Precast Novex Tris-glycine polyacrylamide gels (e.g., 10% polyacrylamide, 15-well, 1.5-mm thick; available from Invitrogen, Carlsbad, CA, cat. no. EC60785).
15. Sodium dodecyl sulfate-polyacrylamide gel electophoresis (SDS-PAGE) and Western blot transfer equipment (e.g., Novex brand XCell SureLock Mini-Cell and XCell II Blot Module, available from Invitrogen, cat. no. EI0002).
16. PVDF membrane (Immobilon-P transfer membrane, Millipore Corporation, Bedford, MA, cat. no. IPVH 00010).
17. Ponceau S solution: prepare an initial 10 × stock solution by mixing 2% Ponceau S (w:v; Sigma, cat. no. P-3504) in 30% trichloracetic acid until the Ponceus S fully dissolves. Dilute this 1:10 in H$_2$O prior to use. Both the stock solution and diluted solution are stable at room temperature.
18. Phosphate buffered saline (PBS).
19. 5% Non-fat milk powder (Bio-Rad, Hercules, CA, cat. no. 170-6404) in PBS, prepared fresh.
20. 1% BSA in PBS, prepared fresh.
21. Appropriate primary and secondary antibodies.

3. Methods

The value of the extraction protocol presented here is its flexibility: a homogenate that is compatible with many biochemical analyses is prepared from a tissue sample. The methods that follow describe (1) the preparation of a sucrose homogenate from a mouse hemibrain and illustrate some of the biochemical analyses, such as (2) Western blot analysis of total proteins, and (3) the separation and analysis of secreted and membrane-associated forms of APP, that can be used to characterize protein expression and processing in the sample.

3.1. Tissue Homogenization

After sacrifice, the two hemispheres of a mouse brain are separated and processed differently: one is homogenized in sucrose for biochemical analyses, while the other is immersion-fixed for histological examinations (*see* **Fig. 1**). Other tissue types or tissue from patients can also be used as the starting material.

3.1.1. Mouse Brain Dissection

Rapidly sacrifice the mouse in accordance with current Institutional Animal Care and Use Committee procedures. With scissors, decapitate the mouse and cut along the midline through the skin over the skull (*see* **Fig. 2**). Expose the skull and, using clean, small scissors, carefully cut through the skull along the midline working rostrally from the base of the skull. Avoid cutting into the brain

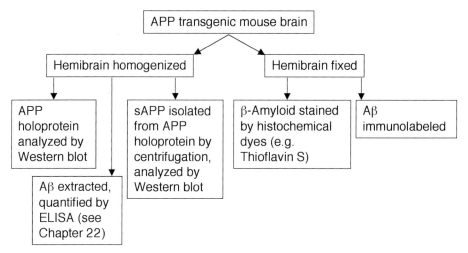

Fig. 1. Flowchart of procedures possible when tissue is processed as described in this chapter.

with the scissors; chipping away at the skull with small cuts of the scissors is a useful technique to avoid damaging the underlying brain. At the level of the eye orbits, cut laterally though the skull in both directions. With forceps, carefully lift and pull laterally the two pieces of skull to expose the brain. Additional cuts may be required to fully expose the skull and some practice may be required— on a wild-type animal—to develop a rapid technique that does not damage the brain.

Once the brain is fully exposed, cut through the spinal cord at the base of the brain, carefully insert forceps under the brain, and gently lift it from the brain case. The olfactory bulbs at the anterior end of the brain often remain within the skull unless special care is taken to free them from the bone. Place the brain on a clean, nonstick surface, and use a razor blade or scalpel to cut through the midline at the corpus callosum, such that the brain is separated into two equal hemispheres (**Fig. 2**). Specific brain regions may be further isolated. (We routinely remove the olfactory bulbs but retain the cerebellum. Rapidly freeze one hemisphere on a piece of aluminum foil placed on dry ice. The other hemisphere can be used for other purposes, such as histological analysis following immersion fixation (*see* **Subheading 3.1.2.**) or stored frozen as reserve tissue. Transfer the frozen hemisphere to a labeled microcentrifuge tube that has been pre-chilled on dry ice and store at −80°C. Tissue can be stored frozen at −80°C for an extended period if necessary. Only a single mouse should be sacrificed at a time and the tissue properly frozen before a second animal is dissected (*see* **Note 2**).

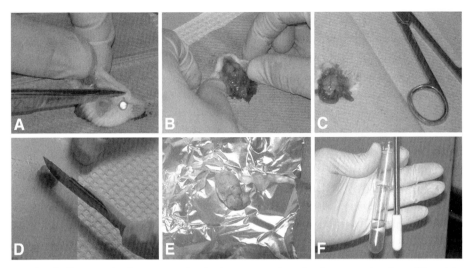

Fig. 2. Removal of a mouse brain in preparation for homogenization. The skin covering the skull is cut with scissors (**A**) and spread apart to expose the skull (**B**). The top of the skull is removed, exposing the brain (**C**). The brain is cut through the midline at the corpus callosum (**D**) and frozen as individual hemispheres on aluminum foil on dry ice (**E**). A Teflon-coated pestle and glass vessel used for homogenization (**F**).

3.1.2. Fixation of the Second Brain Hemisphere for Neuropathology and/or Immunohistochemistry

Immediately drop the remaining brain hemisphere into 15 mL of neutral pH, phosphate-buffered 10% formalin chilled to 4°C. Allow the tissue to immersion fix at 4°C for a minimum of 24 h prior to sectioning and subsequent analysis (*see* Chapter 23).

3.1.3. Preparation of the Sucrose Homogenate

This section of the protocol describes the preparation of a 10% sucrose homogenate (wet weight of tissue/volume) from the frozen mouse brain hemisphere. Such a homogenate can similarly be prepared from human brain or other tissues obtained from patients or animals by dissecting the tissue of interest, weighing it, and adjusting volumes as needed. It is important to work quickly and to keep samples cold throughout.

1. Determine the weight of the tissue (*see* **Note 3**): use an empty microcentrifuge tube to tare the balance and weigh the frozen hemibrains to the nearest 10 mg. This should be done quickly to prevent thawing and samples should be stored on dry ice until homogenized.
2. Cool on ice the glass vessel to be used for homogenization.

3. Calculate the amount of tissue homogenization buffer (THB) that will be needed for a given group of samples: 1 mL of buffer is needed per 100 mg of tissue. Immediately prior to beginning homogenizations, mix protease inhibitors into the THB (1/100th volume of 100 m*M* PMSF and 1/1000th volume 1000 x LAP; *see* **Note 4**). If a precipitate is seen in the 100 m*M* PMSF solution, dissolve it by warming to 37°C. Prepare slightly more (~10% additional) of the THB + inhibitors than the minimum needed. Keep on ice.

4. Pipet the necessary amount of cold THB + inhibitors (e.g., 1.9 mL for a 190 mg tissue sample, the approximate wet-weight of a mouse brain hemisphere) into the glass vessel to be used for homogenization. Drop the frozen brain hemisphere into the cold THB + inhibitors, using a small spatula to pry the tissue from the microcentrifuge tube if necessary.

5. Fully homogenize the tissue using 20 complete up-and-down strokes of the glass vessel while the pestle is rapidly spinning (*see* **Note 1**). Care should be taken not to cause excess foaming by removing the pestle completely from the homogenate, as this denatures proteins.

6. Generate multiple aliquots of the homogenate in pre-labeled microcentrifuge tubes to minimize subsequent refreezing of aliquots following a particular analysis. A useful scheme for subsequent formic acid extraction of Aβ is to dispense the homogenate into several aliquots of 350–500 μL, including a single 5-μL aliquot to be used for protein quantitation (*see* **Subheading 3.1.4.**, *see also* **Note 5**). For the diethylamine extraction procedure described in Chapter 22, make a single 1.3-mL aliquot, with the remaining homogenate distributed into smaller volumes.

7. Freeze aliquots on dry ice immediately.

8. Rinse with water and dry the pestle. Rinse the vessel three times, invert for several seconds, and shake out excess water before homogenizing the next sample. Store aliquoted homogenates at −80°C. Discard any remaining THB to which protease inhibitors have been added.

3.1.4. Determining Protein Concentration in the Homogenate

The protein concentration in each homogenate is determined so that equal amounts of protein from each sample can be analyzed by SDS-PAGE. This may be done using the DC Protein Assay Reagents Package or a similar kit. Diluted aliquots of homogenates and standards are dispensed into duplicate wells of a 96-well plate, followed by the reagents from the kit as described below.

1. Add 20 μL of H$_2$O to the 5 μL aliquot of homogenate set aside for protein quantitation.

2. Dilute the 10 mg/mL BSA stock solution to 0.2, 0.4, 1.0, 2.0, and 4.0 mg/mL and load 5 μL of the diluted standards and samples and a 5 μL H$_2$O blank into duplicate wells.

3. Proceed with the manufacturer's protocol. The 15-min incubation specified in the protocol may be eliminated and the optical density may be read immediately after the addition of reagent B. Typically, the mouse hemibrain homgenization described here yields approx 10 mg/mL total protein.

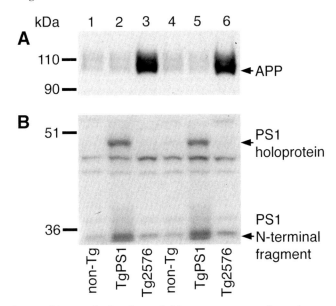

Fig. 3. Western blot analysis of amyloid precursor protein and presenilin 1 from transgenic mouse brain homogenates. Homogenates were prepared from hemibrains isolated from non-transgenic mice (non-Tg), and mice overexpressing a mutant form of human presenilin 1 (TgPS1; *[14]*), or Swedish mutant APP (Tg2576; *[15]*) and subjected to SDS-PAGE as described. Following transfer, one membrane (**A**) was probed with an antibody that binds to the C-terminal of APP and recognizes both the murine and human proteins (C1/6.1; *[11]*) and a second (**B**) was probed with an anti-PS1 antibody (NT1; *[13]*). The increased expression of APP is seen in the APP transgenic mice (lanes 3 and 6) when compared to the endogenous APP expression in the other animals. Similarly, an increase in PS1 expression is seen in the PS1 transgenic mice (lanes 2 and 5).

3.2. Western Blot Analysis of Total Proteins

SDS-PAGE and Western blot have become a routine laboratory procedure and the specifics of each gel and transfer apparatus are typically well-described by the manufacturer's protocol. Choices of polyacrylamide percentage and whether isocratic or gradient gels are more appropriate depend on the molecular weight of the protein being characterized. One source for useful information and pre-cast SDS-polyacrylamide gels is Invitrogen (Carslbad, CA).

Figure 3 shows a representative analysis of mouse brain homogenates by Western blot and the specific method used for these Western blots is described here.

1. Pipet 50 µL of 2 × SB into microcentrifuge tubes (*see* **Note 6**).

2. Pipet the volume of each homogenate that equals 200 μg of protein into the tubes containing 2 × SB (e.g., 20 μL of a 10 mg/mL homogenate).
3. Add H_2O to bring each volume to 100 μL (e.g., 30 μL).
4. Heat tubes to 56°C for 10 min. Vortex each tube for approx 15 s to shear DNA.
5. Load 20 μL of the prepared sample into the wells of a 10% polyacrylamide Tris-glycine minigel. Electrophorese according to manufacturer's protocol.
6. Transfer sized proteins onto PVDF membrane (e.g., Immobilon-P) according to manufacturer's protocol.
7. Confirm the uniformity of protein loading and transfer by Ponceau S staining. Transfer the wet membrane to a small volume (~25 mL) of Ponceau S solution and incubate at room temperature with gentle agitation for 30 s. Wash with multiple changes of PBS until the red-stained protein bands in each lane are visible. The membrane can be dried at this point and stored prior to antibody labeling. The water soluble Ponceau S can be completely removed by washing for approx 15 min.
8. Rewet the membrane according to the manufacturer's protocol if necessary.
9. Block excess protein binding sites by incubating the membrane in 5% non-fat milk powder in PBS for 1 h at room temperature with gentle rocking.
10. Bind primary antibody in 1% BSA in PBS followed by secondary antibody. Choice of antibody binding conditions, type of secondary antibody, and detection technique will need to be guided by the specifics of the primary antibody used.

3.3. Fractionation of Membrane Proteins

A number of the precursor proteins from which amyloids are derived are membrane-associated proteins (e.g., β-amyloid precursor protein [APP], prion protein, and the familial British dementia precursor protein BRI). One advantage of preparing, as the first step in protein characterization and isolation, a sucrose homogenate lacking detergents is that membranes are preserved and fractionation techniques can be used to differentiate membrane-associated proteins from soluble cytosolic or extracellular proteins. We have applied this technique to differentiate the membrane-associated APP holoprotein from the soluble, N-terminal proteolytic fragments of APP that are released from the cell into the extracellular space, the so-called sAPP fragments *(7–10)*. Western blot analysis of APP holoprotein and sAPP isolated from mouse brain by this technique is shown in **Fig. 4** (*see also* **refs.** *11,12* for further examples). Samples should be kept from warming above 4°C throughout the procedure, and rotors and tubes should be pre-chilled (*see* **Note 7**).

1. Thaw on ice one of the smaller volume aliquots of the 10% sucrose homogenate.
2. Pipet 0.2 mL into a 0.5-mL thick-walled untracentrifuge tube.
3. Centrifuge for 1 h at 100,000g (48,000 rpm using a TLA-100.1 rotor) at 4°C.
4. Transfer the supernatant to a new tube and place on ice if SDS-PAGE is to be done immediately, otherwise freeze the supernatant on dry ice and store at −80°C. Multiple aliquots can be prepared prior to freezing to avoid unnecessary freeze-

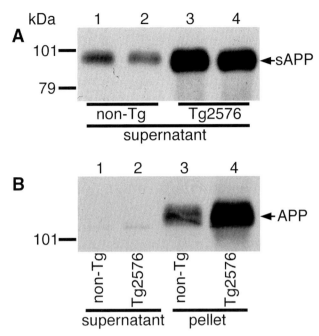

Fig. 4. Detection of sAPP in mouse brain. Sucrose homogenates were prepared from non-transgenic (non-Tg) and Swedish APP transgenic mice (Tg2576) mice and a portion of this homogenate centrifuged at 100,000*g* for 1 h. In (**A**), a portion of the supernatant was examined by Western blot using an antibody that recognizes the N-terminal region of APP (22C11; *[16]*; available from Chemicon, Temecula, CA). The 22C11 epitope is common to both APP holoprotein and the extracellular, secreted fragment of APP (sAPP) generated following α- or β-cleavage of the parental molecule (*see* **ref. 11**). The soluble sAPP fragment remains in the supernatant fraction while the membrane-associated APP holoprotein pellets during the 100,000*g* centrifugation. Much more sAPP was detected in the APP transgenic (lanes 3 and 4) than in the non-transgenic (lanes 1 and 2) mice. In (**B**), both supernatant (lanes 1 and 2) and membrane-pellet-associated (lanes 3 and 4) proteins were electrophoresed and subjected to Western blot analysis with an antibody (C1/6.1; *[11]*) that recognizes APP holoprotein but does not bind to sAPP. No APP holoprotein was detected in the supernatant. APP holoprotein, however, was readily detected in both the membrane pellet prepared from the non-transgenic mouse (lane 3) and, at a higher level, in the membrane pellet from the APP transgenic animal (lane 4).

thaw cycles. Soluble, nonmembrane-associated proteins are contained within this supernatant.
5. Directly resuspend the pellet in 0.4 mL of 1X Sample Buffer containing urea (prepared by diluting 2X SB in H_2O) (*see* **Note 8**). Vortex to fully resuspend the

pellet and spin briefly to collect droplets within the tube. The membrane-associated, nonsoluble proteins within this resuspended pellet can be immediately analyzed by SDS-PAGE and Western blot (*see* **Subheading 3.2.**) or the samples can be stored at −80°C for future analysis.

6. Prior to running the gel add an equal volume of 2 × SB to the supernatant samples and heat these and the resuspended pellets to 56°C for 10 min.

7. Load equal volumes of the supernatant and pellet samples into the gel and electrophorese (*see* **Note 8**). Following migration, transfer to PVDF membrane and probe the membrane with an antibody of interest.

4. Notes

1. While many tools for tissue homogenization can be used to prepare a homogenate from a mouse brain, a Teflon pestle against a glass vessel produces a uniform and consistent homogenate with intact membrane vesicles. In general, the more gentle the technique the better, although the production of a homogenate of uniform consistency is critical. If a motorized stirrer is not available to spin the pestle while moving the vessel up and down, the entire procedure can be done by hand, although more up-and-down strokes will be required. If some debris remains after homogenization, a brief, low-speed spin (~5000g for 5 min) can be used to remove remaining tissue fragments.

2. The quality of the tissue rapidly deteriorates after the animal is sacrificed, so it is critical to dissect rapidly and progress to the point at which tissue is frozen or in fixative before pausing. For this reason, we always freeze the hemibrains one mouse at a time, even if only a small number of hemibrain homogenates will be prepared the same day.

3. In order to obtain equal protein amounts for various assays (such as ELISA, *see* Chapter 22), it is critical that the hemibrain is accurately weighed and that the volume of THB used is adjusted to reflect the weight of each sample. This greatly improves the uniformity across samples in all subsequent analyses.

4. The protease inhibitors have a limited stability in an aqueous solution and should be diluted immediately prior to use. Other inhibitors and inhibitor cocktails can be substituted. Maintaining samples on ice is as critical to preventing proteolysis in the homogenate as are the inhibitors. It is normal for the PMSF to form cloudy precipitate when diluted into THB.

5. The volume of aliquots prepared from the homogenate should reflect the anticipated downstream uses of each sample and should be done to minimize the need to refreeze and reuse an aliquot. The volumes suggested here are ideal for the downstream Aβ ELISA analyses as described in Chapter 22.

6. DNA in the tissue sample interferes with protein migration on an SDS polyacrylamide gel. Urea in the sample buffer denatures the DNA more fully and allows for better protein resolution *(13)*. However, this solution can be viscous and should be vortexed sufficiently to shear DNA and to allow for accurate pipetting of small volumes. Alternatively, nuclei may be spun out of the homogenate at 5000g for 5 min and the post-nuclear supernatant used for subsequent Western blot analysis.

7. Whenever working with the homogenates, they must be kept cold to prevent unwanted proteolysis. Because of the long centrifugation time it is essential that the rotor and centrifuge be pre-chilled to 4°C. Samples should be refrozen immediately if necessary. Once sample buffer has been added to samples, they are significantly more stable.

8. The amount of protein in the supernatant will be much greater than that in the pellet, which contains membrane-associated proteins (~10% of the total protein). However, the volume of sample buffer used to resuspend the pellet is twice the volume of homogenate used for centrifugation. This is done so that when equal volumes of the supernatant plus sample buffer and the pellet resuspended in sample buffer are analyzed by Western blot, the relative distribution of proteins in these two fractions can be directly compared. In order to determine the volume of these fractions to be used for SDS-PAGE, first calculate the volume of homogenate for a given sample that will contain the amount of protein necessary to give a signal by Western blot. For example, in a homogenate containing 10 mg/mL of protein and where 40 µg of protein is typically loaded on the gel, this amount of homogenate would be 4 µL. Remembering that the supernatant has been diluted with an equal volume of SB and pellet resuspend in the same total volume, one would load 8 µL of the supernatant in SB and 8 µL of the resuspended pellet.

Acknowledgment

This work has been supported by the NIH (AG17617 and NS045357). We thank Dr. Karen Duff (Nathan Kline Institute, New York University School of Medicine) for the transgenic mouse brains used in the sample data.

References

1. Gravina, S. A., Ho, L., Eckman, C. B., et al. (1995) Amyloid β protein (Aβ) in Alzheimer's disease brain. Biochemical and immunocytochemical analysis with antibodies specific for forms ending at Aβ40 or Aβ42(43). *J. Biol. Chem.* **270,** 7013–7016.

2. Janus, C., Pearson, J., McLaurin, J., et al. (2000) Aβ peptide immunization reduces behavioural impairment and plaques in a model of Alzheimer's disease. *Nature* **408,** 979–982.

3. Meyer-Luehmann, M., Stalder, M., Herzig, M. C., et al. (2003) Extracellular amyloid formation and associated pathology in neural grafts. *Nat. Neurosci.* **6,** 370–377.

4. Pfeifer, M., Boncristiano, S., Bondolfi, L., et al. (2002) Cerebral hemorrhage after passive anti-β immunotherapy. *Science* **298,** 1379.

5. Rozmahel, R., Huang, J., Chen, F., et al. (2002) Normal brain development in PS1 hypomorphic mice with markedly reduced γ-secretase cleavage of βAPP. *Neurobiol. Aging* **23,** 187–194.

6. Rozmahel, R., Mount, H. T., Chen, F., et al. (2002) Alleles at the Nicastrin locus modify presenilin 1-deficiency phenotype. *Proc. Natl. Acad. Sci. USA* **99,** 14452–14457.

7. Weidemann, A., Konig, G., Bunke, D., et al. (1989) Identification, biogenesis, and localization of precursors of Alzheimer's disease A4 amyloid protein. *Cell* **57**, 115–126.

8. Golde, T. E., Estus, S., Younkin, L. H., Selkoe, D. J., and Younkin, S. G. (1992) Processing of the amyloid protein precursor to potentially amyloidogenic derivatives. *Science* **255**, 728–730.

9. Seubert, P., Oltersdorf, T., Lee, M. G., et al. (1993) Secretion of β-amyloid precursor protein cleaved at the amino terminus of the β-amyloid peptide. *Nature* **361**, 260–263.

10. Caporaso, G. L., Gandy, S. E., Buxbaum, J. D., Ramabhadran, T. V., and Greengard, P. (1992) Protein phosphorylation regulates secretion of Alzheimer β/A4 amyloid precursor protein. *Proc. Natl. Acad. Sci. USA* **89**, 3055–3059.

11. Mathews, P. M., Jiang, Y., Schmidt, S. D., Grbovic, O. M., Mercken, M., and Nixon, R. A. (2002) Calpain activity regulates the cell surface distribution of amyloid precursor protein: inhibition of calpains enhances endosomal generation of β-cleaved C-terminal APP fragments. *J. Biol. Chem.* **277**, 36415–36424.

12. Mathews, P. M., Guerra, C. B., Jiang, Y., et al. (2002) Alzheimer's disease-related overexpression of the cation-dependent mannose 6-phosphate receptor increases Aβ secretion: role for altered lysosomal hydrolase distribution in β-amyloidogenesis. *J. Biol. Chem.* **277**, 5299–5307.

13. Mathews, P. M., Cataldo, A. M., Kao, B. H., et al. (2000) Brain expression of presenilins in sporadic and early-onset, familial Alzheimer's disease. *Mol. Med.* **6**, 878–891.

14. Duff, K., Eckman, C., Zehr, C., et al. (1996) Increased amyloid-β42(43) in brains of mice expressing mutant presenilin 1. *Nature* **383**, 710–713.

15. Hsiao, K., Chapman, P., Nilsen, S., et al. (1996) Correlative memory deficits, Aβ elevation, and amyloid plaques in transgenic mice. *Science* **274**, 99–102.

16. Hilbich, C., Monning, U., Grund, C., Masters, C. L., and Beyreuther, K. (1993) Amyloid-like properties of peptides flanking the epitope of amyloid precursor protein-specific monoclonal antibody 22C11. *J. Biol. Chem.* **268**, 26571–26577.

22

ELISA Method for Measurement of Amyloid-β Levels

Stephen D. Schmidt, Ralph A. Nixon, and Paul M. Mathews

Summary

The neuritic plaque in the brain of Alzheimer's disease (AD) patients consists of an amyloid composed primarily of Aβ, an approx 4-kDa peptide derived from the amyloid precursor protein. Multiple lines of evidence suggest that Aβ plays a key role in the pathogenesis of the disease, and potential treatments that target Aβ production and/or Aβ accumulation in the brain as β-amyloid are being aggressively pursued. Methods to quantitate the Aβ peptide are, therefore, invaluable to most studies aimed at a better understanding of the molecular etiology of the disease and in assessing potential therapeutics. Although other techniques have been used to measure Aβ in the brains of AD patients and β-amyloid-depositing transgenic mice, the enzyme-linked immunosorbent assay (ELISA) is one of the most commonly used, reliable, and sensitive methods for quantitating the Aβ peptide. Here we describe methods for the recovery of both soluble and deposited Aβ from brain tissue and the subsequent quantitation of the peptide by sandwich ELISA.

Key Words: Aβ; amyloid; enzyme-linked immunosorbent assay (ELISA); extraction; quantification; quantitation; Alzheimer's disease; amyloid precursor protein (APP); formic acid; brain.

1. Introduction

In 1985, Masters and colleagues determined that neuritic plaques, one of the defining features of Alzheimer's disease (AD), consist primarily of the small Aβ peptide *(1)*. This peptide, most commonly consisting of 40 or 42 amino acid residues, varies in length at its C-terminus and is derived from the amyloid precursor protein (APP) by specific proteolytic steps *(2)*. The accumulation of Aβ and its deposition as insoluble β-amyloid plaque in the brain parenchyma is generally thought to be central to the pathogenesis of the disease. Much support for this amyloid-cascade model of AD pathobiology has come from the identification of mutations within the APP *(3–6)* and the presenilin proteins (PS1 and PS2; *[7–10]*) that cause early-onset familial AD (FAD) in humans. These mutations result in either more Aβ generation *(11,12)* or increased production

From: *Methods in Molecular Biology, vol. 299: Amyloid Proteins: Methods and Protocols*
Edited by: E. M. Sigurdsson © Humana Press Inc., Totowa, NJ

of particularly pathogenic forms of Aβ (e.g., Aβ42 *[13–15]*). Expression in mice of FAD-mutant human APP transgenes either alone *(16–18)*, or in combination with mutant presenilin transgenes *(13,19,20)*, has resulted in animals that deposit substantial β-amyloid. These amyloid-depositing transgenic mice are now perhaps the most important experimental system in which to evaluate potential AD therapies, such as inhibitors of Aβ generation and Aβ immunotherapy (*see* e.g., **ref. 21**).

Enzyme-linked immunosorbent assay (ELISA) is one of the most commonly employed biochemical techniques for the quantification of the Aβ peptide in the brain of humans and transgenic mice. Following the homogenization of the appropriate tissue (*see* Chapter 21) and a subsequent extraction to dissociate Aβ from the β-amyloid plaque, ELISA is used to precisely and rapidly measure levels of the peptide. Additionally, C-terminal epitope-specific antibodies are frequently used to differentiate Aβ40 from Aβ42, allowing the investigator to separately quantitate these two Aβ species. In this chapter, we describe the methodology used in our laboratory *(21–25)* for extracting Aβ from brain tissue and the subsequent quantitation of Aβ40 and Aβ42 by sandwich ELISA.

2. Materials

1. Ultracentrifuge and rotors (e.g., Beckman Coulter Optima TLX or MAX-E ultracentrifuge, TLA-100.1 and TLA 100.3 rotors).
2. Microplate spectrophotometer capable of reading the optical density at 450 nm of samples in a 96-well plate.
3. Platform rocker.
4. Sonic dismembrator with a probe capable of processing samples of approx 600 μL.
5. 7-mL Dounce glass tissue grinder (Kontes brand; available from Fisher Scientific, Pittsburgh, PA, cat. no. K885300-0007).
6. 8-Channel pipet, with a minimum range of 50–200 μL.
7. 0.5-mL Thickwall polycarbonate ultracentrifuge tubes (8 × 34 mm, Beckman Instruments, Palo Alto, CA, cat. no. 343776).
8. 3.5-mL Thickwall polycarbonate ultracentrifuge tubes (13 × 51 mm, Beckman Instruments, cat. no. 349622).
9. 96-Well high-binding microtiter plates (Nunc-Immuno 96 MicroWell Maxisorp plates recommended, Nalge Nunc International, Rochester, NY, cat. no. 439454; available from Fisher Scientific, cat. no. 12-565-135).
10. Adhesive sealing film for microplates (non-sterile, SealPlate brand recommended, cat. no. 100-SEAL-PLT, Excel Scientific, Wrightwood, CA; available from Sigma-Aldrich, St. Louis, MO, cat. no. Z36,965-9).
11. Multichannel pipetter basins (V-shaped bottom, non-sterile; available from Fisher Scientific, cat. no. 13-681-100).
12. Centricon Centrifugal Filter, 10 kDa cut-off (Centricon YM-10, Millipore, Bedford, MA, cat. no. 4241).

13. Peroxidase Labeling Kit (Roche Diagnostics Corp., Indianapolis, IN, cat. no. 1 829 696).
14. 0.5 mg Human $A\beta_{1-40}$ (lyophilized peptide, American Peptide Co., Sunnyvale, CA, cat. no. 62-0-78A).
15. 0.5 mg human $A\beta_{1-42}$ (lyophilized peptide, American Peptide Co., cat. no. 62-0-80A).
16. Formic acid (minimum 95% purity).
17. FA neutralization solution: 1 M Tris base, 0.5 M Na_2HPO_4, 0.05% NaN_3. Store at room temperature; stable for months.
18. 0.4% Diethylamine, 100 mM NaCl. Store at 4°C; stable for months.
19. 0.5 M Tris base, pH 6.8. Store at 4°C; stable for months.
20. 10 mM EDTA in phosphate-buffered saline (PBS).
21. Dimethyl sulfoxide (DMSO).
22. Purified Aβ40- and Aβ42-specific capture antibodies (*see* **Note 1**).
23. Purified anti-Aβ antibody with an epitope not overlapping with the epitopes of the capture antibodies (*see* **Note 2**).
24. 10% NaN_3. Store at room temperature. Very toxic, so be careful when preparing stock solution from powder.
25. Coating buffer: 30 mM $NaHCO_3$, 70 mM Na_2CO_3, 0.05% NaN_3, pH 9.6 (2.52 g $NaHCO_3$, 7.42 g Na_2CO_3, 5 mL 10% NaN_3, ddH_2O to 1 L; adjust pH to 9.6). Store at 4°C; stable for months.
26. 10X PBS: 1369 mM NaCl, 27 mM KCl, 43 mM Na_2HPO_4, 15 mM KH_2PO_4. (80 g NaCl, 2 g KCl, 11.5 g Na_2HPO_4·$7H_2O$, 2 g KH_2PO_4, ddH_2O to 1 L; adjust pH to 7.4). Store at room temperature. Make fresh monthly in a sterile container.
27. Blocking solution: 1% Block Ace, 0.05% NaN_3, in PBS, pH 7.4. (4 g Block Ace powder, 2 mL 10% NaN_3, 1X PBS to 400 mL; adjust pH to 7.4). Store at 4°C; stable for months (*see* **Notes 3** and **4**).
28. ELISA capture (EC) buffer: 5 mM NaH_2PO_4, 15 mM Na_2HPO_4, 2 mM EDTA, 400 mM NaCl, 0.2% bovine albumin, 0.05% CHAPS, 0.4% Block Ace, 0.05% NaN_3, pH 7.0. (0.69 g NaH_2PO_4·H_2O, 2.13 g Na_2HPO_4, 0.74 g EDTA disodium salt, 23.3 g NaCl, 2.0 g bovine albumin, 0.5 g CHAPS, 4 g Block Ace powder, 5 mL 10% NaN_3, ddH_2O to 1 L; adjust pH to 7.0.) Store at 4°C; stable for months (*see* **Note 4**).
29. PBST: 0.05% Tween®-20 in PBS. Store at room temperature. Make fresh monthly in a sterile container.
30. Detection antibody buffer: 3 mM NaH_2PO_4, 17 mM Na_2HPO_4, 2 mM EDTA, 400 mM NaCl, 1% BSA, pH 7.0. (0.41 g NaH_2PO_4·H_2O, 2.41 g Na_2HPO_4, 0.74 g EDTA disodium salt, 23.3 g NaCl, 10.0 g bovine albumin, ddH_2O to 1 L; adjust pH to 7.0). Sterile filter and make approx 11 mL aliquots under sterile conditions in a tissue culture hood. Store at 4°C; stable for months (*see* **Notes 4** and **5**).
31. TMB Microwell Peroxidase Substrate System (Kirkegaard & Perry Laboratories, Gaithersburg, MD, cat. no. 507600).
32. Stop solution: 5.7% *o*-phosphoric acid. Store at room temperature; stable for months (*see* **Note 6**).

3. Methods

3.1. Designing an ELISA and Preparing Solutions and Standards

3.1.1. Designing an ELISA

A sandwich ELISA consists of a "capture" antibody and a "detection" antibody, both of which bind to the peptide one wants to quantitate, but at distinct and non-overlapping epitopes. Initially, the capture antibody is coated onto the plastic of the microtiter plate. When a complex mixture of proteins (such as those in a brain homogenate) is then incubated in this microtiter plate well, those peptides that are recognized by the capture antibody are bound and "captured" onto the ELISA plate. This reaction is driven by the high concentration and high affinity of the capture antibody, which allows for very small amounts of a peptide, such as Aβ, to be efficiently tethered to the plate via the antibody. Extraneous proteins are then washed away and a second antibody, conjugated to a reporter molecule or enzyme such as horseradish peroxidase, is used to detect the Aβ now bound to the initial capture antibody. Using such a "detection" antibody confers a number of advantages. First, this sandwich method greatly increases the sensitivity of the ELISA. Second, the ELISA gains specificity for a particular protein or peptide by combining the specificities of the two antibodies. For example, we commonly employ antibodies with specificity for the C-terminus of Aβ (e.g., an anti-Aβ40 antibody) to specifically capture Aβ40 from the complex mixture of proteins found in a brain extract *(21–26)*. Importantly, this step eliminates Aβ42, APP, and other APP metabolites that may also be recognized by the detection antibody (*see* **Note 7**). The captured Aβ40 is then detected using an antibody that binds to the N-terminal region of Aβ (as well as other APP metabolites, which, if present, would potentially confound the specific quantitation of Aβ40). The following methodology will focus on the use of sandwich ELISAs to quantitate human Aβ40 and Aβ42 from β-amyloid-depositing human APP transgenic mouse brain (*see* **Fig. 1**). These protocols are also applicable to human brain from AD patients. In these cases, insoluble Aβ that has been deposited into β-amyloid plaques is first solubilized in formic acid (*see* **Subheading 3.2.**). Additionally, sensitive sandwich ELISAs can also quantitate Aβ from non-β-amyloid containing tissue, such as APP transgenic mouse brain prior to β-amyloid deposition and non-transgenic mouse brain (*see* **Subheading 3.3.1.**) or from human or mouse plasma (*see* **Subheading 3.3.2.**).

The specificity and sensitivity of an ELISA is dependent upon the purity, affinity, and epitope specificity of the antibodies used. In general, good monoclonal antibodies make for successful ELISAs; polyclonal antibodies tend to be more variable, and much effort needs to be put into the affinity-purification

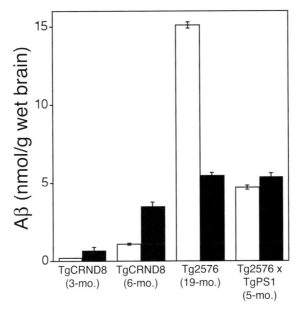

Fig. 1. Brain Aβ levels in transgenic mice determined by sandwich ELISA. Aβ was quantitated from formic acid-extracted 10% brain homogenates as described in this chapter. Data from three different β-amyloid-depositing transgenic mouse lines analyzed at the indicated ages (mo) are presented: TgCRND8 mice *(31; see also* **ref. 21**) and Tg2576 mice *(16)* overexpress mutant forms of human APP while the Tg2576 × TgPS1 is a cross with a mutant presenilin 1-expressing line *(19,20)*. The increase in the amount of deposited Aβ40 and Aβ42 that occurs as the TgCRND8 mice age from 3 to 6 mo is apparent. Additionally, both the TgCRND8 and Tg2576 × TgPS1 lines deposit more Aβ42 relative to Aβ40 than does the Tg2576 line, which is evident from this ELISA analysis. The sandwich ELISAs have a high specificity for Aβ40 and Aβ42 owing to the use of C-terminal specific anti-Aβ40 and Aβ42 antibodies (*see* **Fig. 2**).

of the antiserum prior to its use in an ELISA. Thus, for success in any ELISA, the first step must be the identification of appropriate antibodies, preferably in quantities that will permit the investigator to expend the necessary amounts of antibody in the critical task of ELISA development and quality control. To date, we have primarily used monoclonal antibodies to detect Aβ by sandwich ELISA, antibodies which where the kind gift of Dr. Marc Mercken (Johnson and Johnson Pharmaceutical Research and Development/Janssen Pharmaceutica; *see* e.g., **refs. 21–27**). These, and any particular set of antibodies used by others, however, may not be available to a given investigator and much preliminary effort may need to be made to obtain an antibody repertoire that will be suitable. In **Notes 1** and **2**, we describe commercially available antibodies that have been successfully used in our laboratory, and would therefore allow most laboratories to develop a sandwich ELISA to detect human Aβ. The selection of antibodies is best guided by extensive conversations with laboratories and/or commercial suppliers that have had success detecting the protein of interest, prior to initiating any study. The user should be prepared to try a number of antibodies and antibody combinations during the development of an in-house ELISA. The less abundant the protein or peptide of interest, the more will rest on the pairing of the best antibodies in the sandwich ELISA. Commercial ELISA kits are also available from a number of sources for the quantitation of Aβ, and much of the preparation of samples and standards described in this chapter applies to the use of these kits.

3.1.2. Preparing Solutions

In our laboratory, quantitating a peptide at low fmol/mL levels seems to be remarkably dependent upon the quality of the H_2O used to make solutions; contamination of solutions or the H_2O source with fungus or other microorganisms can be problematic. We therefore recommend periodically disinfecting water storage containers by autoclaving or other means. For these reasons, PBS, including the 10X stock solution, and PBST should also frequently be made fresh, in sterile containers. Several of the solutions contain 0.05% NaN_3 (sodium azide) as a preservative (1/200th volume of a 10% NaN_3 stock solution).

3.1.3. Preparing Aβ Standards

Aβ standards for the ELISA are prepared from high quality synthetic peptides. The peptides are dissolved in dimethyl sulfoxide (DMSO), diluted in ELISA capture (EC) buffer, divided into aliquots, and stored at −80°C. To minimize the possibility of aggregation of the Aβ standard, carry out the following steps quickly and keep the EC buffer on ice. Allow vials of lyophilized peptide to equilibrate to room temperature before opening.

1. Preparation of 1 mg/mL master stock solutions: dissolve 0.5 mg of peptide in 500 μL room temperature DMSO. Make 25 μL aliquots, freeze on dry ice, and store at −80°C. This stock solution is 231 nmol/mL for human Aβ1–40 and 222 nmol/mL human Aβ1–42.
2. Preparation of 1 nmol/mL stock solutions: thaw an aliquot of the 1 mg/mL master stock solution from earlier and dilute 8 μL into 1840 μL EC buffer for human Aβ1–40 and 1764 μL for human Aβ1–42. Divide into 250 μL aliquots, freeze on dry ice, and store at −80°C.
3. Preparation of 5 pmol/mL working stock solutions: thaw an aliquot of 1 nmol/mL solution and dilute 200 μL into 40 mL EC buffer. Divide into 100 μL aliquots, freeze on dry ice, and store at −80°C. You will need about 400 small tubes to make

these aliquots, so label and arrange the tubes before thawing the 1 nmol/mL stock solution.

3.1.4. Preparation of an HRP-Coupled Detection Antibody

Our experience is that the linear range of the sandwich ELISA is best if the detection antibody is directly coupled to horseradish peroxidase (HRP), rather than when coupled to biotin and then detected using streptavidin-HRP. Although suitable biotinylated anti-Aβ detection antibodies can be purchased (*see* **Note 2**) and may suffice for some applications, a group that is investing in the development of highly sensitive ELISAs is better off using HRP-coupled antibodies, which they will likely need to produce themselves. This section describes the preparation of the antibody, the HRP coupling, quality control, and storage of an HRP-coupled detection antibody.

1. Start with the appropriate purified antibody (*see* **Note 2**). The antibody cannot contain azide in subsequent steps. Dialyze against PBS if an azide-free antibody source cannot be found.
2. The coupling reaction requires that the antibody concentration be approx 4 mg/ mL in a volume of 300 μL. Most antibody solutions are more dilute than this, and if so, the antibody first needs to be concentrated using a 10 kDa cut-off Centricon Centrifugal Filter according to the manufacturer's protocol. Apply a sufficient amount of antibody to the column to obtain the needed amount of concentrated antibody. Ensure that the column never completely dries during centrifugation.
3. Follow the Roche Peroxidase Labeling Kit's protocol for the coupling of HRP to an antibody. Per the protocol, remember that the PBS-antibody solution must be initially alkalinized before coupling.
4. The HRP-coupled antibody should not be frozen, but stored at 4°C as freezing can inactivate the peroxidase. Do not add azide as this also inactivates the peroxidase. For long-term storage, thimerosal can be added to 0.002% of the coupled antibody. We find that with proper storage, HRP-coupled antibodies can remain active for more than 1 yr at 4°C.

The efficiency of the coupling reaction must be determined empirically for each batch of HRP-coupled detection antibody. This is done by testing the ability of different dilutions of the newly coupled antibody to detect Aβ standards. When prepared according to the Roche kit protocol, which gives a final antibody volume of approx 1 mL, an HRP-coupled anti-Aβ antibody can typically be diluted 1:1000 or more for the ELISA (*see* **Note 8**).

3.2. Solubilization of Aβ From β-Amyloid Plaques Prior to ELISA

This procedure dissociates the aggregated Aβ in β-amyloid plaques so that it is possible to quantitate the peptide by ELISA. A small volume of homogenized β-amyloid-containing tissue (prepared as described in Chapter 21) is first sonicated in the presence of formic acid, which solubilizes Aβ from the densely aggregated β-amyloid plaque, and then subjected to high-speed centrifugation. Three layers result: a thin, upper lipid layer that is not collected, the predominant intermediate phase that contains the Aβ peptides, and a barely noticeable pellet *(28)*. The intermediate phase is recovered and neutralized for the ELISA; the upper lipid layer and the pellet are discarded.

1. Mix 200 μL of a 10% (w:v) brain homogenate (*see* Chapter 21) into 440 μL of cold formic acid (minimum 95% purity) in a 1.7-mL microcentrifuge tube.
2. Sonicate each sample individually for 1 min on ice: immerse the tip of the probe in the sample and move the tube up and down over the probe while sonicating. Keep the tube on ice during this process by holding it in a small (e.g., 50 mL) beaker filled with ice. Rinse the probe and wipe it dry before processing the next sample.
3. Centrifuge 400 μL of the sonicated mixture at 100,000*g* for 1 h at 4°C (48,000 rpm using a Beckman Coulter TLA-100.1 rotor), in a 0.5-mL thickwall polycarbonate ultracentrifuge tube.
4. Dilute 210 μL of the intermediate phase into 4 mL of room temperature FA neutralization solution. Vortex briefly.
5. Divide the neutralized solution into six 0.5-mL aliquots. Freeze the aliquots on dry ice. Each aliquot is more than sufficient for 4 ELISA-well readings (typically Aβ40 and Aβ42 measurements, each in duplicate). These samples can be loaded onto the ELISA plate neat or, more typically, diluted in EC buffer as necessary (*see* **Subheading 3.4.**).

3.3. Preparation of Non-β-Amyloid Containing Tissue and Blood Plasma Prior to ELISA

3.3.1. Extraction of Aβ From Tissue Without Plaque Pathology

This section describes the extraction of Aβ from homogenates of tissue without β-amyloid plaque pathology. Although the Aβ present in a tissue that lacks β-amyloid is generally soluble, Aβ is a "sticky" peptide and the uniformity of its recovery and subsequent ELISA quantitation is improved if the Aβ is first extracted in diethylamine *(23,24,29,30)*. Other agents, such as Triton X-100, can be used, but we find that the presence of non-ionic detergents in the sample frequently increases the non-specific background of the ELISA. Additionally, diethylamine extraction has the advantage of eliminating lipids (which are particularly abundant in brain and can non-specifically increase background) and membrane-associated proteins such as APP, which remain in the pellet fraction following extraction *(29)*. Diethylamine does, however, disrupt membrane vesicles so that intracellular as well as extracellular Aβ is recovered. In contrast to the formic acid extraction technique described in **Subheading 3.2.**, the diethylamine-extracted sample is much less dilute prior to applying to the ELISA plate (*see* **Subheading 3.5.4.**). This allows for the quantitation of the typically low levels of Aβ found in tissue without plaques (e.g., young APP transgenic mice prior to β-amyloid deposition and non-transgenic mice [*see* **Note 9**]) *(23, 24,30)*. Tissue homogenates prepared as described in Chapter 21 are first mixed with diethylamine using a dounce. Following a high-speed centrifugation, Aβ is recovered in the supernatant, which is pH-neutralized and applied to the ELISA plate. Plaque-associated Aβ, however, is not completely extracted with this method and therefore formic acid extraction should be used in tissues containing β-amyloid.

1. Mix 1 mL of a 10% (w:v) brain homogenate (*see* Chapter 21) with 1 mL of cold 0.4% diethylamine (DEA), 100 m*M* NaCl with six up and down strokes of the glass pestle in a dounce glass tissue grinder. Keep the thawed homogenates, 0.4% DEA, 100 m*M* NaCl, and the dounce on ice. Transfer 1.9 mL of the homogenate/DEA mixture to a 3.5-mL thickwall polycarbonate ultracentrifuge tube.
2. Between samples, rinse the pestle and dounce with H_2O and dry.
3. Centrifuge the tube containing the homogenate/DEA mixture at 100,000*g* for 1 h at 4°C (43,000 rpm in a TLA 100.3 rotor).
4. Add 1.7 mL of the supernatant to a tube containing 170 μL of 0.5 *M* Tris base, pH 6.8 and vortex briefly. Divide into four 440 μL aliquots, freeze on dry ice, and store at –80°C. (440 μL is sufficient to run ELISAs for both Aβ40 and Aβ42 when loading 100 μL neat into duplicate wells for each assay.) These samples can be loaded onto the ELISA plate neat or diluted in EC buffer as necessary (*see* **Subheading 3.4.**).
5. Discard the pellet.

3.3.2. Preparation of Blood Plasma Prior to ELISA

Aβ levels in blood plasma can also be determined by ELISA. The best method to prevent clotting is to mix the blood immediately with EDTA to chelate Ca^{++} as heparin may bind Aβ and interfere with the ELISA measurement. Plasma prepared this way can be directly applied to an ELISA plate following a freeze–thaw cycle to inactivate endogenous peroxidase. Without this freeze–thaw cycle, the peroxidase found in blood can give a high background when the ELISA is developed using a HRP-coupled detection antibody.

1. Immediately mix the blood sample as it is being collected with an equal volume of 10 mM EDTA in PBS.
2. Gently mix the sample and centrifuge at 10,000*g* for 5 min at room temperature.
3. Collect the supernatant and divide the plasma into aliquots prior to freezing at –80°C. These samples can be loaded onto the ELISA plate neat or diluted in EC buffer as necessary (*see* **Subheading 3.4.**).

3.4. Quantification of Ab by Sandwich ELISA

Solutions should be kept on ice during the following steps with the exception of the PBS, PBST, and 5.7% *o*-phosphoric acid, which are stored and used at room temperature. Plates should be tightly sealed with sealing film during all incubations to prevent drying. The sandwich ELISA takes 3 d to complete.

3.4.1. Day 1

1. The capture antibody is coated onto the necessary number of wells of a 96-well high-binding microtiter plate by adding 100 µL/well of antibody diluted in coating buffer. Incubate overnight at 4°C with rocking. Typically, the capture antibodies are diluted to 2.5–10 µg/mL for coating. The antibody diluted in coating buffer must be free of any other proteins (*see* **Note 1**).

3.4.2. Day 2

2. Wash wells twice with PBS. This step can be done by various methods, including a wash bottle containing PBS that is used to spray PBS into the wells. Invert the plate quickly over a sink to discard the solution between washes. Residual wash solution can be removed by inverting the plate and patting on a paper towel.
3. Block non-specific binding sites on the plastic by adding 200 µL/well of blocking solution and incubating for 4 h at room temperature with gentle rocking. This blocking step can be extended for significantly longer than 4 h, and plates may even be left at this step for up to 1 wk if stored at 4°C.
4. Prepare by thawing and diluting as necessary all samples and standards. Immediately before loading or diluting formic acid extracts, they should be incubated at 37°C for 5 min to solubilize any precipitate and not placed on ice before being diluted or a precipitate will quickly reform. Samples may need to be diluted with EC buffer to generate readings within the linear range of the standards. The approximate dilution needs to be determined empirically, and may require that multiple dilutions be tested in a trial ELISA and/or that assays are repeated with additional dilutions. Typically, formic-acid extracted samples from human tissue or transgenic mouse models with significant β-amyloid deposition will need to be diluted from 1:10 to 1:100 for an ELISA with a linear range of approx 25–400 fmol/mL. DEA extracts prepared from non-β-amyloid containing tissue and blood plasma is typically loaded onto the ELISA plate neat or up to a 1:10 dilution. Aβ standards are similarly diluted in EC buffer immediate prior to the ELISA. As previously noted, the concentration of the standards used will depend on the inherent sensitivity of the ELISA and must match the range of the Aβ concentrations found within the samples. As described in **Subheading 3.5.**, care needs to be taken that the values for all samples fall between values obtained for the standards and within the linear range of the standard curve. As an example, standards at 400, 200, 100, 50, 25, 12.5, and 6.25 fmol/mL would be used for samples optimally diluted to give readings of approx 25–200 fmol/mL.
5. Immediately prior to proceeding with the ELISA and after samples are fully prepared for loading, dump the blocking solution. Quickly add 50 µL/well of EC buffer to prevent the wells from drying while the individual samples are being added.

6. Add 100 μL of standards and samples to wells containing 50 μL of EC buffer (final volume in each well is now 150 μL; *see* **Note 10**). For a measurement of background signal, blank wells should contain 150 μL EC buffer.
7. Incubate overnight at 4°C with rocking.

3.4.3. Day 3

8. Wash wells twice with PBST, then once with PBS.
9. Add 100 μL of HRP-conjugated detection antibody, diluted in detection antibody buffer (*see* **Note 8**), to each well. Incubate for 4 h at room temperature with rocking.
10. Wash wells twice with PBST, then once with PBS.
11. Add to the wells 100 μL of a 1:1 mixture of the two solutions (TMB peroxidase substrate and peroxidase substrate solution B) of the TMB microwell peroxidase substrate system (*see* **Note 11**). Allow plates to develop until the second to the least concentrated standard has a slight blue color change, and then stop the reaction by adding 100 μL of stop solution to each well (*see* **Note 6**). If the samples change color rapidly, the reaction may be stopped more quickly. If the samples have a low signal, however, a longer reaction time may produce more useful results. Regardless, samples and standards need to be stopped as simultaneously as possible. If the plates are allowed to overdevelop, the linear range of the ELISA will be compromised.
12. Read the OD_{450} with a microplate spectrophotometer. Aβ concentrations in the sample are interpolated from the OD_{450} using a standard curve generated from the known Aβ amounts in the standards (*see* **Fig. 2**).

3.5. Assessing Antibody Specificity, Discarding Below Sensitivity and Saturated Points From Standard Curves, Accounting for Dilutions

3.5.1. Assessing Antibody Specificity

Once a combination of capture and detection antibodies has been obtained, pilot assays should be performed to determine the specificity of the antibody combinations for Aβ40 or Aβ42. This is accomplished by coating a microtiter plate with the putative Aβ40- or Aβ42-specific antibody. If an antibody has good specificity for Aβ40 vs Aβ42, it should not recognize Aβ42 standards (**Fig. 2A**). Similarly, an Aβ42 antibody should not recognize Aβ40 standards (**Fig. 2B**). Such specificity assays are very useful in pointing out unexpected cross-reactivity in an ELISA.

3.5.2. Determining Sensitivity and Discarding Measurements Below a Meaningful Sensitivity Limit

The majority of the serially diluted standards in an Aβ ELISA should show a linear relationship, but at some point a "bottoming-out" effect will be observed in which there is no longer a linear change in optical density with successively lower concentration standards (**Fig. 3**). The lowest concentration at which a

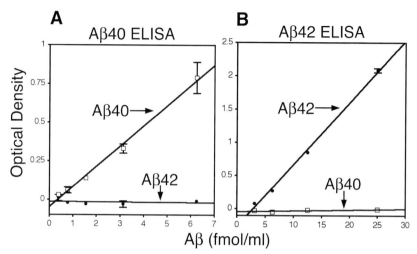

Fig. 2. Assessing antibody specificity. Affinity-purified polyclonal antibodies from a commercial source (Signet Laboratories, Dedham, MA; *see* **Note 1**) were coated on microtiter plates at 5 µg/mL and allowed to capture human Aβ standards. One of the antibodies, made against an Aβ40 epitope, showed sensitivity to sub-femtomole/mL levels to Aβ40 (**A**, open boxes) while exhibiting no crossreactivity to Aβ42 standards included on the same plate (**A**, closed circles). The second antibody, made against an Aβ42 epitope, showed similar specificity to Aβ42 but with a higher linear range and lower sensitivity (**B**). Both Aβ40 and Aβ42 were detected with the human Aβ-specific monoclonal antibody JRF/Aβtot/17, which binds to the N-terminus of Aβ, as previously described *(26)*.

point is still linear with the other points on the curve, and for which duplicate readings are in close agreement, thus becomes a given assay's "sensitivity limit."

Several different concentrations of capture and detection antibodies should be tested in initial pilot assays to determine which antibody concentrations produce the optimal sensitivity for the concentration of Aβ expected in the samples. Since sensitivity will vary somewhat from assay to assay it should be determined for each ELISA, even after pilot assays have demonstrated a general sensitivity limit for a given combination and concentration of antibodies. OD values obtained for standards below this sensitivity limit are meaningless and may skew the standard curve, particularly in the region relevant to the OD measurements obtained from the samples (**Fig. 3**, dotted line). These "below sensitivity" values should be excluded in a consistent fashion before a curve fit is applied (**Fig. 3**, solid line).

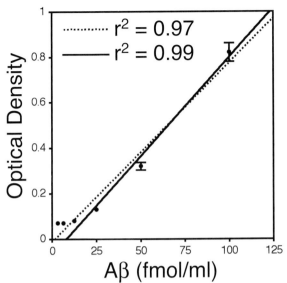

Fig. 3. Determining sensitivity and discarding standards that are below sensitivity. The optical density representing various concentrations of human Aβ40 standard was determined by sandwich ELISA. The bottom two points, representing the lowest concentrations of Aβ40, were below sensitivity and skew a linear curve fit away from the other points (dotted line, $r^2 = 0.97$). Discarding these two points from the analysis improves the curve fit (solid line, $r^2 = 0.99$) and would be more appropriate for calculating the amount of Aβ40 in samples containing approx 25–100 fmol/mL Aβ40.

3.5.3. Determining Linear Range and Discarding Saturated Standards

Frequently, the highest concentration standards develop rapidly, become saturated, and are nonlinear relative to lower concentration standards. Interpolating from a saturated region of the standard curve may cause significant differences between samples to be underestimated. Standard points that show signs of saturation should therefore be discarded and interpolation should only be from within the linear range of the standards (**Fig. 4**, solid line). This will often require additional dilution of the samples in a subsequent assay to ensure that all sample measurements are within the linear range of the ELISA.

Indeed, an ELISA with a broad linear range is desirable, in that this will increase the likelihood that a given sample will be within range on a first ELISA run and that the assay will not have to be repeated. As with optimizing antibody concentrations to produce good sensitivities, pilot assays may be used to determine which antibody concentrations produce the broadest linear range. Interassay variability dictates that the linear range be determined for every assay, however, and that saturated standards be discarded wherever necessary.

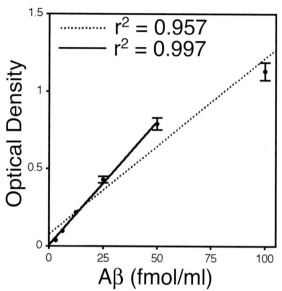

Fig. 4. Discarding a saturated point to improve the linear curve fit. Human Aβ42 standards of various concentrations were analyzed by sandwich ELISA to produce a standard curve. The 100 fmol/mL Aβ42 standard was beginning to saturate when this ELISA was developed, and the linear relationship between peptide concentration and optical density is not optimal. This produces a relatively poor linear curve fit (dotted line, $r^2 = 0.957$); removing 100 fmol/mL Aβ42 standard data point generates a better linear curve (solid line, $r^2 = 0.997$), which would be more appropriate for calculating Aβ42 concentrations in samples from approx 5 to 50 fmol/mL.

3.5.4. Accounting for Dilutions Made During the Assay

Multiple steps throughout this protocol introduce dilution factors that must be taken into account. Aβ amounts are often reported as femto-, pico-, or nanomole per gram of wet tissue. If the neutralized formic acid extract of a 10% tissue homogenate is loaded directly into the ELISA well as described in this chapter, the Aβ value determined in fmol/mL by the ELISA needs to be multiplied by 704 to convert to fmol/g of wet brain tissue (*see* **Note 12**). If the neutralized DEA extract of a 10% tissue homogenate is loaded directly onto the ELISA plate, the Aβ value determined in fmol/mL by the ELISA needs to be multiplied by 24.2 to convert to fmol/g of wet brain tissue (*see* **Note 12**).

4. Notes

1. Capture antibodies must be purified in IgG and PBS or a similar protein-free buffer. Any other proteins will compete with the antibody for binding to the microtiter plate. Thus, ascites fluid, antiserum, etc.—which contain many other proteins—are

not suitable and only purified IgG or affinity-purified antibody from these sources can be used. Typically, capture antibodies are coated at 2.5 –10 μg/mL. Increasing the coating antibody concentration more than this is often a waste of expensive antibody and will not improve the ELISA's sensitivity. Using the minimal amount of antibody for coating is cost-effective, but this amount must be determined empirically and a useful linear range of standards must be maintained. Signet Laboratories (Dedham, MA) sells high-affinity Aβ40 and Aβ42 affinity-purified polyclonal antibodies; these antibodies were used for the assays shown in **Fig. 2**. Additional sources of Aβ antibodies are available and others will undoubtedly become available, so the investigator is advised to search for the most cost-effective source of high-affinity antibodies during the development phase of an ELISA.

2. As with the capture antibodies, the detection antibody must be purified IgG and in PBS or a similar protein-free buffer. A commonly used monoclonal antibody that binds to an epitope within the N-terminus of human Aβ is 6E10 (available from Signet Laboratories, Dedham, MA). A very serviceable sandwich ELISA for Aβ40 and Aβ42 can be constructed from the C-terminal-specific capture antibodies available from Signet and 6E10. Biotinylated 6E10 can also be purchased, and HRP-streptavidin used to detect the 6E10 bound during the ELISA, but this often results in a narrower linear range and therefore a less useful ELISA than one using an antibody directly coupled to horseradish peroxidase (HRP). Regardless, if biotinylated antibody followed by HRP-streptavidin is used, much attention needs to be given to ensuring that the ELISA remains linear in the range of all samples (*see* **Subheading 3.5.**). In our laboratory, we have not found chemiluminescent or fluorescent detection systems to improve sensitivity over directly HRP-coupled antibodies and colorimetric systems.

   An alternative detection strategy is to use a coating antibody of one species (e.g., mouse) and a detection antibody from another (e.g., rabbit). In this case, one could use an HRP-coupled anti-rabbit secondary antibody prior to developing the ELISA. However, we have found that this rarely works in practice, particularly if a high-sensitivity ELISA is required, as crossreactivity of the anti-rabbit secondary antibody with murine IgG leads to overwhelming background.

3. Block Ace powder is a product of Dainippon Pharmaceutical Co. (Osaka, Japan), and is available from Serotec (Raleigh, NC, cat. no. BUF029). Other protein blocking reagents are often used for ELISAs (BSA, casein). Our laboratory's experience has been that while these reagents work, Block Ace gives the lowest non-specific background and the most consistent assays over time, and we therefore recommend it. However, 3% BSA in PBS with azide can be substituted in many applications.

4. Preparing the solutions containing protein (e.g., blocking solution, EC buffer, and detection antibody buffer) is easiest if all of the dry components are added to half of the final volume of the solution and stirred very vigorously—denaturation of these blocking proteins by foaming is not a problem and may even be advantageous. After the protein is dissolved, the solution is brought up to the appropriate final volume.

5. Do not add sodium azide to detection antibody buffer as the azide will inactivate the HRP conjugated to the detection antibodies in later steps.

6. In addition to 5.7% o-phosphoric acid, other acids such as H_2SO_4 and HCl are sometimes used to stop the TMB reaction. *O*-phosphoric acid, however, gives more consistent results and is recommended by the manufacturer.

7. Aβ sandwich ELISAs are typically done using C-terminal Aβ40- or Aβ42-specific antibodies as the capture antibody. This offers some advantages as N-terminal-directed anti-Aβ antibodies bind all Aβ species and usually recognize APP and other APP metabolites. If sufficiently abundant, these may compete with the Aβ of interest during the initial binding to the capture antibody. In practical application, however, this is not of significant concern when detecting the abundant Aβ extracted from a β-amyloid-depositing mouse, for example. Therefore, in some applications, the capture antibody can be N-terminal-directed with other antibodies, recognizing other epitopes either within the Aβ sequence or at the C-terminus, used for detection.

8. The Roche Peroxidase Labeling Kit generally gives very uniform HRP coupling reactions if the starting antibody concentration is carefully adjusted to 4 mg/mL. Nevertheless, the efficiency of the coupling must always be determined by trying each new batch of detection antibody in a trial ELISA. A straightforward way to do this is to test the newly coupled antibody at various dilutions in detection antibody buffer on a series of Aβ standards. A typical scheme would include the detection antibody diluted 1:1000, 1:2000, 1:5000 and perhaps 1:10,000 against Aβ standards ranging from approx 3 fmol/mL to 200 fmol/mL. Frequently, using the detection antibody at higher concentrations will increase the sensitivity of the ELISA, although eventually higher concentrations will non-specifically increase the background.

9. Murine and human Aβ differ in their N-terminal region at three residues. Thus, most antibodies that recognize the N-terminus of Aβ are species specific, with the vast majority of the available antibodies recognizing the human and not the murine peptide (e.g., 6E10). The limited availability of highly sensitive anti-murine Aβ antibodies may make the development of an ELISA that detects endogenous murine Aβ in wild-type mice problematic. Nevertheless, additional anti-murine Aβ antibodies may become available, so the investigator is advised to search both commercial sources and the recent literature during the development phase of such an ELISA.

10. As both the samples and standards are diluted equally 2:3 in EC buffer at this point, no arithmetic adjustment for this dilution is necessary when calculating Aβ concentrations from the standard curve.

11. Warm the two bottles of TMB Microwell Peroxidase Substrate System kit to 37°C, then prepare the developing solution immediately before use by mixing an equal volume from each bottle. Change pipets between bottles to prevent cross-contamination. For every set of two plates, this mixture should be made fresh and used immediately.

12. This correction factor of 704 for the formic-acid extracted homogenate takes into account the 1:11 dilution when the tissue is initially homogenized (*see* Chapter 21), the further 1:3.2 dilution in formic acid, and the 1:20 dilution when the formic

acid extract is neutralized. The correction factor of 24.2 for the DEA-extracted homogenate takes into account the 1:11 dilution when the tissue is initially homogenized (*see* Chapter 21), the further 1:2 dilution in DEA, and the 1:1.1 dilution when the DEA is neutralized.

Acknowledgment

This work has been supported by the NIH (AG17617 and NS045357). We thank Drs. Karen Duff (Nathan Kline Institute, New York University School of Medicine) and David Westaway (Centre for Research in Neurodegenerative Diseases, University of Toronto) for the transgenic mouse brains used in the sample data. We are very indebted to Dr. Marc Mercken (Johnson and Johnson Pharmaceutical Research and Development/Janssen Pharmaceutica) for the use of his anti-Aβ antibodies.

References

1. Masters, C. L., Simms, G., Weinman, N. A., Multhaup, G., McDonald, B. L., and Beyreuther, K. (1985) Amyloid plaque core protein in Alzheimer disease and Down syndrome. *Proc. Natl. Acad. Sci. USA* **82,** 4245–4249.
2. De Strooper, B. and Annaert, W. (2000) Proteolytic processing and cell biological functions of the amyloid precursor protein. *J. Cell Sci.* **113,** 1857–1870.
3. Goate, A., Chartier-Harlin, M. C., Mullan, M., et al. (1991) Segregation of a missense mutation in the amyloid precursor protein gene with familial Alzheimer's disease. *Nature* **349,** 704–706.
4. Chartier-Harlin, M. C., Crawford, F., Houlden, H., et al. (1991) Early-onset Alzheimer's disease caused by mutations at codon 717 of the β-amyloid precursor protein gene. *Nature* **353,** 844–846.
5. Levy, E., Carman, M. D., Fernandez-Madrid, I. J., et al. (1990) Mutation of the Alzheimer's disease amyloid gene in hereditary cerebral hemorrhage, Dutch type. *Science* **248,** 1124–1126.
6. Tanzi, R. E. and Bertram, L. (2001) New frontiers in Alzheimer's disease genetics. *Neuron* **32,** 181–184.
7. Mullan, M., Houlden, H., Windelspecht, M., et al. (1992) A locus for familial early-onset Alzheimer's disease on the long arm of chromosome 14, proximal to the a1-antichymotrypsin gene. *Nat. Genet.* **2,** 340–342.
8. Levy-Lahad, E., Wasco, W., Poorkaj, P., et al. (1995) Candidate gene for the chromosome 1 familial Alzheimer's disease locus. *Science* **269,** 973–977.
9. Rogaev, E. I., Sherrington, R., Rogaeva, E. A., et al. (1995) Familial Alzheimer's disease in kindreds with missense mutations in a gene on chromosome 1 related to the Alzheimer's disease type 3 gene. *Nature* **376,** 775–778.
10. Sherrington, R., Rogaev, E. I., Liang, Y., et al. (1995) Cloning of a gene bearing missense mutations in early-onset familial Alzheimer's disease. *Nature* **375,** 754–760.

11. Cai, X. D., Golde, T. E., and Younkin, S. G. (1993) Release of excess amyloid β protein from a mutant amyloid beta protein precursor. *Science* **259,** 514–516.
12. Citron, M., Oltersdorf, T., Haass, C., et al. (1992) Mutation of the β-amyloid precursor protein in familial Alzheimer's disease increases β-protein production. *Nature* **360,** 672–674.
13. Borchelt, D. R., Thinakaran, G., Eckman, C. B., et al. (1996) Familial Alzheimer's disease-linked presenilin 1 variants elevate Aβ1-42/1-40 ratio in vitro and in vivo. *Neuron* **17,** 1005–1013.
14. Citron, M., Westaway, D., Xia, W., et al. (1997) Mutant presenilins of Alzheimer's disease increase production of 42-residue amyloid β-protein in both transfected cells and transgenic mice. *Nat. Med.* **3,** 67–72.
15. Lemere, C. A., Lopera, F., Kosik, K. S., et al. (1996) The E280A presenilin 1 Alzheimer mutation produces increased Aβ42 deposition and severe cerebellar pathology. *Nat. Med.* **2,** 1146–1150.
16. Hsiao, K., Chapman, P., Nilsen, S., et al. (1996) Correlative memory deficits, Aβ elevation, and amyloid plaques in transgenic mice. *Science* **274,** 99–102.
17. Games, D., Adams, D., Alessandrini, R., et al. (1995) Alzheimer-type neuropathology in transgenic mice overexpressing V717F β-amyloid precursor protein. *Nature* **373,** 523–527.
18. Sturchler-Pierrat, C., Abramowski, D., Duke, M., et al. (1997) Two amyloid precursor protein transgenic mouse models with Alzheimer disease-like pathology. *Proc. Natl. Acad. Sci. USA* **94,** 13287–13292.
19. Duff, K., Eckman, C., Zehr, C., et al. (1996) Increased amyloid-β42(43) in brains of mice expressing mutant presenilin 1. *Nature* **383,** 710–713.
20. Holcomb, L., Gordon, M. N., McGowan, E., et al. (1998) Accelerated Alzheimer-type phenotype in transgenic mice carrying both mutant amyloid precursor protein and presenilin 1 transgenes. *Nat. Med.* **4,** 97–100.
21. Janus, C., Pearson, J., McLaurin, J., et al. (2000) Aβ peptide immunization reduces behavioural impairment and plaques in a model of Alzheimer's disease. *Nature* **408,** 979–982.
22. Meyer-Luehmann, M., Stalder, M., Herzig, M. C., et al. (2003) Extracellular amyloid formation and associated pathology in neural grafts. *Nat. Neurosci.* **6,** 370–377.
23. Rozmahel, R., Huang, J., Chen, F., et al. (2002) Normal brain development in PS1 hypomorphic mice with markedly reduced γ-secretase cleavage of βAPP. *Neurobiol. Aging* **23,** 187–194.
24. Rozmahel, R., Mount, H. T., Chen, F., et al. (2002) Alleles at the Nicastrin locus modify presenilin 1-deficiency phenotype. *Proc. Natl. Acad. Sci. USA* **99,** 14452–14457.
25. Pfeifer, M., Boncristiano, S., Bondolfi, L., et al. (2002) Cerebral hemorrhage after passive anti-Aβ immunotherapy. *Science* **298,** 1379.
26. Mathews, P. M., Jiang, Y., Schmidt, S. D., Grbovic, O. M., Mercken, M., and Nixon, R. A. (2002) Calpain activity regulates the cell surface distribution of amy-

loid precursor protein: inhibition of calpains enhances endosomal generation of β-cleaved C-terminal APP fragments. *J. Biol. Chem.* **277,** 36415–36424.

27. Grbovic, O. M., Mathews, P. M., Jiang, Y., et al. (2003) Rab5-stimulated up-regulation of the endocytic pathway increases intracellular β-cleaved amyloid precursor protein carboxyl-terminal fragment levels and Aβ production. *J. Biol. Chem.* **278,** 31261–31268.

28. Gravina, S. A., Ho, L., Eckman, C. B., et al. (1995) Amyloid β protein (Aβ) in Alzheimer's disease brain. Biochemical and immunocytochemical analysis with antibodies specific for forms ending at Aβ40 or Aβ42(43). *J. Biol. Chem.* **270,** 7013–7016.

29. Savage, M. J., Trusko, S. P., Howland, D. S., et al. (1998) Turnover of amyloid β-protein in mouse brain and acute reduction of its level by phorbol ester. *J. Neurosci.* **18,** 1743–1752.

30. Phinney, A. L., Drisaldi, B., Schmidt, S. D., et al. (2003) In vivo reduction of amyloid-β by a mutant copper transporter. *Proc. Natl. Acad. Sci. USA* **100,** 14193–14198.

31. Chishti, M. A., Yang, D. S., Janus, C., et al. (2001) Early-onset amyloid deposition and cognitive deficits in transgenic mice expressing a double mutant form of amyloid precursor protein 695. *J. Biol. Chem.* **276,** 21562–21570.

23

Histological Staining of Amyloid-β in Mouse Brains

Einar M. Sigurdsson

Summary

The increased availability of transgenic mouse models for studying human diseases is shifting the focus of many laboratories from in vitro to in vivo assays. The purpose of this chapter is to provide investigators with methods that will allow them to obtain well-preserved mouse brain sections to be stained with the standard histological dyes for amyloid, Congo red and thioflavin-S. These sections can as well be used for immunohistological procedures that allow detection of amyloid-β plaques as well as pre-amyloid deposits.

Key Words: Mouse; perfusion; brain; fixation; histology; Congo red; thioflavin-S; immunohistochemistry; amyloid-β; plaques.

1. Introduction

Over the last decade, numerous transgenic mouse models have been developed for Alzheimer's disease and other amyloid-related disorders. The increased availability of these models continues to shift the focus of many laboratories from in vitro and cell-based assays to animal models. Several tissue preparation methods, histological stains and immunohistochemical procedures can be used for amyloid and pre-amyloid detection, and have been described in varying details by others *(1–10)*. Here, a combined protocol is provided *(11–13)*, which focuses on these methods applied to fixed mouse-brain sections. First of all, appropriate tissue preservation is fundamental for any following steps; systematic sectioning and storage of tissue sections is important for subsequent analysis as well. Congo red and thioflavin-S are the two major histological stains used to detect any form of amyloid. These dyes bind to the characteristic β-pleated sheet conformation of amyloid. Additionally, monoclonal mouse antibodies are increasingly employed to detect pre-amyloid and amyloid deposits and their use has led to the development of kits that reduce nonspecific binding

From: *Methods in Molecular Biology, vol. 299: Amyloid Proteins: Methods and Protocols*
Edited by: E. M. Sigurdsson © Humana Press Inc., Totowa, NJ

of these antibodies to mouse tissue. Although these kits come with detailed protocols, the exact procedure often has to be adjusted depending on the tissue preparation, the antigen, and the monoclonal antibodies.

2. Materials

1. Sodium pentobarbital.
2. Surgical instruments: scissors, various forceps, feeding needles for newborn mice or birds.
3. Perfusion pump with variable flow (Mini-Pump, Variable flow, Fisher, cat. no. 138761), with tubing (inner diameter: 0.38 mm).
4. 0.1 M Sodium/potassium phosphate buffer, pH 7.4: Dissolve 4.03 g KH_2PO_4 monobasic and 10 g Na_2HPO_4 dibasic, anhydrous in distilled water, pH to 7.4 with dilute NaOH or HCl, fill to 1 L with distilled water.
5. Heparin (1000 USP units/mL).
6. 4% Paraformaldehyde in 0.1 M sodium/potassium phosphate buffer: Prepare 0.2 M sodium/potassium phosphate buffer (8.06 g of KH_2PO_4 monobasic and 20 g of Na_2HPO_4 dibasic, anhydrous dissolved in distilled water, pH to 7.4 and adjust volume to 1 L). A portion of this solution can be diluted to 0.1 M for use in mouse perfusions. The following steps should be performed in a chemical hood.
   a. Weigh 40 g of prilled paraformaldehyde.
   b. Heat 450 mL of distilled water to 55°C, remove from heat, and add paraformaldehyde while stirring.
   c. Add 10 N NaOH until the solution clears, which occurs when the fixative dissolves (*see* **Note 1**).
   d. Allow to cool to room temperature and then filter through Whatman no. 1 (*see* **Note 2**). Add water to 500 mL and then add an equal volume of 0.2 M phosphate buffer to obtain 1 L of the fixative. Use freshly prepared and/or store refrigerated (*see* **Note 3**).
7. 15 mL Conical tubes.
8. 20% Glycerol/2% dimethylsulfoxide in 0.1 M sodium/potassium phosphate buffer, pH 7.4: Prepare buffer as listed earlier, then add 200 mL of glycerol and 20 mL of dimethylsulfoxide to a total volume of 1 L.
9. Single-edge razor blades No. 9.
10. Tissue Freezing Medium® for frozen tissue specimens.
11. Paint brushes, size 0 and 1.
12. 24-Well tissue culture plates.
13. Ethylene glycol-based cryoprotectant: 30% sucrose, 30% ethylene glycol in 0.1 M sodium/potassium phosphate buffer, pH 7.4. Prepare the buffer as listed before, then add 300 g of sucrose and 300 mL ethylene glycol to a total volume of 1 L.
14. Microscope slides (25 × 75 × 1.0 mm).
15. 95% Ethanol : 1.0 M acetic acid (1:1).
16. Gelatin.
17. Chromium potassium sulfate.

18. 1% Congo red solution in 50% ethanol: Stir 2.5 g of Congo red into 250 mL of 50% ethanol.
19. 100% Ethanol.
20. Xylene or CitriSolv® (Fisher Scientific).
21. Saturated lithium carbonate solution: Stir 2.6 g of lithium carbonate into 200 mL of distilled water.
22. 100-μm nylon cell strainers (Falcon, No. 352360) or equivalent.
23. 6-Well culture dishes.
24. MOM® kit (Vector Laboratories) or equivalent.
25. Vectastain ABC Elite kit to detect rabbit IgG primary antibodies (Vector) or equivalent.
26. Primary antibody diluent in phosphate buffered saline for polyclonal antibodies: 2% Triton X-100, 0.1% sodium azide, 0.01% bacitracin, 2% bovine serum albumin, and 10% normal serum from the same species that the secondary antibody is raised in.
27. Phosphate buffered saline (PBS): 0.01 *M* phosphate buffer, 0.0027 *M* potassium chloride and 0.137 *M* sodium chloride (PBS tablets [Sigma]: dissolve 1 tablet in 200 mL of water).
28. 0.3% hydrogen peroxide in PBS made fresh from 30% solution.
29. 6E10 (Signet) or other anti-Aβ antibodies.
30. 0.2 *M* Sodium acetate buffer, pH 6.0, adjust pH with 1 *M* acetic acid.
31. Diaminobenzidine tetrahydrochloride.
32. Nickel ammonium sulfate.
33. Filter disks (0.45-μm).

3. Methods

The methods described here outline (1) a mouse perfusion technique and brain sectioning, (2) preparation of gelatin-coated slides, (3) two amyloid histology procedures, and (4) an immunohistological method for staining of pre-amyloid and amyloid deposits in mouse brain.

3.1. Mouse Perfusion and Brain Sectioning

1. The mouse has to be deeply anesthetized by intraperitoneal injection of sodium pentobarbital (150 mg/kg).
2. The mouse is subsequently placed over a collection container on a tray or grid that will allow the perfusate to drain into the container (*see* **Note 4**).
3. The chest cavity is then exposed and the left ventricle is punctured with pointed forceps.
4. A perfusion cannula is then placed into the ventricle and held in place with hemostats or with forceps that can be secured with a paperclip (*see* **Note 5**). The right atrium is then cut and the perfusion pump started. Cutting the atrium allows the perfusate to exit the circulatory system (*see* **Note 6**).
5. The initial perfusion should be performed with a physiological solution such as 0.1 *M* sodium/potassium phosphate buffer or PBS, pH 7.4.

6. Heparin (1 U/g body weight) may be added to the perfusion solution to eliminate blood clotting (*see* **Note 7**). This is especially helpful for investigators who are still learning the perfusion technique.

7. If biochemical analysis is needed on the tissue, the brain can be removed from the skull following perfusion with 25–50 mL of phosphate buffer (5–10 min or until the perfusate runs clear). The whole brain or its portion can then be snap frozen on dry ice (*see* **Note 8**). The remaining portion can then be immersion fixed in 4% paraformaldehyde overnight. If the whole brain is to be used for histology, the perfusion has to be temporarily halted while the tubing is transferred to a 4% paraformaldehyde solution.

8. The perfusion is then resumed with about 50–100 mL of fixative. The fixative will shrink the brain, which will aid in removing it from the skull, although with experience, an unfixed brain can easily be removed intact.

9. The brain can then be postfixed overnight at 4°C in the same fixative (*see* **Note 9**).

10. The brain is subsequently transferred to 20% glycerol/2% dimethylsulfoxide in 0.1 M phosphate buffer and kept refrigerated overnight (*see* **Note 10**).

11. The brain can then be sectioned at 40 μm on a freezing microtome or in a cryostat (*see* **Note 11**).

12. It is often necessary to save all the sections; this can be done conveniently by placing them into series in 24-well cell culture tray filled with ethylene glycol cryoprotectant. We usually save sections from each brain into five series at 200 μm intervals. The sections are then stored at −20°C until used (*see* **Note 12**).

3.2. Preparation of Gelatin Coated Slides

For immunohistochemical procedures, commercially available coated slides are sufficient but for Congo red staining of slide-mounted sections, it is preferable to use gelatin-coated slides.

1. Use commercially available pre-cleaned slides (75 × 25 mm).
2. Place in a slide rack (*see* **Note 13**).
3. Rinse in distilled water.
4. Dip for 1 min in 1:1 solution of 95% ethanol and 1.0 M acetic acid.
5. Rinse in distilled water until water sheets off slides (15 to 30 s).
6. Dip for 2 min in freshly prepared filtered gelatin solution. Prepare by heating 300 mL of distilled water to 60°C, then add 3 g of gelatin and stir. When fully dissolved, add 0.3 g of chromium potassium sulfate to harden the gelatin and stir until dissolved. Filter through Whatman no. 1 or equivalent filter paper before use.
7. Dry at 30–50°C overnight.
8. Place in slide containers until used.

3.3. Histology

3.3.1. Congo Red Staining

1. Prepare a saturated Congo red solution by adding 2.5 g of Congo red to a total volume of 250 mL of 50% ethyl alcohol. Undissolved material should be filtered

through a Whatman no. 1 or equivalent filter paper (*see* **Note 14**). It is preferable to perform this staining on sections that have been mounted onto slides (*see* **Note 15**).

2. After defatting the tissue in xylene or CitriSolv®, it is hydrated by taking it through a series of ethyl alcohol solutions (100%, 95%, 80% and 70%, 1–2 min in each) before staining in the Congo red solution for 1 h.
3. The slides are subsequently dipped into a saturated lithium carbonate solution for 10–20 s and then rinsed for a similar period in tap water or distilled water (*see* **Note 16**).
4. Dip the slides for 5–60 s in 70% ethanol to get rid of some of the background staining, and then transfer the slides quickly through 80%, 95%, and two sets of 100% ethanol (15 s to 2 min in each) before placing them in xylene or CitriSolv for 5–10 min (*see* **Note 17**).
5. The slides can now be cover-slipped with mounting media such as DePex (*see* **Note 18**).
6. Congo red staining can then be viewed under plan polarized light. Amyloid plaques should give apple-green birefringence, usually as a Maltese cross, whereas the neurons that are non-specifically stained will not emit birefringence (*see* **Note 19**).

3.3.2. Thioflavin-S Staining

1. As with the Congo red staining method, it is preferable to stain mounted sections with thioflavin-S.
2. Prepare 1% thioflavin-S solution in distilled water, and filter it through Whatman no. 1 before use (*see* **Note 20**).
3. Sections are defatted in xylene or CitriSolv (5–10 min) and then hydrated through a series of ethyl alcohol solutions (100%, 95%, 80%, 70%, 1–2 min in each one), placed for a few seconds in water, and then stained with 1% thioflavin-S for 30–60 min.
4. The sections are subsequently dehydrated through a series of ethyl alcohol solutions (70%, 80%, 95%, 100%, 100%, 1–2 min in each one) and then placed in xylene or CitriSolv® (5–10 min) before being cover-slipped with DePex mounting media.
5. The slides are then cleaned as described previously for the Congo red method before being viewed under fluorescence using the filter set recommended for that particular microscope. Thioflavin-S bound to amyloid will emit fluorescence.
6. The stained sections should be stored in a cool, dark place such as a refrigerator.

3.3.3. Immunohistochemistry: Staining for Amyloid Plaques and Pre-Amyloid Deposits

The following protocols are for free-floating mouse brain sections (*see* **Note 21**). Initially, certain controls should be included such as pre-adsorption of the primary antibody with the antigen and omitting of the primary and secondary antibodies as well as the avidin-peroxidase complex. For mouse-on-mouse detection it is appropriate to include omission of the primary antibody in each run. For optimal immunodetection of plaques it is not necessary to pretreat the sections with heat or formic acid as is preferable for human sections.

3.3.3.1. Mouse Monoclonal Antibodies

The following description applies to the use of mouse monoclonal antibodies on mouse brain sections. To eliminate background staining, it is recommended to use MOM® kit (Vector Laboratories) or equivalent reagents. We have modified the kit instructions to our conditions. It is likely that further adjustments can be performed to save reagents.

1. The sections are treated for 15 min in 0.3% hydrogen peroxide, freshly prepared from a 30% solution, to quench endogenous peroxidase activity.
2. Following a 5-min wash in PBS, the sections are placed in MOM blocking reagent (Vector Laboratories) for 1 h (5 drops per 10 mL of PBS; *see* **Note 22**), and then washed twice for 5 min in PBS.
3. The sections are then placed for 1 h or longer in mouse anti-Aβ antibody (primary antibody) in MOM diluent (*see* **Note 23**). The diluent is prepared by adding 1200 μL of protein concentrate from the MOM kit into 15 mL of PBS. Control sections are placed in MOM diluent alone.
4. After two 5-min washes in PBS, the sections are incubated for 30 min in biotinylated anti-mouse IgG (secondary antibody) solution that is prepared by adding 20 μL of stock solution into 15 mL of MOM diluent (*see* **Note 24**).
5. Following two 5 min washes in PBS, the sections are placed for 30 min in avidin-peroxidase solution that is prepared 30 min in advance by adding two drops of Reagent A and two drops of Reagent B into 15 mL of PBS.
6. Subsequently, the sections are washed for 5–10 min in 0.2 *M* sodium acetate solution, pH 6.0 (*see* **Note 25**), and then reacted with the chromogen of choice. We usually use diaminobenzidine tetrahydrochloride (DAB) as a chromogen and nickel ammonium sulfate to intensify the reaction. This solution can be prepared by dissolving 35 mg of DAB in 100 mL of 0.2 *M* sodium acetate buffer, pH 6.0. Nickel ammonium sulfate (2.5 g) is subsequently added and once dissolved, the solution is filtered through 0.45-μm filter disk by using a 20–60 mL disposable syringe (*see* **Note 26**).

3.3.3.2. Polyclonal Antibodies

When rabbit polyclonal antibodies are used on mouse brain sections, background staining owing to crossreactivity is less of an issue than with mouse antibodies. Regular secondary antibody kits such as the Vector Elite kit for rabbit IgG antibodies can be used. This protocol is the same as the mouse monoclonal protocol except that the primary antibody diluent is different (as detailed in **Subheading 3.**) and the secondary antibody is diluted into PBS (*see* **Note 27**).

4. Notes

1. Stop adding NaOH when it does not enhance the clarity of the solution. Powdered paraformaldehyde is more soluble than the prilled version but is not recommended because of enhanced risk of exposure to airborne particles. Although the prilled fixative may not dissolve completely, undissolved material can be filtered away.

2. It is recommended to use double filter paper or to allow the solution to cool down. Warm solution may rupture the filter paper.

3. It is usually recommended to use freshly prepared fixative, but we have not noticed any differences in the quality of tissue sections or amyloid staining when the fixative has been refrigerated for up to several weeks prior to use.

4. It is convenient to perform this procedure over a sink, but the perfusate should be disposed of according to institutional guidelines.

5. As a perfusion cannula, 22 gage feeding needle for newborn mice or birds is recommended. Blunted needles are less likely to tear the heart during perfusion. The cannula will fit onto a regular intravenous tubing extension set that also has an injection site for the heparin solution. Alternatively, heparin can be added directly to the perfusion solution.

6. The lungs will expand if there is a buildup of fluids within the circulatory system. The cause is usually that efflux of the perfusate out of the right atrium is blocked and/or the cannula has accidentally been placed in the right ventricle so the perfusion liquid will enter the lungs first instead of the main circulation.

7. Blood clots will give non-specific staining in immunohistochemistry.

8. If one hemisphere needs to be saved for biochemical analysis, it is convenient to place the brain on a cold surface and cut the brain slightly off the midline with a single-edged razor blade. The larger portion should be immersion fixed and will contain a fraction of the contralateral hemisphere. However, if the brain is cut through the midline between the hemispheres, the hippocampus will often separate from the brain section during staining and/or mounting onto the slides.

9. If the brain is fixed by perfusion, post-fixation for 6–8 h may be sufficient, but overnight fixation does not affect amyloid staining.

10. Without this step, the brain may crack while being frozen for sectioning.

11. By sectioning the brain at 40 μm, the sections will withstand free-floating staining procedures. Paint brush no. 1 is convenient to use to move the sections from the blade into the ethylene glycol cryoprotectant solution. Tissue Freezing Medium® or equivalent is used to secure the brain to the specimen holder. It usually takes less time to section the brain on a freezing microtome than in a cryostat.

12. Prolonged storage at −20°C does not seem to affect staining of amyloid.

13. The slide rack should have a handle and an open bottom so that the rack can be moved quickly between solutions (equivalent to Wheaton No. 900200 that includes a staining dish that fits the rack). Nineteen slides will fit in each rack if placed in a zig-zag way. Larger racks can also be used.

14. If the staining will be performed in regular glass slide container, a total volume of 200–250 mL will be sufficient to cover the slides. The solution should be stirred for a few hours or even overnight to ensure that most of the chemical is dissolved. It is preferable to use freshly made solution. After sitting at room temperature for several days or weeks, the stain will not work properly.

15. Commercially available coated slides, such as Fisherbrand® Superfrost®*/Plus work fine for immunohistochemistry, but for histological stains such as Congo red or Cresyl Violet, it is preferable to use gelatin-coated slides for improved adherence of the sections to the slide.

16. Some protocols suggest washing the sections in running tap water for 15 min. This may be appropriate for human sections, but in our experience, mouse and rat sections partially come off the slides when left in water for too long.

17. Protocols for Congo red staining usually involve counterstains such as Harris hematoxylin but this procedure can be avoided by not washing off the nonspecific Congo red neuronal staining. Most of this staining comes off in the 70% ethanol, so care should be taken to keep the slides only briefly in this solution. The exact time varies but 5 to 60 s is usually sufficient to reduce the background staining. The remaining intensity should be similar to Cresyl Violet staining. The sections can always be restained in the Congo red solution if they become too destained in the alcohol solutions. Leave the slides for at least 2 min in the final set of 100% ethanol. Water will turn the xylene or Citrosolv® cloudy and can interfere with microscopic analysis of the tissue sections.

18. Various mounting media can be used. We are particularly fond of DePex mounting media. It does not shrink much while setting, which reduces bubble formation, and it does not come off in sharp flakes when the slides are cleaned with a razor blade after it has set overnight. It is best to use coverslips (22 × 60 mm) that are slightly smaller than the slide (25 × 75 mm). The slides will be easier to clean and bubbles are less likely to form.

19. Birefringence is often seen in larger blood vessels and in threads associated with the surface of the brain. This observation does not confirm the presence of amyloid and can be owing to Congo red binding to collagen which is known to result in birefringence.

20. As with Congo red, a volume of 200–250 mL is sufficient if standard glass staining containers are used.

21. A similar protocol can be used for slide-mounted sections, although incubation periods may have to be increased to allow sufficient penetration of the antibodies. It is most convenient to use cell strainers, with 100-μm nylon mesh, to move the sections between solutions. The strainers can be placed in 6-well culture dishes for all the washes. To save material for all other incubations, the strainers can be placed in small individual petri dishes of a similar diameter as the wells. Less volume is needed in these dishes to immerse the sections.

22. The MOM blocking solution can be used at least few times.

23. 6E10 (Signet; 1:1000-fold dilution) binds to human Aβ but does not crossreact with mouse Aβ. The proper dilution of other monoclonal antibodies will have to be assessed.

24. We have also diluted the secondary antibody into PBS with similar results.

25. It is convenient to prepare 1.0 M sodium acetate, pH 6.0 stock solution. pH can be adjusted with 1.0 M acetic acid. This solution is best kept refrigerated. The 0.2 M solution can be prepared from the stock in 500 mL aliquots and stored refrigerated.

26. The syringes can be used repeatedly and discarded when it becomes difficult to push the plunger through the syringe, which will occur over time.

27. For general immunohistochemistry, we have often used 0.3% Triton-X in PBS as the wash solution and to dilute the secondary antibody as well as the A+B solu-

tion. The detergent may enhance epitope exposure, which facilitates detection of certain antigens, but is not necessary for staining amyloid plaques in mouse section. The primary antibody diluent can be stored refrigerated prior to use for at least several months.

Acknowledgments

This work was supported by NIH grant AG20197 and the Alzheimer's Association. These protocols were adapted in part from methods obtained from Stanley A. Lorens and Debra Magnuson at Loyola University Chicago. I thank Drs. Fernando Goni and Ayodeji Asuni for their comments on the manuscript.

References

1. Kitamoto, T., Ogomori, K., Tateishi, J., and Prusiner, S. B. (1987) Formic acid pretreatment enhances immunostaining of cerebral and systemic amyloids. *Lab. Invest.* **57**, 230–236.
2. Davies, L., Wolska, B., Hilbich, C., et al. (1988) A4 amyloid protein deposition and the diagnosis of Alzheimer's disease: prevalence in aged brains determined by immunocytochemistry compared with conventional neuropathologic techniques. *Neurology* **38**, 1688–1693.
3. Gentleman, S. M., Bruton, C., Allsop, D., Lewis, S. J., Polak, J. M., and Roberts, G. W. (1989) A demonstration of the advantages of immunostaining in the quantification of amyloid plaque deposits. *Histochemistry* **92**, 355–358.
4. Lamy, C., Duyckaerts, C., Delaere, P., et al. (1989) Comparison of seven staining methods for senile plaques and neurofibrillary tangles in a prospective series of 15 elderly patients. *Neuropathol. Appl. Neurobiol.* **15**, 563–578.
5. Wisniewski, H. M., Wen, G. Y., and Kim, K. S. (1989) Comparison of four staining methods on the detection of neuritic plaques. *Acta Neuropathol. (Berl)* **78**, 22–27.
6. Vallet, P. G., Guntern, R., Hof, P. R., et al. (1992) A comparative study of histological and immunohistochemical methods for neurofibrillary tangles and senile plaques in Alzheimer's disease. *Acta Neuropathol. (Berl)* **83**, 170–178.
7. Raskin, L. S., Applegate, M. D., Price, D. L., Troncoso, J. C., and Hedreen, J. C. (1995) Comparison of new and traditional methods for detection of senile plaques in Alzheimer's disease. *J. Geriatr. Psychiatry Neurol.* **8**, 125–131.
8. Cullen, K. M., Halliday, G. M., Cartwright, H., and Kril, J. J. (1996) Improved selectivity and sensitivity in the visualization of neurofibrillary tangles, plaques and neuropil threads. *Neurodegeneration* **5**, 177–187.
9. Shiurba, R. A., Spooner, E. T., Ishiguro, K., et al. (1998) Immunocytochemistry of formalin-fixed human brain tissues: microwave irradiation of free-floating sections. *Brain Res. Brain Res. Protoc.* **2**, 109–119.
10. Cummings, B. J., Mason, A. J., Kim, R. C., Sheu, P. C., and Anderson, A. J. (2002) Optimization of techniques for the maximal detection and quantification of Alzheimer's-related neuropathology with digital imaging. *Neurobiol. Aging* **23**, 161–170.

11. Sigurdsson, E. M., Lorens, S. A., Hejna, M. J., Dong, X. W., and Lee, J. M. (1996) Local and distant histopathological effects of unilateral amyloid-β 25-35 injections into the amygdala of young F344 rats. *Neurobiol. Aging* **17,** 893–901.
12. Sigurdsson, E. M., Lee, J. M., Dong, X. W., Hejna, M. J., and Lorens, S. A. (1997) Bilateral injections of amyloid-β 25-35 into the amygdala of young Fischer rats: Behavioral, neurochemical, and time dependent histopathological effects. *Neurobiol. Aging* **18,** 591–608.
13. Sigurdsson, E. M., Scholtzova, H., Mehta, P. D., Frangione, B., and Wisniewski, T. (2001) Immunization with a non-toxic/non-fibrillar amyloid-β homologous peptide reduces Alzheimer's disease associated pathology in transgenic mice. *Am. J. Pathol.* **159,** 439–447.

24

The Mouse Model for Scrapie

Inoculation, Clinical Scoring,
and Histopathological Techniques

Harry C. Meeker, Xuemin Ye, and Richard I. Carp

Summary

The mouse is a popular and versatile model for the study of scrapie and other transmissible spongiform encephalopathies. In this chapter, information is given for preparation of infectious material for inoculation and a method of clinical scoring that yields accurate and reproducible quantification of the scrapie incubation period. With the help of histopathological and immunopathological techniques, we can detect brain pathological changes in scrapie-infected animals at the cellular and molecular level. We will also describe the histological and immunocytochemistry methods we use for scrapie research, outline step-by-step procedures, discuss tissue preparation, fixation, and processing of specimens, and provide special hints to achieve successful staining. We also include our results of PrP^{Sc} and GFAP immunostaining in scrapie research. In conclusion, immunopathological staining is an important and useful tool in the research of scrapie pathology.

Key Words: Mouse; scrapie; inoculation; brain; histopathology; immunohistopathology; PrP^{Sc}; GFAP.

1. Introduction

The mouse offers many advantages as an experimental model for the study of transmissible spongiform encephalopathies (TSEs). Mice are easy to handle, inexpensive to maintain and readily available in a variety of well-characterized inbred and outbred strains. Transgenic mice, carrying *PrP* genes from a variety of species, allow for the study of TSEs beyond the mouse-adapted scrapie strains *(1–5)*. Knockout mice, with the *PrP* gene inactivated, have been used for a variety of scrapie studies *(5)*. In addition, knockout mice for genes other than the *PrP* gene have afforded an opportunity to explore the interaction of these genes with the TSE disease process *(6,7)*.

From: *Methods in Molecular Biology, vol. 299: Amyloid Proteins: Methods and Protocols*
Edited by: E. M. Sigurdsson © Humana Press Inc., Totowa, NJ

With the recent appearance of diseases such as bovine spongiform encephalopathy (BSE), variant Creutzfeldt-Jakob Disease (vCJD) and chronic wasting disease (CWD), interest in the field of TSEs has increased. Researchers new to the field, or those wishing to expand into different areas of the TSE arena, may be in need of information about the hands-on aspects of TSE in the mouse.

Observation of scrapie-infected mice for clinical signs of disease provides an accurate and reproducible method for defining the scrapie incubation period. In addition, appearance of the ataxia associated with clinical scrapie coincides with maximal brain titers of infectious agent and the most extensive signs of brain pathology *(8–10)*.

Scrapie in the mouse produces well-defined pathological signs in the brain. The hallmarks of scrapie pathology in the brain are vacuolation and astrocytosis *(11)*. In addition, some scrapie strain–mouse strain combinations (i.e., 87V in MB mice, 22A in C57BL/6J mice) produce amyloid plaques in the brain *(12, 13)*. Different scrapie strains have been found to produce characteristic patterns of neuropathology in specific areas of the brain, termed lesion profiles *(11,14)*.

The focus of this chapter is to review techniques involved in production and diagnosis of scrapie in the mouse: injection, clinical scoring and histopathological techniques for diagnosis and study of scrapie.

2. Materials

2.1. Inoculation and Clinical Scoring

1. Insulin syringes, 0.5 cc, 28-gauge integral needle (Becton-Dickinson, Franklin Lakes, NJ).
2. Phosphate-buffered saline (PBS), 0.01 *M*, pH 7.4.
3. Ground-glass Tenbroeck tissue homogenizers, 2 mL–15 mL. (Fisher Scientific, Pittsburgh, PA).
4. Analytical balance.
5. Anesthetic (Avertin or Isoflurane).
6. Scoring grid.

2.2. Tissue Preparation

1. Sodium pentobarbital (Nembutal, Abbott Laboratories, North Chicago, IL).
2. Normal saline (0.9% sodium chloride: 9 g NaCl in 1000 mL distilled water).
3. 4% Paraformaldehyde.
4. Xylene.
5. Paraffin.
6. Alcohol (30%, 70%, 95%, and 100%) should be prepared with distilled water.
7. 0.1 *M* PBS: Add the following chemicals to a 1000 mL volumetric flask: 6.63 g monobasic sodium phosphate; 40.4 g dibasic sodium phosphate; 8.78 g sodium chloride; add distilled water to make 1000 mL. The pH should be 7.4. Use 1 *N*

hydrochloric acid, if too alkaline, or 1 *N* sodium hydroxide, if too acid, and adjust the pH to proper value.

2.3. Histopathological Staining

1. Graded alcohols (50%, 70%, 95%, and 100%).
2. Xylene.
3. HEMO DE (Fisher).
4. Hematoxylin and eosin (H&E) staining solution.
5. Periodic acid-Schiff (PAS) staining solution.
6. Thioflavin-S fluorescence stain solution.
7. Permount, DPX mount (Aldrich Chem. Co. Milwaukee, WI).

Permount, xylene, and most of the chemicals and reagents listed can be obtained from Sigma Chemical Company (St. Louis, MO), or Fisher Scientific (Pittsburgh, PA).

2.4. Immunocytochemistry for Astrocytosis and PrP^{Sc}

1. Oven or incubator (capable of maintaining 37°C).
2. Staining jars.
3. Timers.
4. Absorbent wipes.
5. Microscope slides.
6. Coverslips for slides.
7. Light microscope with final magnification ×100 and ×400.
8. Distilled or deionized water, reagent grade.
9. Xylene or HEMO DE.
10. Ethanol (50%, 70%, 95%, 100%).
11. Absolute methanol.
12. 30% Hydrogen peroxide.
13. PBS, commercially available: Dulbecco's Phosphate Buffered Saline (DPBS) (Sigma).
14. 0.01 *M* PBS or Tris-buffer (TB) containing 0.05% Tween-20, pH 7.5.
15. Tris-buffered saline, pH 7.6.
16. 3,3'-Diaminobenzidine (DAB) tetrahydrochloride or DAB tablets (Sigma) (store DAB below 0°C).

Normal serum, primary antibody solution, secondary antibodies, chromogen substrate solutions and levamisole solution should all be stored at 2–8°C.

3. Methods

3.1. Inoculation

Scrapie agent can be found in a number of tissues in affected animals, including brain, spleen, lymph nodes and, to a much lesser extent, in other tissues *(8)*.

The scrapie titer in brain at the time of clinical manifestations of disease is at least 10-fold higher than in spleen *(15,16)*, thus brain is the organ of choice for obtaining scrapie material for purification of PrP^{Sc} and for propagation of the infectious agent.

For preparation of brain homogenate for injection into mice, brains are weighed, suspended at 10% (w/v) in 0.01 *M* PBS and homogenized by 20 strokes in a ground-glass tissue homogenizer *(17)*. Homogenates are kept frozen at –20°C to –70°C.

The route of injection for scrapie is dictated by the experimental design (*see* **Note 1**). For production of infectious material, the route of choice is intracerebral (ic) as this route produces the shortest incubation times. Typically, for production of infectious brain material in mice, we inject 0.025 mL/mouse ic of a 1% (w/v) scrapie brain homogenate into the right temporal area of the brain. This is done using a 0.5-cc insulin syringe with an integral 28-gauge needle which eliminates the air space typically encountered with luer-tip syringes.

Weanling mice 4–6 wk of age are preferred for ic injection because of the relative softness of the skull at this age (*see* **Note 2**). Although we have not kept detailed records on ic injection mortality, one should expect over 95% of mice to survive the injection. For inhalation anesthesia, we have used isoflurane or ether. With these, it is critical not to overdose the animal and kill it. We have also used Avertin, 2.5% (0.015 mL/g body wt), injected intraperitoneally, from which the animals recover quickly. Intraperitoneal injection of scrapie can be accomplished without anesthesia.

3.2. Clinical Scoring to Determine Scrapie Incubation Period

Quantification of the scrapie incubation period involves determining the appearance of clinical symptoms. The appearance of clinical symptoms coincides with maximal titers of infectious agent in the brain *(8–10)*.

The most reliable method that we have found of determining the onset of clinical scrapie in mice consists of an apparatus we call a scoring grid with 3 mm-wide parallel bars set 7 mm apart *(17,18)* (**Fig. 1**). The mice are placed on the scoring grid and observed for ataxia. Signs to look for include increased stumbling, ragged or wobbly gait, and lack of coordination (*see* **Note 3**). As the disease progresses, mice will stumble on the grid, extending a limb between the bars of the grid and have difficulty righting themselves (**Fig. 1**). Three consecutive positive weekly scores conclude the scrapie incubation period.

We start to observe mice on the grid shortly before the expected onset of symptoms. As such, it is imperative to know the approximate incubation period for the scrapie strain–mouse strain combination under study (**Table 1**). The incubation period of a given scrapie strain–mouse strain combination depends on the interplay of a number of factors including dosage, route of infection and

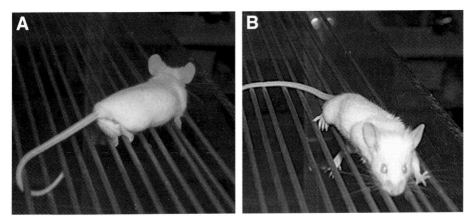

Fig. 1. Clinically positive scrapie-injected mouse on scoring grid. (**A**) Mouse does not support itself with rear legs. (**B**) Front limb extends below grid. Mouse does not support itself on grid.

the genetic makeup of the mouse. The *Sinc* (*19,20*) or *Prn-i* (*21,22*) gene, the product of which is PrP, is the genetic factor with the greatest influence on incubation period. In the mouse, there are two alleles for *Sinc*, which can be characterized by the incubation period for the ME7 scrapie strain. Homozygous mice, termed *s7s7*, will have a relatively short incubation time for the ME7 strain, while homozygous strains, termed *p7p7*, will have a prolonged ME7 incubation period. Heterozygotes will generally have an incubation period for ME7 that is intermediate, however, with some scrapie strains such as 139A, heterozygotes may have a longer incubation period than either *s7s7* or *p7p7* parents, a phenomenon termed overdominance (*23,24*). Most scrapie strains have a shorter incubation period in *s7s7* mice than in *p7p7* mice, but there are several scrapie strains in which the pattern is reversed with a long incubation period in *s7s7* mice and a much shorter incubation period in *p7p7* mice; examples include the 22A and 87V scrapie strains (**Table 1**).

3.3. Tissue Preparation

All animals are sacrificed with an intraperitoneal injection of sodium pentobarbital (Nembutal, intraperitoneally; 70 mg/kg body wt). The choice of whether to fix tissue by sample immersion in fixative or by perfusion depends largely upon the size of the tissue and the ability of the fixative to penetrate the tissue (*see* **Note 4**). Herein we give the procedure for perfusion, which is best for producing well-fixed brain (*25*).

1. Mice are perfused via the heart by making a small incision in the right atrium and inserting a canula connected to the perfusion media into the left ventricle. The

Table 1
Scrapie Strain-Mouse Strain Incubation Times (Mean days ± SEM).
Intracerebral Inoculation With 0.025 mL of 1% (w/v) Scrapie Brain Homogenate

Mouse strain	Sinc genotype	Scrapie strain									
		n	ME7	n	139A	n	22L	n	22A	n	87V
C57Bl	s7s7	60	156 ± 1	10	145 ± 3	8	146 ± 2	8	365 ± 4		>600[a]
CW[b]	s7s7	43	146 ± 2	28	136 ± 1	16	142 ± 2	8	405 ± 6		N.D.[c]
MB/IM	p7p7	12	252 ± 3[d]	13	244 ± 5[d]		N.D.	11	179 ± 1[a]	27	282 ± 4[e]
SJL	s7s7	8	145 ± 1	9	120 ± 3	10	145 ± 2		N.D.		N.D.

[a]See ref. 24.
[b]CW = Compton White.
[c]N.D. = Not done.
[d]MB mouse strain.
[e]IM mouse strain.

mice are perfused with normal saline (5 mL/min) at room temperature for 1 min, followed by perfusing with 4% paraformaldehyde or 10% neutral buffered formalin in 0.1 M PBS (pH 7.4, 5 mL/min) for 10 min at room temperature.

2. The brains and other organs are removed immediately and immersion fixed in the same fixative (4% paraformaldehyde or 10% neutral buffered formalin) for 24 h at 4°C.
3. The tissues are then placed in 0.1 *M* PBS (pH 7.4) for 2–3 d and rinsed thoroughly.
4. Specimens are then cut into small blocks, which are put into labeled carriers.
5. Tissue blocks are dehydrated in graded alcohols (30%, 70%, 95%, 100%, for 5 min, two times each) and three times in xylene and are then processed into paraffin blocks and sectioned into 7-μm thick histological sections.

3.4. Slide Preparation

To prevent sections from detaching from the glass slide, slides can be treated with Vectabond reagent (Vector Lab., Burlingame, CA), a non-protein tissue section adhesive. Alternatively, 5% gelatin can be used to coat the slide before mounting tissue sections.

1. Put slides into the slide holder, and clear in acidified 70% alcohol (add a few drops of glacial acetic acid into 300 mL 70% alcohol) for 2 h.
2. Rinse well with distilled water, put slides into 5% gelatin solution for 30 min and then air dry. Slide can also be precoated with 0.1% polylysine in water and then air-dried.

3.5. Histopathological Staining

1. The sections are dewaxed in xylene, or HEMO DE (Fisher), then placed in 100% ethanol, and then are rehydrated through graded ethanol solutions (95%, 70%, 50%) prior to being rinsed in deionized water (*see* **Note 5**).
2. The sections can then be stained by a variety of techniques including H&E for general histopathology, PAS (**Fig. 2A**) for abnormal amyloid-like deposits or thioflavin-S fluorescence stain for amyloid deposits (**Fig. 2B**).

3.6. Immunocytochemistry for Astrocytosis and PrP^{Sc}

The demonstration of antigen in tissue and cells by immunostaining is a two-step process involving first, the binding of an antibody to the antigen of interest, and second, the detection and visualization of bound antibody by one of a variety of enzyme chromogenic systems. With any of these methods, appropriate controls are essential (*see* **Note 6**).

1. Fill three containers (covered with lids) with 300 mL xylene or HEMO DE (so the tray is covered with liquid). The sections are dewaxed in xylene or HEMO DE, then placed in 100% ethanol, and then are rehydrated through graded ethanol solutions (95%, 70%, 50%) prior to being rinsed in PBS.

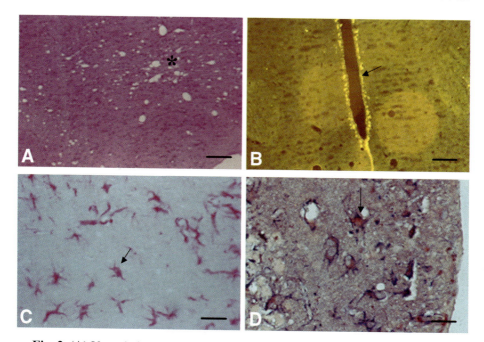

Fig. 2. (A) Vacuolation (*) in the brainstem in a scrapie ME7-infected CD1 mouse; magnification bar = 200 μm, PAS stain. **(B)** Prion plaques around the ventricles appear bright yellow against a dark background under UV light (arrow) in a scrapie 263K-infected hamster; magnification bar = 200 μm, thioflavin-S stain. **(C)** Astrocytosis (arrow) in a brain of scrapie 263K-infected hamster; magnification bar = 50 μm, GFAP immunostaining (red color). **(D)** Double immunostaining: PrP^{Sc} (brown color) and GFAP (blue color) immunostainings were co-localized in astrocytes (arrow) in the area of hippocampus of a 263K-infected hamster; magnification bar = 50 μm.

2. If using peroxidase as the labeling enzyme, endogenous peroxidases are quenched with 0.3% H$_2$O$_2$ in 100% methanol (add 3 mL of 30% hydrogen peroxide into 297 mL of methanol) for 15 min.
3. Nonspecific sites are blocked with normal sheep serum (Cappel Organon Teknika Co., Durham, NC), diluted 1 in 50 in PBS and the section incubated for 30 min at room temperature.
4. The sections are then incubated overnight at 4°C in a moist chamber with 100 μL primary antibody, e.g., monoclonal antisera are reacted with GFAP at 1/500 dilution (BioGenex Laboratories, San Ramon, CA) (**Fig. 2C**), or antibody to PrP^{Sc}. Antibody that reacts with mouse PrP^{Sc} is diluted with PBS. To unmask epitopes of the aggregated abnormal isoform of the scrapie protein, sections are treated with 88% formic acid for 15 min prior to reaction with the primary antiserum for PrP^{Sc}. Alternatively an antigen retrieval method may be used (*see* **Note 7**).

5. After incubation with primary antibody, slides are washed or rinsed 3 times with PBS followed by incubation with 100 µL secondary antibody (*see* **Note 8**) at optimum dilution for 30–60 min at room temperature. Rinse three times with PBS.
6. Blot slides around sections. Too much rinse buffer left on slides may cause excessive dilution of reagents. Tissue that dries during staining will not be well-stained.
7. Stain sections using several drops of the appropriate substrate-chromogen solution (*see* **Note 9**). If possible, monitor color development under a microscope. Tap off excess reagent, rinse in distilled water and mount section (*see* **Note 10**).

3.7. Double Labeling

Double staining is a technique used to reveal two distinct antigens in a single tissue. Different antigens such as PrP^{Sc} and GFAP can be detected together in tissue sections by using double-labeling procedures (**Fig. 2D**) (ZYMED Laboratories, South San Francisco, CA). To double-immunostain for PrP^{Sc} and GFAP, the section incubated in the following order:

1. 10% Normal serum in 0.01 M PBS for 1 h.
2. Polyclonal antibody against PrP^{Sc} is diluted and applied for 24 h at 4°C.
3. After rinsing three times in 0.01 M PBS, apply two drops or 100 µL of biotinylated second antibody to each section.
4. Incubate 10–30 min. Rinse well three times in PBS.
5. Apply two drops of streptavidin-alkaline phosphatase to each section. Incubate 10–30 min. Rinse well three times in PBS.
6. Apply two drops of substrate-chromogen mixture to each section. Incubate 5–10 min. Rinse well in distilled water.
7. Apply two drops of double staining enhancer to each section. Incubate 30 min. Wash well with distilled water, then rinse with PBS.
8. Add two drops of serum blocking solution to each section. Incubate 10 min. Blot off the solution. Do not rinse.
9. Apply two drops of polyclonal antibody against GFAP (The Jackson Laboratories, Bar Harbor, ME) diluted 1/400 and leave for 24 h at 4°C. Rinse well in PBS and apply two drops of biotinylated second antibody to each section. Incubate 10–30 min. Rinse well in PBS.
10. Apply two drops of streptavidin-peroxidase to each section. Incubate 10–30 min. Rinse well with PBS.
11. Apply two drops of substrate-chromogen mixture to each section. Incubate and monitor under a microscope to assess color development. Rinse well with tap water.
12. Mount slide with mounting medium and dry.

4. Notes

1. The intraperitoneal, or intravenous route, may be used to study the role of the spleen, lymphoreticular system, or peripheral nervous system, in scrapie infections (*26– 28*). Other routes such as oral, intranasal, and subcutaneous, may be used to mimic naturally acquired infections (*29–33*).

2. After about 6 wk of age, the skull becomes more difficult to penetrate. In such cases, and sometimes even with young mice, it is helpful to twist the syringe back and forth as pressure is applied, in effect to "drill" through the skull. While needles larger than 28 gauge may be used, this may increase mortality.

3. We have found that different mouse strains have different levels of competence on the scoring grid. Therefore, it is helpful to observe the mice under study on the scoring grid either before the appearance of clinical symptoms or to test uninjected mice of the same strain on the grid.

4. The most widely used fixing agent for pathologic histology is formaldehyde. Formaldehyde is commercially available at 37% to 40% gas dissolved in water, sold as formalin. Formalin fixes cytoplasm and nuclear sap but hardens tissue and prevents paraffin from easily penetrating the tissue. It makes cytoplasm basophilic so that acid dyes do not work well and chromosomes are poorly fixed, therefore, prolonged formalin fixes should be avoided. At least 15 to 20 volumes of fixative should be used for every volume of tissue. Most tissues should remain in fixative for 24 h at room temperature (25°C) or 48 to 72 h at 4°C and should be then stored in 70% alcohol. For better PrP^{Sc} immunocytochemistry staining, animals should be perfused with 4% paraformaldehyde, 0.34% L-lysine, 0.05% sodium *m*-periodate (4% PLP).

5. The deparaffinization process functions to remove the paraffin wax from tissues. All steps of deparaffinization should be sufficiently long to completely remove the paraffin from the sections.

   Xylene or HEMO DE (Fisher) is the usual reagent used for this purpose and three changes over a 10-min period are desirable. Slides should not be allowed to dry from the time hydration is begun until cover slip is applied. Dewax slides in xylene or HEMO DE for 3 × 5 min, immerse in 100% alcohol for 2 × 3 min, and hydrate in 95% alcohol for 2 × 3 min, hydrate in 70% alcohol for 2 × 3 min, hydrate in 50% alcohol 2 × 3 min. Immerse in pure water for 2 × 3 min.

6. Three control slides are necessary for the interpretation of results: (1) positive control, a specimen processed in the same way as the unknown, which contains the antigen to be stained; (2) negative control, a specimen processed in the same way as the unknown, which does not contain the antigen to be stained; and (3) reagent control, an additional slide that is treated with a non-immune serum instead of primary antibody. Any staining observed on this slide is probably owing to non-specific protein binding, or non-specific binding of other reagents.

   A major problem investigators have faced in attempts to use immunohistochemical techniques with mouse primary antibodies on mouse tissues is the inability of the anti-mouse secondary antibody to distinguish between the mouse primary antibody and endogenous mouse immunoglobulins in the tissue. Investigators can use MOM, an immunodetection kit (Vector Lab), to localize mouse primary monoclonal and polyclonal antibodies on mouse tissues.

7. In some cases, an antigen retrieval method should be used. Antigen retrieval citra solution (BioGenex, San Ramon, CA) may be used to recover antigenicity in formalin-fixed, paraffin-embedded tissue. This citric acid-based formula, using a

high temperature treatment procedure, is highly effective at revealing antigens in tissue sections that are formalin-fixed and paraffin-embedded. It consists of heating sections in a microwave oven in the presence of an antigen retrieval solution. If the antigens are incompletely retrieved, the staining is light and the background may be high.If an antigen retrieval method is necessary, use the following protocol as suggested by BioGenex: (1) Rinse slides in deionized water. Place slides in a plastic staining holder with any empty slots filled with blank slides. Place the holder in a plastic slide bath containing 250–300 mL of working strength antigen retrieval citra solution. Place a lid loosely on the bath and center it inside a microwave oven on a paper towel to adsorb any liquid runover. (2) Turn the oven on high power (500–1000 W) and closely watch the solution until it comes to a rapid boil, and then turn off the oven. It is very important that a rapid boil is reached before turning off the oven. 3) Set oven power to approx 50% level and heat for 10–15 min. The power setting should be adjusted so that the oven cycles on and off every 20–30 s and the solution boils about 5–10 s each cycle. (4) Remove the slide bath from the microwave oven. Allow slides to cool for 20–30 min at room temperature. Rinse with several changes of deionized water. Place slides in PBS and continue with the immunostaining procedure.

8. Immunostaining is accomplished using the peroxidase–antiperoxidase (PAP) system, the avidin-biotin complex (ABC) system (Vector Laboratories), or the supersensitive alkaline phosphatase (Biogenex Laboratories, San Ramon, CA) technique. One advantage of the ABC system is that it utilizes the same peroxidase complex for all primary antisera irrespective of its origin in different animal species.

   PAP is the most widely used enzyme and in combination with the most favored chromogen, i.e., 3,3'-diaminobenzidine tetrahydrochloride (DAB), it yields a crisp, insoluble, stable, dark brown reaction end product which is capable of enhancement.

   For the supersensitive alkaline phosphatase method, after reaction with the primary antibody and rinses, the sections are incubated sequentially at room temperature with biotinylated secondary antibody (1/100), alkaline phosphatase-conjugated streptavidin (1/100) and chromogen substrate (BioGenex Laboratories). Levamisole solution is used to reduce background staining of alkaline phosphatase-based detection systems by inhibiting endogenous alkaline phosphatase in most tissues.

   The use of alkaline phosphatase enzyme labels for immunocytochemical staining has been developed as an alternative to immunoperoxidase methods, in part, owing to the problem of endogenous peroxidase activity present in many tissues. However, endogenous alkaline phosphatase activity in tissues such as liver, bone, intestine, placenta, and tumor, can be a significant problem. Levamisole (BioGenex Laboratories) is a potent inhibitor of most of these forms of the enzyme.

   For the ABC system, following a 10-min wash in PBS, the excess buffer is removed from around the sections, which are then incubated with avidin-biotin complex at a dilution of 1/400 or 1/800 for 60 min at room temperature. After treatment with ABC solution, the sections are washed with PBS for 10 min. The slides are then treated for 10 min in freshly prepared DAB-hydrogen peroxide-peroxidase substrate solution. After incubation with the substrate, the slides are washed in Tris-HCl buffer and PBS for 5 min each, dehydrated through ascending grades of ethanol, cleared in two changes of xylene for 3 min each, and mounted with Permount or DPX mount and a cover glass.

9. DAB is a widely used chromogen for peroxidase staining. To make DAB-peroxide solution, dissolve 1 tablet of DAB (Sigma) in 15 mL of Tris-buffered saline pH 7.6. Add 12 μL of fresh 30% hydrogen peroxide prior to use. Solutions should be filtered through a 0.2-μ filter immediately prior to use. DAB is insoluble in alcohol and, therefore, is suitable for permanent mounting with Permount. Another chromogen for peroxidase staining is AEC (3-amino-9-ethylcarbazole). It forms a reddish-brown end product that is alcohol-soluble and, therefore, alcohol-containing solutions for counterstaining or dehydration cannot be used, AEC requires an aqueous counterstain (e.g., Mayer's hematoxylin) and aqueous mounting medium (available from BioGenex). 5-bromo-4-chloro-3-indolyl phosphate/nitro blue tetrazolium (BCIP/NBT) is for use with the alkaline phosphatase enzyme system. After application of the alkaline phosphatase solution, tissue sections should be washed in 0.1 M Tris-wash buffer pH 9.5 for 5–10 min at room temperature. Carefully wipe excess liquid from around the tissue section. Apply one to two drops of BCIP/NBT substrate solution directly to the tissue section, and incubate at room temperature for approx 10 min, or until blue-color development is complete. Tap off excess reagent, rinse section in two to three changes of distilled water, counterstain if possible and mount section. Since BCIP/NBT substrate is insoluble in alcohol or xylene, permanent mounting medium can be used. BCIP/NBT, AEC and DAB (3,3'-diaminobenzidine), tetrahydrochloride can be bought from BioGenex, Sigma Chemicals, or Polysciences, Inc. Store these reagents between 0 and 4°C. AEC and DAB, like other benzidine-containing compounds, are carcinogenic. BCIP/NBT, AEC, and DAB can also cause skin irritation upon contact and should be handled with appropriate care. Gloves need to be used for BCIP/NBT, AEC, and DAB staining. If there is contact, flush immediately with copious amounts of water. Do not pour AEC or DAB solutions down the drain. Contact your safety laboratory department for appropriate disposal.

10. Crystal/Mount (Biomeda Corp) is an aqueous-based, mounting medium designed especially for the permanent preservation of immunoperoxidase and alkaline phosphatase stained tissue section. (1) Place all slides to be mounted on a horizontal surface and apply three drops of Crystal/Mount to the tissue sections. (2) Rotate the slides covered by the Crystal/Mount so that it spreads to cover an area approximately the size of a quarter. Do not apply a coverslip on top of the Crysal/Mount. (3) Place the slides horizontally in an oven set at 70–80°C for at least 10 min. (4) Remove the slides from the oven and allow them to reach room temperature. The slides are now ready for microscopic visualization.

Gel/Mount (Biomeda Corp) is an aqueous mounting medium designed for preserving the fluorescence in the tissue sections. (1) Rinse stained tissue sections with distilled water. (2) Remove excess water around the slides. (3) Place slides on a horizontal surface and apply two drops of Gel/Mount on the tissue section. (4) Place coverslip carefully on the top of the mounting fluid. Allow mounted slides to air dry overnight before placing them in a permanent storage box.

Acknowledgments

The authors wish to thank Ms. Joanne Lopez for her secretarial assistance in preparation of the manuscript.

References

1. Prusiner, S. B., Scott, M., Foster, D., et al. (1990) Transgenetic studies implicate interactions between homologous PrP isoforms in scrapie prion replication. *Cell* **63,** 673–686.
2. Collinge, J., Palmer, M. S., Sidle, K. C. L., et al. (1995) Unaltered susceptibility to BSE in transgenic mice expressing human prion protein. *Nature* **378,** 21–28.
3. Vilotte, J. L., Soulier, S., Essalmani, R., et al. (2001) Markedly increased susceptibility to natural sheep scrapie of transgenic mice expressing ovine PrP. *J. Virol.* **75,** 5977–5984.
4. Chiesa, R., Piccardo, P., Ghetti, B., and Harris, D. A. (1998) Neurological illness in transgenic mice expressing a prion protein with an insertional mutation. *Neuron* **21,** 1339–1351.
5. Raeber, A. J., Brandnar, S., Klein, M. A., et al. (1998) Transgenic and knockout mice in research on prion diseases. *Brain Pathol.* **8,** 715–733.
6. Oldstone, M. B., Race, R., Thomas, D., et al. (2002) Lymphotoxin-α- and lymphotoxin-β-deficient mice differ in susceptibility to scrapie: evidence against dendritic cell involvement in neuroinvasion. *J. Virol.* **76,** 4357–4363.
7. Prinz, M., Heikenwalder, M., Schwarz, P., Takeda, K., Akira, S., and Aguzzi, A. (2003) Prion pathogenesis in the absence of Toll-like receptor signaling. *EMBO Reports* **4,** 195–199.
8. Outram, G. W. (1976) The pathogenesis of scrapie in mice, in *Slow Virus Diseases of Animals and Man* (Kimberlin, R. H., ed.), Elsevier, North Holland, Amsterdam, pp. 325–358.
9. Kimberlin, R. H. and Walker, C. A. (1979) Pathogenesis of mouse scrapie: dynamics of agent replication in spleen, spinal cord and brain after infection by different routes. *J. Comp. Pathol.* **89,** 551–562.
10. Bruce, M. E. and Fraser, H. (1981) Effect of route of injection on the frequency and distribution of cerebral amyloid plaques in scrapie mice. *Neuropathol. Appl. Neurobiol.* **7,** 289–298.
11. Fraser, H. (1979) Neuropathology of scrapie: the precision of the lesions and their diversity, in *Slow Transmissible Diseases of the Central Nervous System*, Vol. 1 (Prusiner, S. B. and Hadlow, W. J., eds.), Academic Press, NY, pp. 387–406.
12. Fraser, H. and Bruce, M. E. (1973) Argyrophilic plaques in mice inoculated with scrapie from particular sources. *Lancet* **I,** 617–618.

13. Wisniewski, H. M., Bruce, M. E., and Fraser, H. (1975) Infectious etiology of neuritic (senile) plaques in mice. *Science* **190,** 1108–1110.

14. Kim, Y. S., Carp, R. I., Callahan, S. M., Natelli, M., and Wisniewski, H. M. (1990) Vacuolization, incubation period and survival time in three mouse genotypes injected stereotactically in three brain regions with the 22L scrapie strain. *J. Neuropathol. Exp. Neurol.* **49,** 106–113.

15. Rubenstein, R., Merz, P. A., Kascsak, R. J., et al. (1991) Scrapie-infected spleens: analysis of infectivity, scrapie-associated fibrils and protease-resistant proteins. *J. Infect. Dis.* **164,** 24–35.

16. Eklund, C. M., Kennedy, R. C., and Hadlow, W. J. (1967) Pathogenesis of scrapie virus infection in the mouse. *J. Infect. Dis.* **117,** 15–22.

17. Carp, R. I., Kim, Y. S., and Callahan, S. M. (1989) Scrapie-induced alterations in glucose tolerance in mice. *J. Gen. Virol.* **70,** 827–835.

18. Carp, R. I., Callahan, S. M., Sersen, E. A., and Moretz, R. C. (1984) Preclinical changes in weight of scrapie-infected mice as a function of scrapie agent-mouse strain combination. *Intervirol.* **21,** 61–69.

19. Dickinson, A. G. and Meikle, V. M. (1969) A comparison of some biological characteristics of the mouse-passaged scrapie agents, 22A and ME7. *Genet. Res.* **13,** 213–225.

20. Dickinson, A. G., Meikle, V. M. H., and Fraser, H. (1968) Identification of a gene which controls the incubation period of some strains of scrapie agent in mice. *J. Comp. Pathol.* **78,** 293–299.

21. Carlson, G. A., Kingsbury, D. T., Goodman, P. A., et al. (1986) Linkage of prion protein and scrapie incubation time genes. *Cell* **46,** 503–511.

22. Hunter, N. J., Hope, J., McConnell, I., and Dickinson, A. G. (1987) Linkage of the scrapie-associated fibril protein (*PrP*) gene and Sinc using congenic mice and restriction fragment length polymorphism analysis. *J. Gen. Virol.* **68,** 2711–2716.

23. Dickinson, A. G. and Fraser, H. (1979) The assessment of the genetics of scrapie in sheep and mice, in *Slow Transmissible Diseases of the Central Nervous System*, Vol. 1 (Prusiner, S. B. and Hadlow, W. J., eds.), Academic Press, NY, pp. 13–32.

24. Carp, R. I., Moretz, R. C., Natelli, M., and Dickinson, A. G. (1987) Genetic control of scrapie: incubation period and plaque formation in I mice. *J. Gen. Virol.* **68,** 401–407.

25. Vaughn, J. E., Barber, R. P., Ribak, C. E., and Houser, C. R. (1981) Methods for the immunocytochemical localization of protein and peptide involved in neurotransmission, in *Current Trends in Morphological Techniques* (Johnson, J. E., ed.), CRC Press, Boca Raton, FL, pp. 33–70.

26. Fraser, H. and Dickinson, A. G. (1970) Pathogenesis of scrapie in the mouse: the role of the spleen. *Nature* **226,** 462–463.

27. Clarke, M. C. and Haig, D. A. (1971) Multiplication of scrapie agent in mouse spleen. *Res. Vet. Sci.* **12,** 195–197.

28. Aucouturier, P., Geissman, F., Damotte, D., et al. (2002) Infected splenic dendritic cells are sufficient for prion transmission to the CNS in mouse scrapie. *J. Clin. Invest.* **108,** 703–708.

29. Eklund, C. M., Hadlow, W. J., and Kennedy, R. C. (1963) Some properties of the scrapie agent and its behavior in mice. *Proc. Soc. Exp. Biol. Med.* **112,** 974–979.
30. Carp, R. I. (1982) Transmission of scrapie by the oral route: effect of gingival scarification. *Lancet* **1,** 170–171.
31. Thackray, A. M., Klein, M. A., and Bujdoso, R. (2003) Subclinical prion disease induced by oral inoculation. *J. Virol.* **77,** 7991–7998.
32. Glatzel, M. and Aguzzi, A. (2000) Peripheral pathogenesis of prion diseases. *Microbes Infect.* **2,** 613–619.
33. Baier, M., Norley, S., Schultz, J., Burminkel, M. K., Schwarz, A., and Reimer, C. (2003) Prion diseases: infectious and lethal doses following oral challenges. *J. Gen. Virol.* **84,** 1927–1929.

25

Radiolabeling of Amyloid-β Peptides

Miguel Calero and Jorge Ghiso

Summary

Nowadays, a wide variety of protocols for labeling proteins is available. However, radiolabeling remains one of the most powerful, sensitive and accurate methods to trace and quantitate proteins. Additionally, radiolabeling techniques are steadily gaining importance for diagnosis and treatment in nuclear medicine. There is a considerable number of radioisotopes, but only some are commonly used for basic biomedical research. Among them, the iodine radioisotopes (γ-emitters) have several advantages for the labeling of proteins. This chapter focuses on radioiodination protocols for amyloidogenic peptides, using the Aβ peptides as a paradigm. The chloramine T, Iodo-Gen®, and lactoperoxidase methods can be successfully applied to radioiodination of different amyloid peptides as long as free tyrosyl (or histidyl) groups are available. However, these methods differ in their yield and the degree of oxidative damage conferred to labile peptides. When no tyrosines are available, the Bolton-Hunter methodology can be used. The labeling by the tyramine-cellobiose ligand trapping method is applicable to the study of cellular uptake and catabolism of amyloid peptides.

Key Words: Amyloids; radiolabeling; chloramine T; lactoperoxidase; iodination; Bolton-Hunter; Iodo-Gen; oxidation; tyramine-cellobiose; ligand-trapping.

1. Introduction

The process of radioactive decay is a consequence of the instability of some combinations of protons and neutrons in the atomic nucleus. This nuclear instability manifests as radioactivity by spontaneously converting part of its mass into energy, emitted as particles and electromagnetic radiation. This process is spontaneous and the time when a particular atom will decay cannot be predicted. However, when a large number of radioactive atoms is considered, the fraction of atoms that will decay in a given time span (decay rate) can be accurately estimated as the half-life of the radionuclide. During radioactive decay, principles of conservation of energy, linear and angular momentum, charge

From: *Methods in Molecular Biology, vol. 299: Amyloid Proteins: Methods and Protocols*
Edited by: E. M. Sigurdsson © Humana Press Inc., Totowa, NJ

and nucleon number apply. The emission of radiation by a radionuclide is determine by the type, energy and intensity of the radiation. The major types of radiation are α particles, β particles, and γ rays (*see* **Note 1**). Radioactivity is generally measured in Becquerels (Bq) or Curies (Ci) (*see* **Note 2**). The specific activity of a radioisotope or radiolabeled molecule corresponds to the radioactivity of the material per unit of mass (*see* **Note 3**).

The radiolabels most commonly used for biomedical research are either γ rays or β-particle emitters, namely iodine isotopes (^{125}I, ^{131}I, ^{123}I), carbon-14 (^{14}C), phosphorous-32 (^{32}P), sulfur-35 (^{35}S), and tritium (^{3}H) (*see* **Table 1**). ^{14}C and ^{3}H are low-energy β-particle emitters, whereas ^{35}S is a medium-energy, and ^{32}P is a high-energy β emitter. Diverse biological processes have been studied using radioisotopes. Thus, labeled amino acids, nucleotides, sugars, and different ligands have been used for the experimental analysis of cell proliferation and toxicity, metabolic studies, protein modifications (protein phosphorylation [^{32}P and ^{33}P], acetylation and methylation [^{3}H-*S*-adenosyl methionine]), ligand receptor interactions (^{125}I, ^{3}H), radioimmunoassays (^{125}I), and enzymatic assays (^{3}H,^{14}C), among others *(1–6)*. With the exception of research focused on the effects of radiation on living beings, one of the most important advantages for working with radioisotopes is based on the fact that radioactive materials seldom change their other physical, chemical, or biological properties (*see* **Note 4**).

Within the amyloid field, studies based on metabolic labeling with ^{35}S-methionine have provided some of the most important data on support of the prion hypothesis, by demonstrating the ability of PrP^{res} (prion protein resistant to proteinase K digestion) to induce the conversion of newly synthesized radiolabeled PrP^{sen} (PrP sensible to proteinase K digestion) *(7,8)*. In addition, radioisotopes have been widely used for the characterization of in vivo and in vitro protein amyloid deposition *(9,10)*, as well as for the study of catabolism and clearance of different amyloid peptides *(11,12)*.

1.1. Labeling Strategies

Labeled proteins or peptides suitable for different specifications may be produced by a variety of strategies. Cell-free expression systems and biosynthetic labeling of proteins by pulse-chase or short-term metabolic labeling of cells are techniques commonly used in the study of biochemical properties, synthesis, processing, intracellular transport, secretion, and degradation of proteins. These procedures allow the incorporation of ^{35}S-Cys or ^{35}S-Met, or less commonly tritiated forms of the rest of amino acids to newly synthesized proteins. Chemical synthesis with labeled amino acids (^{3}H, ^{14}C, and ^{35}S) is an effective (although expensive) alternative specially suited for short peptides.

Table 1
Physical Characteristics of Commonly Used Radionuclides[a]

Radionuclide	Atomic mass	Half life	Type of emission	Max decay energy (MeV)	Max range of emission (cm)[b]	Resulting stable nuclide	Target organ
$_1H^3$ (tritium)	3.01605	12.43 yr	β	0.019	0.42 (air)	$_2He^3$	whole body
$_6C^{14}$ (carbon-14)	14.00324	5730 yr	β	0.156	21.8 (air)	$_7N^{14}$	bone, fat
$_{15}P^{32}$ (phosphorous-32)	31.97391	14.2 d	β	1.711	610 (air) 0.8 (water) 0.76 (Plexiglas)	$_{16}S^{33}$	bone
$_{15}P^{33}$ (phosphorous-33)	32.97173	25.4 d	β	0.249	49 (air)	$_{16}S^{33}$	bone
$_{16}S^{35}$ (sulfur-35)	34.96903	87.4 d	β	0.167	24.4 (air)	$_{17}Cl^{35}$	testes
$_{37}Rb^{86}$ (Rubidium-86)	85.91117	18.7 d	β	1.774	640 (air) 0.8 (water) 0.9 (lead)	$_{38}Sr^{86}$	Bone (25%) and rest of the body (75%)
$_{39}Y^{90}$ Yttrium-90	89.90715	64.1 h	γ β	1.077 2.28	900 (air) 1.1 (water)	$_{40}Zr^{90}$	Bone (50%), liver (15%)
$_{43}Tc^{99m}$ Technetium-99	98.90625	2.13×10^5 yr	β	0.294	63 (air)	$_{44}Ru^{99}$	Whole body
$_{49}In^{111}$ (Indium-111)	110.90511	2.83 d	γ	0.245	0.02 (lead)	$_{48}Cd^{111}$	Spleen, red bone marrow, liver
$_{53}I^{123}$ (Iodine-123)	122.90560	13.27 d	β γ	1.242	0.04 (lead)	$_{52}Te^{123}$	thyroid
$_{53}I^{125}$ (Iodine-125)	124.90462	60 d	γ	0.27–0.035	0.02 (lead)	$_{52}Te^{125}$	thyroid
$_{53}I^{131}$ (Iodine-131)	130.90612	8.04 d	β γ	0.606 0.364	165 (air) 2.4 (lead)	$_{54}Xe^{130}$	thyroid

[a]Adapted and modified from **ref. 2**. Other data are from "Guide to the safe handling of radioactive material in research. PerkinElmer Life Sciences (www.perkinelmer.com/lifesciences)."

[b]For γ radiation, these values correspond to the half-value layer for lead shielding.

Other methods for labeling proteins and peptides rely on the modification of the molecule either by direct binding of the radioisotope of interest, or through the attachment to a labeled intermediate molecule. Among the wide variety of radioisotopes now available, radioisotopes of iodine offer several advantages for easy and reliable labeling of proteins and peptides, namely: well-known chemistry, high specific activity, easy availability, and low costs. Moreover, these radioisotopes are γ emitters and can be counted directly in a γ counter without the need for sample preparation. There are three γ-emitting radioisotopes of iodine available that have distinct properties and applications (*see* **Table 1**). Several methods of radioiodination of proteins have been developed, but most rely on the oxidation of the iodine anion (I^-) to the reactive species I_2 or I^+ and the subsequent electrophilic attack of these species to the ortho-positions of the aromatic ring of tyrosine residues. These protocols for radioiodination will be the focus of the methods presented here (*see* **Table 2**).

1.2. Notes for the Safe Use of Radiosotopes

To work safely with radioactive materials, it is necessary to understand the potential hazards they pose. The danger associated with radioactivity is linked to the perturbation of biological processes as a consequence of formation of ions after the interaction of biological materials with α, β, or γ emissions. Limitation of the exposure of people who work with radioactive materials that emit penetrating radiation must be achieved by adjusting several parameters: (1) distance from the source, (2) duration of the exposure, and (3) density of the barrier (air, water, shielding material) between the individual and the source.

As the radiation falls off in proportion to the square of the distance, distance is very useful for protection when handling radioactive sources. Therefore, it is advisable to avoid direct handling of sources of penetrating radiation by using forceps, tongs, holders, and spacers to maintain distance between you and the source (*see* **Note 5**). The time of exposure can be greatly reduced by careful planning. It is of critical importance to (1) review prior to labeling the safety aspects of the operation in detail, (2) carry out trial runs with no radioactivity, and (3) design operations based on simple and feasible steps. Shielding from the radioactive source is also a very effective measure to reduce the exposure to the radiation. Dense materials such as lead should be used as barriers for γ radiation, while lighter materials like glass, water, or Plexiglas are good shielding materials for pure β-particle emitters. In order to minimize the exposure to radiation (1) calculate the shielding needs using half value layers and γ-ray constants (*see* **Table 1**), (2) check shielding in all directions accessible to personnel, and (3) do not look directly at the source, use transparent shields, mirrors, or periscopes instead.

Table 2
Reagents for Radioiodination of Proteins

Reagent (trivial name)	Scientific name Molecular mass	Chemical structure	Advantages	Disadvantages
Chloramine T[a]	sodium p-toluenesulfo-chloramide 228.66 Da.		Simple and, reproducible; High yield	Oxidative damage
Iodo-Gen®[a]	1,3,4,6-tetrachloro-3α,6α-diphenylglycoluril 432.09 Da.		Reaction at the liquid–solid interphase; Milder than Chloramine T	Adsorption to the reagent; Some degree of oxidative damage
Bovine Lactoperoxidase[a] (LPO)	Enzyme (hydrogen peroxide oxidoreductase): EC. 1.11.1.7 78500 Da.	Complex structure	Gentle and specific	Technically demanding Lactoperoxidase itself may be labeled

(continued)

329

Table 2 (Continued)

Reagent (trivial name)	Scientific name Molecular mass	Chemical structure	Advantages	Disadvantages
Bolton–Hunter Reagent[b]	Succinimidyl-3-[4-hydrophenyl]propionate 263.26 Da.		Effective for proteins with no tyrosyl residues.	Unpredictable site of labeling; May alter the structure and function of proteins
Tyramine-cellobiose[b]	N-(4-O-(β-D-Glucopyranosyl)-β-D-glucopyranosyl)-3-hydroxyphenylethylamine 463.49 Da		Intracellularly-trapped ligand; Appropriate for studies on cellular uptake and catabolism	Synthesis technically demanding; No commercial resource

[a]Catalyst molecules involved in the formation of iodine reactive species for radioiodination of proteins.
[b]Derivatizing reagents that modify proteins (mainly at N-terminal α-amino or lysine ε-amino residues). They have to be labeled by oxidative iodination.

This chapter is focused on protocols for labeling amyloid proteins with iodine radioisotopes. Therefore, specific precautions for other radioisotopes are not addressed in detail. Some important data such as half-life, type of emission, energy, shielding requirements, target organs, etc. can be viewed in **Table 1**.

The specific hazards derived from the use of iodine radioisotopes (^{125}I, ^{131}I, and ^{123}I) come both from the direct exposure to the γ radiation and the inhalation of volatile free iodine (**Table 1**). Lead or lead-impregnated Plexiglas shields must be used for protection. During the labeling protocol, where mCi amounts are used, it is also necessary to wear a lead apron to protect most of the body, and a ring badge to measure exposure to the unshielded parts. Any experiment that involves the use of free, unbound iodine should be performed behind a shield in a chemical hood equipped with a charcoal trap to absorb volatile iodine. Inhaled iodine is concentrated in the thyroid (**Table 1**) and scans of the throat and neck area should be performed on a regular basis in accordance with the Radiation Safety Office and local rules at your institution.

Radioiodine isotopes (as all radioactive materials) have to be stored and clearly identified as radioactive by using appropriate labels. Na^{125}I solutions (or other iodine radioisotopes) should be stored at room temperature, since freezing results in subsequent volatilization of radioiodine. Microcurie (or higher) quantities of ^{125}I have to be stored in containers surrounded by at least 3-mm thick lead. Avoid keeping iodide solution in acidic conditions to minimize volatilization. Waste material has to be kept isolated in sealed, clearly labeled, and shielded containers (*see* **Note 6**).

Finally, it is important to remark that any use of radioisotopes presents inherent risks. Nowadays, a wide variety of labels for proteins and other ligands is available, including enzymes, biotin, and fluorophores. Therefore, the advantages and disadvantages of choosing radiolabeling for a particular experimental protocol must be carefully evaluated.

2. Materials

2.1. Peptide, Protein, or Amyloid Source (see Note 7)

1. Amyloid β$_{1–40}$ peptides (*see* **Note 8**).
2. Human Aβ$_{1–40}$ (*see* **Note 9**):
   DAEF**R**HDSGYEV**H**HQKLVFFAEDVGSNKGAIIGLMVGGVV.
3. Rat Aβ$_{1–40}$ (*see* **Note 10**):
   DAEF**G**HDSGF**EV**RHQKLVFFAEDVGSNKGAIIGLMVGGVV.

2.2. Carrier-Free Radioiodine Isotopes (see Note 11)

1. Na^{125}I, 185 MBq (5 mCi) (100 mCi/mL, Amersham Bioscience, Piscataway, NJ).

2.3. Non-Reusable Materials

1. Aerosol filter tips.
2. Yellow hazard tape and labels printed with the international symbol for radio-activity.
3. Disposable absorbent paper sheets.
4. Disposable labcoats and/or sleeves.
5. All-purpose radioactivity decontaminant (e.g., PerkinElmer NEN Count-Off).

2.4. Reagents

1. Chloramine T (sodium *p*-toluenesulfochloramide) (ICN Biomedicals Inc., Aurora, OH).
2. Sodium metabisulfite (Fischer Scientific, Pittsburgh, PA).
3. Iodo-Beads® (chloramine T reagent immobilized on nonporous polystyrene beads) (Pierce Chemical Co., Rockford, IL).
4. Iodo-Gen® pre-coated tubes (1,3,4,6-tetrachloro-3α,6α-diphenylglycoluril) (Pierce Chemical Co.).
5. Bolton-Hunter Reagent (Succinimidyl-3-[4-hydrophenyl]propionate) (Pierce Chemical Co.).
6. Bovine Lactoperoxidase (Sigma Chemical Co., St Louis, MO).
7. Trichloroacetic acid (Sigma Chemical Co.).
8. Trifluoroacetic Acid (Sigma Chemical Co.).
9. Acetonitrile (Merck, Darmstadt, Germany).
10. Sodium Cyanoborohydride (NaBH$_3$CN) (Sigma Chemical Co.).
11. Tyramine (3-hydroxyphenylethylamine) (Sigma Chemical Co.).
12. Cellobiose (4-O-(β-D-Glucopyranosyl)-β-D-glucopyranosyl) (Sigma Chemical Co.).

2.5. Dedicated Equipments and Materials

1. Pipets.
2. Microfuge.
3. Chemical hood equipped with a charcoal filter.
4. HPLC system and C4-reverse-phase column (e.g., no. 214TP52, 250 × 2.1 mm size; 5-μm diameter beads; 300 Å pore size, Vydac The Separations Group, Hesperia, CA).

2.6. Installations

1. "Hot" lab or dedicated space.
2. Storing facilities.

2.7. Safety Equipments and Materials

1. Dosimeter (*see* **Note 12**).
2. Appropriate hand-held radioactivity monitor (*see* **Note 13**).
3. Ring-badge monitor.

4. Protective eyewear.
5. Shielded residue containers.
6. Lead-impregnated Plexiglas transparent shields and boxes (*see* **Note 14**), lead apron.

3. Methods

3.1. Radioiodination of Amyloid Peptides

Different methods for chemical or enzymatic radioiodination of proteins have been developed, but almost all of them rely on the oxidation of the radio-iodide anions (I^-) to the reactive iodous ions I^+ (or ICl), and subsequent electrophilic attack to the ortho positions to the hydroxyl group of the aromatic ring of tyrosine residues *(13)* (*see* **Fig. 1**). The efficiency of the methods for chemical iodination is optimal at pH values between 6.0 and 8.5. At pH 5.0 the iodination is negligible. However, at pH >9.0 the imidazole ring of histidine can be also labeled *(14)*.

Alternative protocols, such as the Bolton-Hunter method *(15)*, that involve the modification of other residues, have to be used when tyrosine residues are not available or their modification is not convenient.

The specific conditions for the production of good quality radiolabeled products at high yields have to be established by experimental approaches, although some basic rules, based on the chemistry of the reaction may be followed. Thus, buffers containing tyrosine, iodide, and reducing agents should be avoided for direct iodination of proteins. Most of these methods are feasible within a relatively wide range of temperatures (0–37°C) and pHs (6.0–8.5). Therefore, the labeling method and the experimental conditions should be chosen to provide a buffer system and temperature that are compatible with the specific sample and biological system. Even under the most suitable conditions, labile proteins may undergo unwanted modifications, such as oxidation of sensitive residues. In this case, adequate purification protocols are required to produce radiolabeled proteins of good quality and high specific activity (*see* **Subheading 3.3.**).

Keeping in mind that most of the amyloidogenic peptides and proteins are prone to aggregation and conformational change, special care should be taken to avoid conditions that adversely affect the solubility and/or conformational stability of the polypeptide chain *(16)*. Here, we present some of the most commonly used protocols for the radioiodination of peptides and proteins by using amyloid β peptides as a paradigm for other amyloidogenic peptides. Any of the radioiodine isotopes (^{125}I, ^{131}I, and ^{123}I) can be used in these labeling protocols.

Because radioactive material is involved, it is very important to carefully lay out the experiments prior to labeling. All the reagents and equipment should be ready for use.

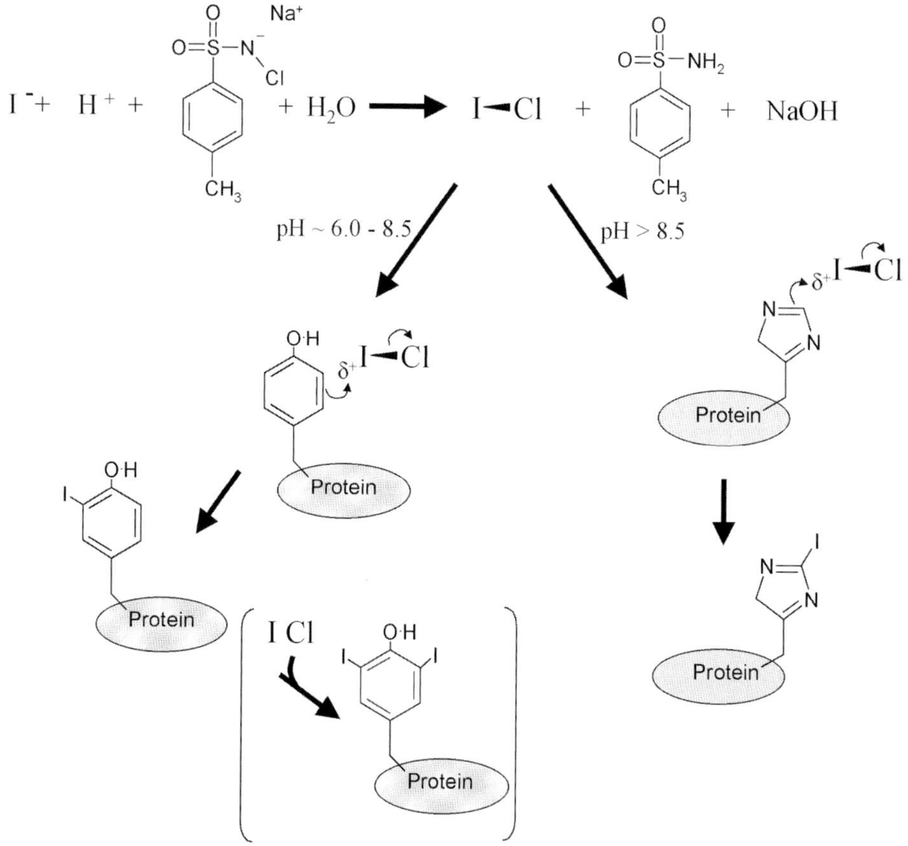

Fig. 1. Protein labeling by the chloramine T method. Molecular mechanism of protein labeling by the chloramine T method of tyrosyl residues at pH 6.0–8.5 (left) and hystidyl residues at pH >8.5 (right). Oxidation of iodide ions (I^-) by the chloramine T generates reactive iodine species (ICl) that mediates the electrophilic attack to the aromatic groups of tyrosine or histidine.

3.1.1. The Chloramine T Method
for Direct Radioiodination of Amyloid Peptides

The chloramine T method was developed by Hunter and Greenwood in 1962 *(3)*, and is the protocol most widely used for protein radioiodination. This method requires a tyrosine residue for labeling, and its main advantages are its simplicity, reproducibility, and good yield *(3,17)*. The major disadvantage is that the protein is directly exposed to strong oxidizing conditions, and therefore oxidation-sensitive proteins can be readily inactivated or denatured, mainly at methionine residues (*see* **Note 15**).

Procedure (*see* **Note 16**):

1. Run a high-performance liquid chromatography (HPLC) blank gradient and equilibrate the system at the initial conditions according to **Subheading 3.3.** or other appropriate conditions.
2. Freshly prepare chloramine T (1 mg/mL) (*see* **Note 17**) and sodium metabisulfite (2.5 mg/mL) solutions in milli-Q water.
3. Dissolve 10–50 μg of Aβ$_{1–40}$ peptide (*see* **Note 18**) in 50 μL of H$_2$O (*see* **Note 19**). Once the peptide is completely dissolved, add 15 μL of 1 *M* phosphate buffer, pH 7.4 (*see* **Notes 20** and **21**). Transfer to a 1.5-mL screw-cap tube. Add 10 μL of Na^{125}I (1 mCi) (*see* **Note 22**). Adjust up to 130 μL total with H$_2$O (*see* **Note 23**). For example:
   - 50 μL Aβ40 solution (0.2–1 μg/μL);
   - 15 μL phosphate buffer 1 *M*, pH 7.4;
   - 55 μL milli-Q H$_2$O;
   - 10 μL Na^{125}I (1 mCi).
4. Initiate the iodination reaction by adding 10 μL of chloramine T solution (1 mg/mL), mix by pipetting up and down, and incubate for one minute (*see* **Note 24**) at room temperature (*see* **Note 25**).
5. Stop the reaction by adding and mixing 10 μL of sodium metabisulfite (2.5 mg/mL). The reaction mix (150 μL of total volume) is then ready for purification by reverse-phase HPLC or other convenient method (*see* **Note 26** and **Subheading 3.3.**).
6. After purification, the radioiodinated peptide can be kept prior to use at –20°C for up to 1 mo (*see* **Note 27**) in a well labeled 3-mm thick lead container.

3.1.2. The Iodo-Gen® Method for Direct Radioiodination of Amyloid Peptides

Iodo-Gen iodination reagent (1,3,4,6-tetrachloro-3α,6α-diphenylglycoluril) was first described as a reagent for the iodination of proteins and cell membranes *(18)*. This reagent appears to be as effective as enzymatic methods for iodination of externally exposed residues and as effective as chloramine T for general protein iodination, but it is milder and may therefore result in less oxidative damage to the protein or peptide *(14,18–20)*. The Iodo-Gen reagent can be supplied as a dry powder for coating your own reaction vessels. However, that procedure involves organic solvent and is cumbersome and may give variable results. Therefore, unless otherwise justified, it may be more convenient to purchase already Iodo-Gen-coated tubes commercially available from Pierce.

The protocol presented here is for direct iodination of amyloid β peptides, according to described methods *(20)*. An alternative protocol, where the ^{125}I is pre-activated directly in the Iodo-Gen tube before it is added to the peptide, potentially eliminates the oxidative damage to labile proteins, and prevents losses from nonspecific binding to the reagent-coated vessel. However, lower yields are normally obtained.

1. Run a HPLC blank gradient and equilibrate the system at the initial conditions according to **Subheading 3.3.** or other appropriate conditions.
2. Wash an Iodo-Gen (10 µg) precoated tube with 1 mL of 0.2 M phosphate buffer, pH 7.4 (*see* **Note 21** and **Note 28**).
3. Dissolve 10–50 µg of Aβ$_{1-40}$ peptide in 120 µL of milli-Q water (*see* **Note 19**). Once the peptide is completely dissolved, and add 15 µL of phosphate buffer 1 M, pH 7.4. Transfer the peptide solution to an Iodo-Gen precoated tube.
4. Immediately add 10 µL of *cold* Na^{125}I (1 mCi) and allow the reaction to proceed for 20 min, with occasional agitation.
5. Stop the reaction by removing the sample from the reaction vessel and add 15 µL of 10 mM NaI, mix and incubate for 1 min. The reaction mixture (150 µL of total volume) is then ready for purification by reverse-phase HPLC or other convenient method (*see* **Subheading 3.3.**).
6. After purification, the radioiodinated peptide can be kept prior to use at –20°C for up to 1 mo (*see* **Note 27**) in a well labeled 3-mm thick lead container.

3.1.3. Lactoperoxidase Method for Enzymatic Radioiodination of Amyloid Peptides

The lactoperoxidase method uses bovine lactoperoxidase in the presence of hydrogen peroxide as an oxidative reagent to specifically produce reactive species that iodinate the tyrosine residues *(21)*. The main advantage of this method is that the protein is subjected to milder conditions compared to the chloramine T or similar methods. Therefore, there is less chance of denaturation and/or oxidation of proteins or peptides *(21–23)*. However, this method is technically demanding and more difficult to optimize.

1. Run a HPLC blank gradient and equilibrate the system at the initial conditions according to **Subheading 3.3.** or other appropriate conditions.
2. Dissolve 10–20 µg of Aβ$_{1-40}$ peptide in 40 µL of milli-Q water (*see* **Note 19**). Once the peptide is completely dissolved, transfer the solution to a 1.5-mL screw-cap tube. Sequentially, add 20 µL of Na^{125}I (2 mCi), 5 µL of lactoperoxidase solution (20 µg/mL) (*see* **Note 29**), and 7 µL of sodium acetate buffer 1 M, pH 5.6 (*see* **Note 30**).
3. Initiate the reaction by adding 10 µL of hydrogen peroxide solution (10 µg/mL) (*see* **Note 31**) and allow the reaction to proceed for 30 min, with occasional agitation.
4. Stop the reaction by adding 100 µL of sodium azide (1% in 0.2 M phosphate buffer, pH 7.4) and incubate for 5 min (*see* **Note 32**).
5. Proceed immediately to the purification step (*see* **Note 33**) or freeze-down the reaction until purification is performed by reverse-phase HPLC or other convenient method (*see* **Subheading 3.3.**).
6. After purification the radioiodinated peptide can be kept prior to use at –20°C for up to 1 mo (*see* **Note 27**) in a well labeled 3-mm thick lead container.

3.1.4. The Bolton-Hunter Method
for Labeling of Amyloid Peptides Devoid of Tyrosine Residues

The Bolton-Hunter method *(15)* for radioiodination of proteins is the procedure of choice when tyrosines are not present or accessible. This method modifies by acylation the N-terminal amino acid and the ε amino groups of lysines, thereby removing the charge from these residues. This modification may alter the functionality and/or structure of the protein *(15,24)* (*see* **Note 4** and **Subheading 2.**).

The following protocol describes iodination by the Bolton-Hunter method of the rat Aβ peptide, which does not contain tyrosyl residues (*see* **Note 10**). In this procedure, the reagent succinimidyl-3-[4-hydrophenyl]propionate is labeled with ^{125}I, then separated from the products of the reaction, and subsequently attached to the primary amines of the peptide to be iodinated.

Labeling of the Bolton-Hunter reagent (*see* Note 34)

1. Freshly dissolve 5 mg of Bolton-Hunter reagent in 1 mL of dimethyl sulfoxide (DMSO) (5 mg/mL).
2. Freshly prepare chloramine T (2 mg/mL) and sodium metabisulfite (12 mg/mL) solutions in 0.25 *M* phosphate buffer, pH 7.5.
3. Add 20 μL of Na^{125}I (2 mCi) carrier-free solution into a screw cap tube.
4. Add to the tube 2 μL (*see* **Note 35**) of the Bolton-Hunter reagent solution and immediately 10 μL of chloramine T solution. Incubate for 15 s with continuous agitation.
5. Stop the reaction by adding 10 μL of bisulfite solution and 10 μL of 10 m*M* carrier cold sodium iodide and incubate for 5 min.
6. Add 5 μL of dimethyl formamide (DMF) to the reaction mixture.
7. Extract the ^{125}I-labeled Bolton-Hunter reagent using 250 μL of redistilled benzene.
8. Remove the aqueous phase and distribute the organic phase containing the ^{125}I-labeled Bolton-Hunter reagent into one or several glass tubes depending on the number of reactions that will be performed.
9. Remove the organic phase by evaporating the benzene with a gentle stream of nitrogen in a chemical hood equipped with charcoal filter.
10. Keep the reagent at −20°C as a dried film in the tubes in a moisture-free environment until used.

Attachment of the ^{125}I-Bolton-Hunter reagent:

1. Dissolve 10–20 μg of Aβ$_{1-40}$ peptide in 40 μL of milli-Q water (*see* **Note 19**). Once the peptide is completely dissolved, add 10 μL of 0.5 *M* sodium borate buffer, pH 8.5. Place the solution in a water/ice bath for 5 min.
2. Transfer the peptide solution to a pre-cooled ^{125}I-Bolton-Hunter reagent coated tube, and allow to react for 30 min with gentle agitation.

3. Add 50 µL of 1 *M* glycine in 0.1 *M* sodium borate buffer, pH 8.5 and incubate for 5 min.
4. Proceed immediately to the purification step or freeze-down the reaction until purification is performed by reverse-phase HPLC or other convenient method (*see* **Subheading 3.3.**).
5. After purification, the radioiodinated peptide can be kept prior to use at –20°C for up to 1 mo (*see* **Note 27**) in a well labeled 3-mm thick lead container.

3.2. The Tyramine-Cellobiose Ligand Trapping Method for the Study of Clearance and Catabolism of Amyloid Peptides

Direct labeling of tyrosine residues of peptides or proteins with iodine isotopes is of limited use for the study of tissue uptake and catabolism, mainly because the label is rapidly excreted from the cells after internalization and intralysosomal proteolytic digestion. This process is mediated by hepatic deiodases that rapidly deiodinate mono- and di-iodotyrosine. However, these physiological deiodases cannot dehalogenate the imidazole ring of histidine. Therefore, for catabolism studies, selective radioiodination of histidyl residues (*see* **Note 21** and **Fig. 1**) may be advantageous compared to radioiodination of tyrosine residues *(13)*.

Alternatively, residualizing forms of radioiodinable labels have been developed, e.g., di-lactitol-tyramine, inulin-tyramine, or tyramine-cellobiose *(25–28)*. Among them, the tyramine-cellobiose adduct has been widely used to determine sites of catabolism of many proteins, including some related to amyloid diseases, i.e., transthyretin (TTR) *(11)* or serum amyloid P-component (SAP) *(12)*. The usefulness of this procedure relies on that the ^{125}I-tyramine-cellobiose adducts cannot be degraded by mammalian cells. These adducts remain trapped intracellularly and, therefore, accumulate in the organs involved in the uptake *(11,12,25)*.

The following procedure is based on the work of Pittman and collaborators, 1983 *(25)* and it is adapted for the labeling of amyloid β peptides by scaling down 50 times all the quantities of reagents and volumes used (for a detailed scheme *see* **Fig. 2**). This type of labeling protocol is valid for the study of clearance, organ uptake, and catabolism, provided that the labeled peptides behaves similarly to the unmodified native peptide. Therefore, ^{125}I-tyramine-cellobiose-peptide and directly ^{125}I-labeled peptide should be studied to assess whether both preparations have similar clearance kinetics in vivo.

Synthesis of the tyramine-cellobiose adduct (*see* Note 36)

1. To a small glass reaction vessel, add 2 mL of 0.2 *M* sodium phosphate buffer, pH 7.5.
2. Add 240 µ*M* of each of the following reagents: 83 mg of Cellobiose (*see* **Note 37**), 33 mg of tyramine, and 16 mg of NaBH$_3$CN. Allow the mixture to react for 6 d at room temperature with continuous stirring (*see* **Note 38**).

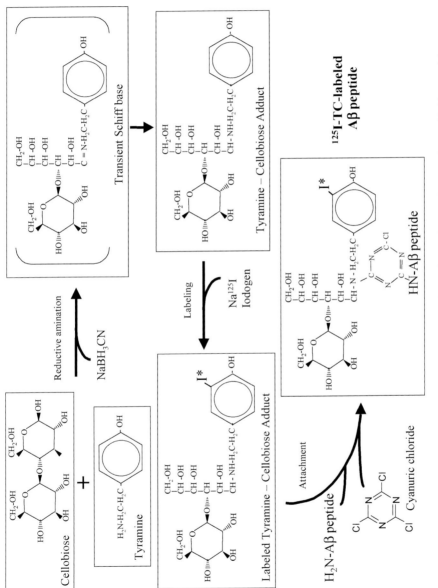

Fig. 2. Chemical reactions involved in the labeling of proteins with tyramine-cellobiose adduct.

339

3. After incubation, adjust the pH of the solution to 5.5 with HCl (*see* **Note 39**) and load the reaction mixture onto a cation-exchange column (0.6 × 18 cm AG-50W, Bio-Rad). Wash the column with 10 volumes of water and elute with 0.5 M NH$_4$OH and freeze-dry the eluted material.
4. Further purify the adduct by silicic acid chromatography on a 0.7 × 26 cm column eluted with butanol/acetic acid/water mix (7:1:2 by vol). The tyramine-cellobiose adduct elutes after the free tyramine at the end of the column volume. Freeze-dry the purified adduct and keep at −20°C until used.

Labeling of the tyramine-cellobiose adduct and attachment to the protein:

1. Weight a small aliquot of the tyramine-cellobiose adduct and dissolve it at a concentration of 0.1 µg/µL in milli-Q water. Transfer 48 µL of adduct solution (4.8 µg, ~10 nmol) to an Iodo-Gen (10 µg)-coated tube. Add 30 µL of Na^{125}I (3 mCi) and allow to react for 30 min at room temperature.
2. Stop the reaction by transferring the solution to a fresh tube containing sodium metabisulfite (10 µL of 0.1 M) and NaI (5 µL of 0.3 M).
3. After labeling, activate the ^{125}I-tyramine-cellobiose adduct by adding 20 µL of cyanuric chloride (0.1 mg/mL in acetone) (*see* **Note 40**) and 5 µL of sodium hydroxide (4 mM). Incubate for 20 s. After activation, quench the reaction by the addition of 3 µL of acetic acid (10 mM).
4. Immediately after activation, attach the activated ^{125}I-TC adduct to the Aβ peptide by adding the ligand to a tube containing 50 µg of the peptide in 50 µL of 0.1 M phosphate buffer, pH 7.5. Allow the reaction to proceed for 1 h at room temperature.
5. Proceed immediately to the purification step or freeze-down the reaction until purification is performed by reverse-phase HPLC, or other convenient method (*see* **Subheading 3.3.**).
6. After purification, the radioiodinated tyramine-cellobiose-peptide adduct can be kept prior to use at −20°C for up to 1 mo (*see* **Note 27**) in a well labeled 3-mm thick lead container.

3.3. Purification of Labeled Aβ Peptides

After completion of the labeling protocol, the reaction mixture contains the labeled protein, together with unlabeled protein, radioiodide, salts, and possibly other components such as reducing agents, carrier proteins, enzymes, etc. For most uses, the labeled protein has to be purified from the reaction mix.

A number of techniques are available for peptide and protein purification. Gel filtration is likely the most widely used of all separation methods following labeling. This is a fast and convenient method to remove unbound radioiodine, as well as inorganic and small organic molecules. It can be accomplished by using home-made disposable columns packed with appropriate resins, such as Sephadex G-25 (Pharmacia). However, this technique does not allow the separation of labeled from unlabeled peptides or proteins, and therefore in most cases, it is unsuitable to obtain labeled material with very high specific activity. Also, this procedure does not allow separation of modified products such as oxidized Aβ peptide. Because the modified protein may have different properties, it is mandatory to use purification methods of higher resolution, capable of separating the different molecular species of the protein.

In the amyloid β peptides used as example in these protocols, the methionine at position 35 (*see* **Fig. 2**) is highly susceptible to oxidation during the exposure to the iodination condition used in most protocols. In our experience and others, oxidation of the peptide modifies its properties (aggregation, neurotoxicity, binding to other proteins, etc). Therefore, it is of primary importance not only to separate the labeled peptide from reagents and unlabeled peptide, but also to separate it from oxidatively damaged forms. This may be achieved by RP-HPLC separation, which has the capacity to resolve single species of the peptide based on the different hydrophobicity of the molecules. The following protocol describes the separation of and purification of amyloid β peptides in a RP-HPLC system.

The actual separation profile for your specific protocol may vary depending on a number of variables, including type of column, solvents, flow, gradient, temperature, etc. Therefore, it is important to establish the best separation conditions for your particular protein before proceeding with the iodination procedure.

A convenient reference for evaluating your separation protocol may come from a "cold" labeling assay using [127]I (non-radioactive) by the chloramine T method. By using equimolar amounts of sodium iodide and peptide, you can expect to observe a mix of species corresponding to the non-iodinated, mono-iodinated and di-iodinated species (or higher depending on the number of tyrosine present) of the peptide, as well as their oxidized counterparts (if any) (**Fig. 3**). This HPLC profile will serve as a guideline for the "hot" experiment, where smaller amounts of radioactive iodine will be used, and therefore the position of the iodinated species will be less apparent. This procedure can be applied regardless of the method employed for the labeling, although the actual profile will vary depending on the condition employed and the residues modified.

Procedure (*see* **Notes 41** and **42**):

1. Before starting the labeling procedure, program the HPLC gradient according to the conditions below (*see* **Note 43**), set the UV detector at 220 nm, run a blank, and equilibrate the HPLC at the initial conditions.
   Phase A: H_2O (0.1% TFA); phase B: acetonitrile (0.1% TFA); flow: 200 μL/min; temperature: 30°C
   Gradient:

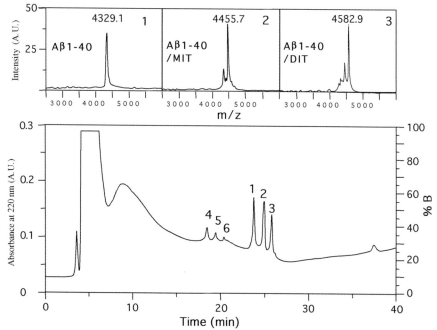

Fig. 3. Separation of Aβ1-40 labeled peptides. An aliquot of 20 μg (~4.5 nmol) of the Aβ1-40 peptide was labeled with 0.6 μg (4 nmol) of cold Na^{127}I with 3 iodobeads for 20 min. The iodinaton mixture was separated under the conditions described in **Subheading 3.3**. The peaks 1, 2, and 3 were analyzed by MALDI-TOF mass spectrometry and found to correspond to the non-iodinated, mono-iodinated (MIT) and diiodinated (DIT) forms of Aβ non-oxidized. Peaks 4, 5, and 6 were identified as the oxidized forms of the same species.

Time (min)	%A	%B
0	75	25
5	75	25
40	60	40
45	20	80
50	20	80
51	75	25
65	75	25

2. After completion of the labeling, centrifuge the iodination reaction mixture for 2 min at 14,000*g* and check that the HPLC system is ready (injection loop, fraction collector, initial conditions, etc).
3. Inject the supernatant (~150 μL) into the RP-HPLC system.

4. Collect first 10 min of gradient (where free iodine elutes) over 5 mL of saturated Tris-base in a 50 mL closed tube (*see* **Note 44**).
5. Using a fraction collector, collect over siliconized tubes (*see* **Note 45**) 1 min-fractions (200 µL) from min 15 to 45.
6. After the HPCL run has finished, take 1-µL aliquots from each fraction and count them in a γ counter using the appropriate channel. Pool the fractions corresponding to the mono-iodinated labeled non-oxidized Aβ40 peptide (*see* **Note 46**), and dry them under nitrogen stream at room temperature on a chemical hood equipped with a charcoal filter.
7. Calculate the specific activity of the labeled peptide (*see* **Note 47**).
8. If higher specific activity is needed, repurify the labeled peptide to homogeneity by using the same gradient as above.

4. Notes

1. An α particle is a large, positively charged particle that corresponds to the nucleus of a helium atom, with two protons and two neutrons. β Particles are essentially particles of nuclear origin with a mass equivalent to the electron that are released when a neutron is converted into a proton. Thus, release of a β particle changes the atomic number and elemental status of the isotope. The γ-radiation is also of nuclear origin and has both particle and wave properties. The release of γ-radiation produces an isotopic change, rather than an elemental one; however, often the resulting nuclei are unstable and they further decay by releasing β particles.
2. 1 Bq= 1 dps; 1 Ci=3.7 × 10^{10} Bq ; 1µCi=3.7 × 10^4 dps = 2.22 × 10^6 dpm.
3. Often, specific activity in biomedical research is expressed as µCi/µg or Ci/mmol.
4. Whenever radiolabeled peptides are used as tracers, it is essential that they behave like their native counterparts. Therefore, labeled peptides should be characterized functionally and structurally by studying a batch of peptide labeled by the same protocol using a "cold" isotope and comparing them to unlabeled peptides.
5. When manipulating directly a radioactive source the distance can be as short as 1 mm. In this case, by using a 10-cm tong to manipulate the source, the exposure to the skin is reduced 10,000 times.
6. Disposal of radioactive animal carcasses generates a difficult problem, because they cannot be disposed of as animal waste, nor as radioactive material. A simple way may be to store the material (conveniently shielded by using lead boxes or lead foil) at –80°C for several disintegration periods until the radioactivity is minimal.
7. As most of the labeling protocols are not selective for specific proteins, labeling of particular peptides or proteins requires preparations of the highest purity. Although purity is not usually a major concern for studies with synthetic peptides, residual organic contaminants (e.g., TFA) and salts may alter the properties of the peptide or interfere with the labeling protocol.
8. Amyloid β_{1–40} peptides homologous to residues 672–711 of human Aβ-precursor protein AβPP770 were synthesized at the W.M. Keck Facility at Yale University using *N*-tert-butyloxycarbonyl chemistry and purified by RP-HPLC. Peptides were

analyzed and structurally characterized via analytical reverse phase HPLC, amino-acid analysis, MALDI-TOF mass spectrometry, N-terminal sequence analysis, and circular dichroism spectroscopy.

9. Depending on the batch of the peptide, solubilization at neutral pH or distilled water may be problematic. In general, it may be convenient to weight an amount of peptide (e.g., 0.5 mg), dissolve it in 0.02% ammonia solution (or alternatively, 20 mM carbonate buffer, pH 9.6) at 1 mg/mL and distribute the volume in several small aliquots (5–50 µg) as convenient, and proceed to lyophilized them immediately.

10. The amino acid sequence of rat Aβ_{1-40} differs at positions 5, 10, and 13 (in bold) with respect to human Aβ_{1-40}. The change at position 10 (Y to F) removes the only tyrosine of the human sequence. This peptide will be used as an example for the Bolton-Hunter protocol.

11. Other suppliers for both ^{125}I and ^{131}I isotopes are Nordion International Inc. (Vancouver, BC, Canada) and DuPont/NEN (Boston, MA).

12. Different badges are sensitive to different types of radiation; be sure to wear the appropriate for the radiation used.

13. Geiger counter with thin NaI(T1) scintillation probe for low- (^{125}I) or medium-energy (^{131}I) γ; Ratemeter with end-window or pancake GM probe for medium-(^{14}C, ^{33}P, ^{35}S) or high energy (^{32}P) β particles emitters.

14. These boxes cannot be kept at low temperatures without deterioration, therefore it may be convenient to use regular storing boxes wrapped with lead foil.

15. Other amino acids sensitive to oxidation are cysteine and tryptophan residues.

16. The same protocol may be applied independent of the radioiodine nuclide (^{125}I, ^{123}I, ^{131}I), however protection measures have to be set up accordingly to the type and energy of the radiation emitted by the isotope.

17. Alternatively, 3 Iodo-beads® (Pierce), which are nonporous polystyrene beads with immobilized chloramine T reagent, can be used as iodination reagent. For this, the incubation time should be prolonged to 15 min. The reaction is then stopped by removing the Iodo-beads from the iodination mixture without the addition of sodium metabisulfite. Buffer compatibility is similar to the chloramine T reagent. For a detailed protocol, consult manufacturer's instruction available at www.pierce net.com.

18. The protocol can be scaled up to 50–150 µg of Aβ_{1-40} using 1–3 mCi of ^{125}I.

19. Do not agitate or vortex. Allow the peptide to dissolve slowly within a few minutes. This procedure will avoid the creation of local high concentrations that may result in peptide aggregation.

20. The chloramine T and Iodo-beads protocols are compatible with most common buffer components including phosphate (recommended), Tris and Hepes, as well as several additives such as detergents (SDS, NP-40, Triton® X-100), urea and high-salt concentration (1 M NaCl). These additives, in some cases, may improve the yield of the reaction by increasing the exposure of tyrosine residues. However, reducing agents or antioxidants cannot be used. In addition, Iodo-beads are not compatible with organic solvents, such as dimethyl sulfoxide or dimethyl formamide that readily dissolve the polystyrene beads.

21. The optimum pH range for iodination of tyrosine residues by chemical methods lies between pH 6.5 and 8.5. Above pH 8.5, the iodination of histidine residues appears to be favored.

22. Depending on the distributor, radioiodine nuclides are usually supplied as the sodium salt of the iodide form in either phosphate buffer or 0.1 M NaOH solution. Sodium hydroxide solution may help to minimize the formation of volatile species such as HI, however it has to be neutralized with an equal volume of 0.1 M HCl before proceeding with the iodination.

23. The total volume of the reaction should be kept as low as possible to achieve a rapid and efficient iodination and so that it can be loaded in the HPLC loop. However, prolonged incubation at a high peptide concentration may result in peptide aggregation.

24. Specific conditions for each peptide or protein have to be adjusted by varying protein concentration, amount of chloramine T used and time of reaction.

25. If the protein is labile or significant aggregation occurs at room temperature during the time scale of the iodination protocol, the reaction can be performed at lower temperatures by increasing the time of incubation with the oxidizing chloramine T reagent.

26. This protocol yields radioiodinated peptide with a very good yield. However, a significant degree of oxidation occurs and it is critical to separate non-oxidized from the oxidized forms by RP-HPLC, or other suitable method (*see* **Subheading 3.3.**).

27. Although the half life of ^{125}I radionuclide is about 60 d, significant radiolysis of the peptide occurs during storing. Therefore it is advisable to discard ^{125}I radiolabeled peptides after approx 1 mo after synthesis.

28. Extremely alkaline solutions will both allow the labeling of histidine residues and increase the solubility of the Iodo-Gen reagent and the oxidative damage to the peptide (*see* **Note 19**).

29. A stock solution of 10 mg/mL in 0.1 M sodium acetate buffer, pH 5.6 can be prepared and stored at −20°C. The working solution at 20 μg/mL has to be prepared freshly from the stock solution. It is critical that any buffer added before iodination does not contain sodium azide, because it inhibits lactoperoxidase activity.

30. Optimum pH for iodination varies greatly with the protein to be labeled. Trial iodination assays using cold ^{127}I may be carried out to determine the best conditions. In the case of amyloidogenic peptides, pH around the isoelectric point of the peptide should be avoided to minimize the risk of aggregation.

31. It is critical that this solution is prepared freshly each time from the stock solution. Commercially available hydrogen peroxide solution are commonly at 30% (w/w). A 10 μg/mL solution can be conveniently prepared by adding 1 μL of concentrated hydrogen peroxide solution (30%) to 30 mL of milli-Q water.

32. At this step it is possible to add cold sodium or potassium iodide to minimize contamination risks.

33. By this method, some of the lactoperoxidase may become radioiodinated and may complicate the purification if the enzyme shows characteristics similar to the labeled peptide or protein in the separation system employed.

34. The Bolton-Hunter reagent in a radioiodinated form is commercially available from Amersham and Dupont NEN.
35. This amount corresponds to approx 27 nmol of the Bolton-Hunter reagent.
36. *See* **Fig. 2.**
37. ^{14}C-labeled cellobiose (50 µCi) may be used as a tracer to facilitate the detection and purification of the tyramine-cellobiose adduct.
38. During this period, particulate material may appear in the solution. This will not affect the yield of the reaction.
39. Hydrogen gas evaporates as the pH of the solution decreases.
40. Cyanuric chloride is very corrosive to the eyes, skin, and respiratory tract. Evaporation at 20°C is negligible, but toxic airborne particles can be produced by inadequate pipetting techniques.
41. Owing to the risk of the generation of volatile iodine molecules at low pH, it is highly recommended that the HPLC system is installed in a chemical hood.
42. For highly hydrophobic peptides (such as most of amyloid forming peptides) and proteins, a C4 column (e.g., 214TP52 from Vydac) is recommended.
43. Amyloid β_{1-42} is more hydrophobic than $A\beta_{1-40}$ and the gradient should be run from 30 to 45% (phase B). Even under these conditions, $A\beta_{1-42}$ species appear as broad poorly resolved peaks. Alternatively, the $A\beta_{1-42}$ species can be separated under basic conditions as described *(29)*. In this case, do not use the chloramine T method for labeling, as chloramine may co-elute with the peptide.
44. This will neutralize the acidic pH and help to minimize the evaporation of iodine.
45. When preparing peptides of high specific activity, significant adsorption to surfaces will occur. To minimize this phenomenon, siliconized tubes are necessary. Additionally, we have found that, under certain conditions, coating with gelatin may help to improve the recovery of the peptide from the tubes.
46. Under these conditions, non-oxidized mono-iodinated peptide will elute at min 29 and oxidized peptide at min 23. However, elution time will fluctuate with the type, age, and usage of column.
47. The specific activity (µCi/µg peptide) can be calculated by estimating the area under the peak and comparing it against a well-established amount of cold peptide run under the same conditions. When the total amount of labeled protein cannot be estimated in this way, other techniques such as immunodetection should be used.
48. Pure mono-iodinated peptide has a specific activity of 2170 µCi/nmol (~ 480 µCi/µg).

References

1. Meisenhelder, J. and Hunter, J. (1988) Radioactive protein-labeling techniques. *Nature* **335**, 120.
2. Meisenhelder, J. and Semba, K. (1995) Safe use of radioisotopes, in *Current Protocols in Protein Science* (Collgan, J. E., Dunn, B. M., Ploegh, H. L., Speicher, D. W., and Wingfield, P. T., eds.). John Wiley & Sons, Inc., New York, pp. A.2B.1–12.
3. Hunter, W. M. and Greenwood, F. C. (1962) Preparation of iodine-131 labeled human growth hormone of high specific activity. *Nature* **194**, 495–496.

4. Alberola-Ila, J., Places, L., De La Calle, O., et al. (1991) Stimulation through the TCE-CD3 complex up-regulates the DC2 surface expression on human T lymphocytes. *J. Immunol.* **146,** 1085–1092.

5. Lederkremer, G. Z. and Lodish, H. F. (1991) An alternatively spliced miniexon alters the subcellular fate of the human asialoglycoprotein receptor H2 subunit. *J. Biol. Chem.* **266,** 1237–1244.

6. Goding, J. W. (1996) Radiolabelling of monoclonal antibodies, in *Monoclonal Antibodies: Principles and Practice* (Goding, J. W., ed.). Academic Press, London, pp. 224–228.

7. Kocisko, D. A., Come, J. H., Priola, S. A., et al. (1994) Cell-free formation of protease-resistant prion protein. *Nature* **370,** 471–474.

8. Caughey, B., Horiuchi, M., Demaimay, R., and Raymond, J. (1999) Assays of Protease-resistant prion protein and its formation. *Methods Enzymol.* **309,** 122–133.

9. Esler, W. P., Stimson, E. R., Mantyh, P. W., and Maggio, J. E. (1999) Deposition of soluble amyloid-β onto amyloid templates: with application for the identification of amyloid fibril extension inhibitors. *Methods Enzymol.* **309,** 350–374.

10. Maggio, J. E., Stimson, E. R., Ghilardi, J. R., et al. (1992) Reversible in vitro growth of Alzheimer disease beta-amyloid plaques by deposition of labeled amyloid peptide. *Proc. Natl. Acad. Sci. USA* **89,** 5462–5466.

11. Makover, A., Moriwaki, H., Ramakrishnan, R., Saraiva, M. J. M., Blaner, W. S., and Goodman, D. S. (1988) Plasma transthyretin. Tissue sites of degradation and turnover in the rat. *J. Biol. Chem.* **263,** 8598–8603.

12. Hutchinson, W. L., Noble, G. E., Hawkins, P. N., and Pepys, M. B. (1994) The pentraxins, C-reactive protein and serum amyloid P component, are cleared and catabolized by hepatocytes in vivo. *J. Clin. Invest.* **94,** 1390–1396.

13. Behr, T. M., Gotthardt, M., Becker, W., and Behe, M. (2002) Radioiodination of monoclonal antibodies, proteins and peptides for diagnosis and therapy. A review of standardized, reliable and safe procedures for clinical grade levels kBq to GBq in the Gottingen/Marburg experience. *Nuklearmedizin* **41,** 71–79.

14. Salacinski, P. R. P., McLean, C., Sykes, J. E. C., Clement-Jones, V. V., and Lowry, P. J. (1981) Iodination of proteins, glycoproteins, and peptides using a solid-phase oxidizing agent, 1,3,4,6-tetrachloro-3α,6α-dipenyl glycoluril (IODO-GEN®). *Anal. Biochem.* **117,** 136–146.

15. Bolton, A. E. and Hunter, W. M. (1973) The labelling of proteins to high specific radioactivities by conjugation to a ^{125}I-containing acylating agent. *Biochem. J.* **133,** 529–539.

16. Zagorski, M. G., Yang, J., Shao, H., Ma, K., Zeng, H., and Hong, A. (1999) Methodological and chemical factors affecting amyloid β peptide amyloidogenicity. *Methods Enzymol.* **309,** 189–204.

17. Bailey, G. S. (1996) The chloramine T method for radiolabeling protein, in *The Protein Protocols* (Walker, J. M., ed.), Humana Press, Totowa, NJ, USA, pp. 665–667.

18. Fraker, P. J. and Speck, J. C. Jr. (1978) Protein and cell membrane iodinations with a sparingly soluble chloroamide, 1,3,4,6-tetrachloro-3α, 6α-diphenylglycoluril. *Biochem. Biophys. Res. Commun.* **80,** 849–857.

19. Markwell, M. A. K. and Fox, C. F. (1978) Surface-specific iodination of membrane proteins of viruses and eucaryotic cells using 1,3,4,6-tetrachloro-3a,6a-diphenylglycoluril. *Biochem.* **17**, 4807–4817.

20. McClard, R. W. (1981) Removal of sulfhydryl groups with 1,3,4,6-tetrachloro-3a, 6a-diphenylglycoluril: application to the assay of protein in the presence of thiol reagents. *Anal. Biochem.* **112**, 278–281.

21. Marchalonis, J. J. (1969) An enzymatic method for trace iodination of immunoglobulins and other proteins. *Biochem. J.* **113**, 299–305.

22. Morrison, M. (1980) Lactoperoxidase catalyzed iododination as a tool for investigation of proteins. *Meth. Enzymol.* **70**, 214–220.

23. Bailey G. S. (1996) The lactoperoxidase method for radiolabeling protein, in *The Protein Protocols* (Walker, J. M., ed.), Humana Press, Totowa, NJ, USA, pp. 668–669.

24. Thompson, J. A., Lau, A. L., and Cunningham, D. D. (1987) Selective radiolabeling of cell surface proteins to a high specific activity. *Biochemistry* **26**, 743–750.

25. Pittman, R. C., Carew, T. E., Glass, C. K., Green, S. R., Taylor, C. A. Jr., and Attie, A. D. (1983) A radioiodinated, intracellularly trapped ligand for determining the sites of plasma protein degradation in vivo. *Biochem. J.* **212**, 791–800.

26. Hysing, J. and Tolleshaug, H. (1986) Quantitative aspects of the uptake and degradation of lysozyme in the rat kidney in vivo *Bioch. Bioph. Acta* **887**, 42–50.

27. Thorpe, S. R. and, Baynes, J. W. (1994) Residualizing glycoconjugates: biologically inert tracers for studies on protein endocytosis and catabolism. *Methods Enzymol.* **242**, 3–17.

28. Maxwell, J. L., Baynes, J. W., and Thorpe, S. R. (1988) Inulin-125I-tyramine, an improved residualizing label for studies on sites of catabolism of circulating proteins. *J. Biol. Chem.* **263**, 14122–14127.

29. Näslund, J., Karlström, A. R., Tjernberg, L. O., Schierhorn, A., Terenius, L., and Nordstedt, C. (1996) High-resolution separation of amyloid β-peptides: structural variants present in Alzheimer's disease amyloid. *J. Neurochem.* **67**, 294–301.

26

In Vivo Imaging of Amyloid-β Deposits in Mouse Brain With Multiphoton Microscopy

Jesse Skoch, Gregory A. Hickey, Stephen T. Kajdasz, Bradley T. Hyman, and Brian J. Bacskai

Summary

With the advent of transgenic mouse models expressing cortical amyloid pathology, the potential to study its progression in an intact brain has been realized. Multiphoton microscopy provides a non-destructive means of imaging with micron resolution up to 500 μm deep into the cortex. We detail a surgical procedure and discuss a multiphoton imaging approach that allows for labeling and chronic visualization of amyloid-β deposits through a cranial window. The ability to monitor these hallmarks of Alzheimer's disease enables studies aimed at evaluating the efficacy of treatment and prevention strategies.

Key Words: Multiphoton microscopy; two-photon; in vivo imaging; craniotomy.

1. Introduction

Amyloidogenic pathology, the hallmark of a number of neurodegenerative diseases, can be detected in living animal models using multiphoton microscopy *(1)*. By utilizing pulsed, low-energy, long wavelength light, multiphoton microscopy *(2)* greatly reduces sample damage and allows visualization of deep tissue structure provided these structures can be specifically labeled with a suitable fluorescent biomarker. Previously, imaging amyloid-β (Aβ), the major component of Alzheimer's disease plaques, in brain tissue was limited to static histological approaches. With a laser-scanning multiphoton microscope, Aβ deposition can be visualized in three dimensions and over time in transgenic mouse models. Preparing and imaging a mouse brain with the potential for post-experimental survival can be accomplished with impressive and informative results that eclipse the limitations of ex vivo techniques. While preparing a mouse for cortical imaging is a delicate procedure, it can be executed with relative ease and consistency thanks to the robust nature of this rodent species and the relatively noninvasive properties of the multiphoton technique.

From: *Methods in Molecular Biology, vol. 299: Amyloid Proteins: Methods and Protocols*
Edited by: E. M. Sigurdsson © Humana Press Inc., Totowa, NJ

By exchanging skull tissue for a glass window, multiphoton excitation and resultant fluorescence may pass freely through the superficial layers of the cortex. Aside from having the power to construct three-dimensional maps of Aβ pathology, this approach allows imaging studies which examine the effects over time of numerous drugs, labels, and other conditions on Aβ deposition, clearance, and morphology. The permanently affixed cranial window also allows multiphoton studies of transgene promoted green fluorescent protein (GFP) variants, endogenous autofluorescence, and other fluorescently tagged epitopes within the cortex.

In order to view Aβ in vivo, we have designed a surgical protocol and several labeling techniques. This chapter will detail anesthesia, surgical preparation, craniotomy, window installation, animal recovery, and imaging procedures, including application of fluorescent probes topically, intraperitoneally, and intravenously for use with transgenic mouse models of Alzheimer's disease *(3)*.

2. Materials

 1. 2,2,2-tribromoethanol (Sigma; St. Louis, MO).
 2. Tertiary amyl alcohol (Sigma).
 3. Fine scissors.
 4. Absorbent wedges (Fine Science Tools, Foster City, CA).
 5. Puralube (J.A. Webster, Sterling, MA).
 6. 1-mL Syringes with 27-gauge needle.
 7. Cotton-tipped applicators.
 8. Hand-held micro-drill.
 9. 0.45-mm Round drill burr (VWR, Willard, OH).
10. 8-mm Round, glass cover slips (Warner Instrument, Holliston, MA).
11. Forceps, angled and straight and ultrafine angled.
12. Homoeothermic blanket with rectal probe.
13. Stereo dissecting microscope—variable magnification (×1–4.5).
14. Stereotaxic apparatus.
15. Xylocaine/Lidocaine (2%) (VWR).
16. Size 100 compressed gel foam.
17. 70% Isopropyl alcohol.
18. Betadyne solution.
19. Dental acrylic powder.
20. Krazy Glue™.
21. Low melting point wax (MP < 52°C).
22. Upright multiphoton microscope.
23. Transgel™.

Our anesthetic of choice is Avertin; a cost-effective, easy to use, and relatively long-lasting tribromoethanol based solution. One caveat to this anesthetic is its high alcohol concentration. While more expensive gas-based anesthetics may be considered as an alternative, they are more difficult to maintain in conjunction with the microscope stage.

2.1. Preparation of Anesthetic

To prepare Avertin anesthetic:

1. Weigh 5 g of 2,2,2-tribromoethanol with 2.5 g of tertiary amyl alcohol (2-methyl-butan-2-ol) into a 50-mL tube. Vortex the solution until the tribromoethanol is dissolved.
2. Heat 225 mL of double distilled water to 55°C and add 2.5 mL of 1 *M* PBS pH 7.4 (w/o Mg^{2+} or Ca^{2+}) and 40 mL of 100% ethanol. Turn off the heat source and add the tribromoethanol/tertiary amyl alcohol solution dropwise until dissolved.
3. Sterile filter the solution, then allow it to cool in a refrigerator.
4. pH the solution to a range of 7.0 to 7.6. (Avertin has an extremely low buffering capacity; use 0.1 N NaOH and add dropwise using a Pasteur pipet. Typically, the solution will require approx 300 µL of NaOH. After adding the NaOH, allow the solution to stand for several minutes as the pH stabilizes.)
5. Protect the Avertin from light and store at 4°C. Sterile filter 10–50 mL aliquots as needed. Avertin stock should last for 6–12 mo.
6. Tribromoethanol solid is unstable and may break down into toxic components. If the mice are adversely affected by a fresh batch of Avertin, order new tribromoethanol.

3. Methods

3.1. Anesthetizing and Handling the Animal

The ideal dosage for each animal will vary primarily based upon the animal's body mass and age. We employ a cautious approach to anesthesia because of the cost and fragility of these transgenic animals. For an average animal (30 g, 16 mo), administer 0.35 mL of Avertin, adjusting by up to 0.10 mL in either direction (*see* **Note 1**). Let the anesthetic take effect for at least 20 min before administering additional boosters. The toe-pinch method is a reliable and easy way to assess the level of sedation; simply apply firm pressure to the animal's toe pads and observe whether or not the animal demonstrates a pain response. If there is a response, administer an additional 0.10-mL dose of anesthetic. Repeat this examination followed by 0.10-mL booster dose of anesthetic every 10 min until the animal fails to exhibit a toe-pinch response. If a cumulative dose of 0.80 mL is reached and the animal is not completely anesthetized, it is best to abort the experiment and to allow the animal to recover for at least 48 h before re-attempting anesthesia.

It is slightly more difficult, yet equally important, to monitor anesthesia during imaging. During extended time-course sessions imaging may be jeopardized by a possible toe-pinch reaction and it may be more appropriate to monitor the animal's breathing and stature.

3.1.1. Animal Handling

The animal should be grasped firmly with one hand, using the thumb and forefingers to pinch the scruff of the neck, while remaining fingers secure the tail in order to minimize body movement. Proceed with the intraperitoneal injection by inserting approx 6–7 mm of a 27-gauge needle into the abdomen just left and anterior to the genitalia at a 30° angle from the mouse body. After placing the needle into the animal, pull back gently on the plunger. If air enters the syringe with minimal resistance, proceed with the injection.

3.2. Surgical Preparation

While waiting for the animal to be fully anesthetized, begin sterilizing the work space and all of the tools intended for use during the procedure. Sterile filter at least 25 mL of phosphate-buffered saline (PBS), and cut gel foam into approx 25 2-mm^3 sections. Allow the sections to saturate in 10 mL of sterile PBS. Maintain the saturated gel foam as well as an additional 10 mL of sterile PBS on ice. Trim the animal's whiskers and the dorsal surface of the head. Apply ophthalmic ointment to protect the animal's eyes. Secure the animal in the stereotax (*see* **Note 2**). Disinfect the shaved area by applying alternating coats of Betadyne and isopropyl alcohol (3 coats Betadyne, 2 coats isopropyl alcohol) (**Fig. 1**).

3.3. Surgery

A subcutaneous local anesthetic should be used prior to incision in addition to general anesthesia. Inject a 0.1-mL bolus of 2.0% Xylocaine under the animal's scalp just prior to making the incision. With scissors and angled forceps, remove the skin from the disinfected region by pinching and lifting it such that a single cut will adequately expose the skull. Use a dry cotton swab to completely remove the periosteum membrane from the exposed skull surface. Following the removal of the periosteum, it is important to keep the skull moist by frequent application of sterile PBS. Prior to drilling, use another dry swab to remove excess PBS from the site. Gently score a circle approx 6 mm in diameter into the skull surface with the drill (**Fig. 2**). Position the drill site such that the posterior end of the circle is just anterior to lambda. Once satisfied with the prospective window site, begin drilling through the bone, exposing the cortical surface (*see* **Note 3**). Drill immediately lateral to the midline on either side of the most anterior portion of the scored site. Following the previously scored contours, drill toward the posterior edge of the site, stopping within 1 mm of the midline. Place several pieces of saturated gel foam over the crevice. Continue

Fig. 1. Anesthetized mouse in stereotax; surgical site sterilized with Betadyne and isopropyl alcohol.

drilling in the same manner over the contralateral hemisphere, leaving bone intact at the anterior and posterior midline of the skull. Carefully sever the posterior bone bridge, being mindful that this is the thickest portion of bone in the drill path. Using a pair of angled forceps, grasp the posterior portion of the skull cap just lateral to either side of the midline. Gently yet firmly peel the skull in the anterior direction, severing the anterior bone bridge (**Fig. 3**). Rapidly apply saturated gel foam to the brain surface. Using an absorbent wedge, soak-up excess PBS and blood while being careful to prevent over-drying. Using a syringe, apply additional PBS to the gel foam as needed to maintain a moist environment. Depending on the extent of bleeding, multiple washes and re-application of gel foam may be necessary (*see* **Note 4**).

3.3.1. Dura Mater Reflection

Dependent upon the requirements of the experiment (*see* **Note 5**), reflect the dural surface from the cortex. Many biomarkers such as thioflavin-S and anti-Aβ antibodies are unable to cross the blood–brain barrier. However, these probes may be applied topically with limited cortical access if the dura mater is reflected. Using ultra-fine forceps, gently pinch the dura as distal to the mid-

Fig. 2. Drill path superimposed onto exposed skull surface. Reference points bregma (**A**) and lambda (**B**) are shown here. A 6-mm circle is cut for installation of an 8-mm cover slip.

line as possible. Carefully peel towards the midline. Do not attempt to excise the dura mater; leave the reflected tissue lying across the midline. Repeat this procedure until sufficient cortical surface is exposed. Reflecting the dura is a delicate procedure; be careful not to insult the exposed cortex and to keep it moist at all times.

3.3.2. Window Installation

Once bleeding has subsided (**Fig. 4**), begin with installation of the cranial window. Make certain that there is no gel foam clinging to the brain surface. Using a syringe, apply sterile PBS copiously. With angled forceps, place a cover-slip over the exposed brain, ensuring that it comes into contact with only the

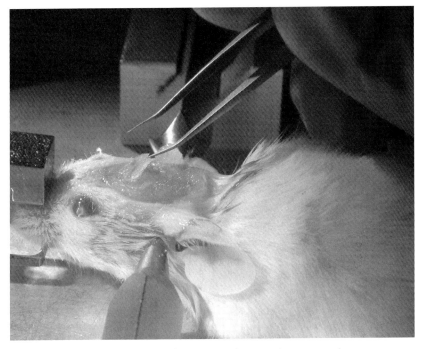

Fig. 3. Skull cap being removed; posterior to anterior.

skull and the protective layer of PBS (*see* **Note 6**). The cover slip should be large enough to cover the exposed brain as well as approx 1 mm of the surrounding bone. Prepare the acrylic mixture by combining approx 150 mg of powdered cement with two to three drops of Krazy Glue™. Mix thoroughly. Before applying the mixture, make sure that there is no air underneath the cover slip; apply additional PBS if necessary. Using the shaft of a cotton swab cut at a 45° angle, apply the mixture to the edge of the cover slip, guiding it away from the brain and onto the skull surface. Allow several minutes for the mixture to set (**Fig. 5**).

3.4. Imaging

3.4.1. Preparation

Water immersion objectives are well-suited for in vivo imaging; particularly high numerical aperture (NA), long working distance, dipping objectives for detecting weak signals and focusing deep into the cortex. We use an upright microscope (*see* **Note 7**) and prefer a 20×, 0.95 NA (Olympus) objective owing to its sensitivity and flexibility. Using software controls to change the scan region, a 3× zoom with this objective is nearly equivalent to a 1× zoom with

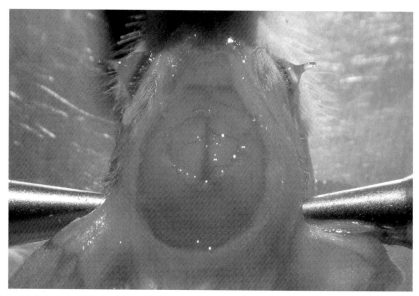

Fig. 4. Exposed brain just prior to cranial window installation. Glare on the cortical surface is indicative of intact dura mater.

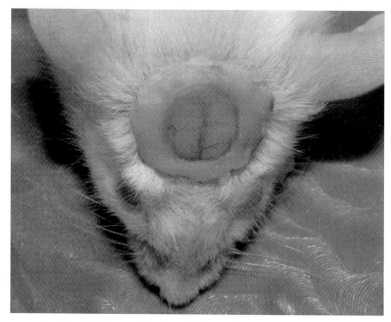

Fig. 5. 11 mo after installation of a cranial window.

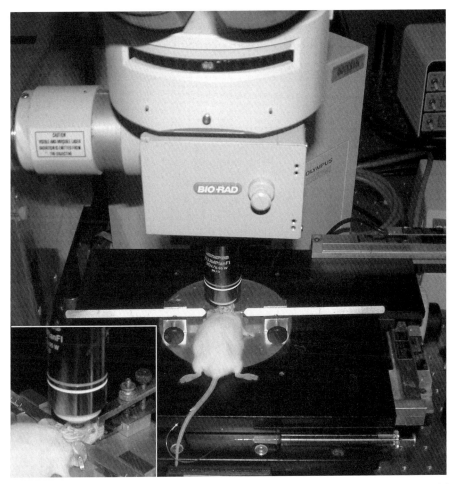

Fig. 6. Mouse under upright multiphoton microscope. Custom stereotaxic assembly fits securely into microscope stage. Inset image shows dipping objective positioned directly above cranial window.

a 60X (Olympus 0.90 NA) dipping objective. It is important to maintain a stable water column above the cranial window. This can be achieved by constructing a restraining ring of wax around the perimeter of the site. Heat a low melting-point wax (MP<52°C, do not exceed 55°) and apply to the perimeter of the cover slip using a blunt implement. When the wax has solidified, the animal is ready to be imaged. Place the entire stereotaxic assembly onto the microscope stage such that the objective is positioned directly above the cranial window (**Fig. 6**). Be careful not to obscure the objective by contact with the wax ring.

3.4.2. Acquisition

In order to determine focus and region of interest, use epi-fluorescence with a standard UV cube and a mercury arc-lamp. It is difficult to visualize all but the most superficial plaques under epi-fluorescence. Cerebral amyloid angiopathy, a difficult pathology to preserve and detect in histological sections, can be seen quite easily even with epi-fluorescence. To maximize the likelihood of locating a plaque in an unlabeled brain, search for small auto-fluorescent (broadband, but primarily green) deposits. Frequently, these lipofuscin deposits will indicate regions of cortical inflammation of which a plaque may be the cause. Limit UV exposure to prevent photodamage and bleaching. Begin scanning with multiphoton excitation at moderate speed with medium laser power and high PMT gain. Gradually increase power if unable to localize an amyloid deposit within the field. Most fluorophores have wide multiphoton absorption spectra, yet it is important to empirically determine the optimal excitation wavelength for maximum signal by initially scanning at a variety of wavelengths. To obtain a z-series of an average-sized plaque (25 µm in diameter) use 1–5 µm z-steps. In order to obtain a z-series of cerebral amyloid angiopathy, be conscious of the size and orientation of the affected vessels. Generally, 10–15 µm steps are suitable for larger vessels, whereas 5–10 µm steps will suffice for smaller arterioles. For kinetic studies, a four-dimensional movie (z-series over time) can be created. Bear in mind, that although multiphoton excitation is relatively benign at low power, long-term repetitive scans can yield tissue damage and photo-bleaching. It is best to calculate an estimated total scan-exposure time, keeping the time between z-series intervals to a maximum; essentially establishing a balance between temporal resolution and photo-damage.

3.4.3. Animal Recovery

If the experiment requires animal survival (*see* **Note 8**), anesthesia may be reversed provided the body temperature of the animal has not dropped too far below the physiological norm. Unfortunately, maintaining body temperature during the imaging process increases the likelihood of motion artifact as the animal begins to awaken. For the most part, the animal should survive the procedure despite the absence of an external heat source. Avertin generally allows 2–3 h of anesthesia.

Immediately after acquisition, remove the animal from the stereotax, place it on the homoeothermic blanket, then lubricate and insert the temperature probe rectally. Make certain the animal is restrained and that it cannot cause harm to itself relative to the probe. When the animal is maintaining its own normal body temperature and has a reflexive response to toe-pinch stimulation, it is ready to be returned to a clean and unoccupied cage (*see* **Note 9**). The usual recovery time for this procedure can range from 24 to 48 h. If the animal has not resumed normal grooming and eating behavior beyond this time frame, it may require additional medical attention or euthanasia.

4. Notes

1. Anesthesia: Prior to the intended date of surgery, it is recommended that precautions be taken to ensure that the animal has not recently experienced stressful circumstances (i.e., recent transport). Not all animals will react consistently to the Avertin, and because it is an ip-administered anesthesia, variation will occur based on the actual injection site. For larger animals, better results may be achieved if the size of the initial dosage is increased.

2. Securing the animal in the stereotax: We use custom-built circular-based stereotaxic assemblies that fit directly into our microscope stage (**Fig. 6**). Any design that can be secured to a microscope stage and incorporates pointed earbars (other styles that may work well in rats do not offer adequate stability in the mouse), a translatable (toward or away from animal) bitebar, and a noseclamp should suffice. In order to effectively immobilize the animal's head, the ear bars must be clamped firmly just anterior to the ears. Gently manipulate the animal's incisors over and around the mouthpiece. Carefully tighten the apparatus in such a manner that it clamps down just posterior to the animal's nose. Once the mouse appears to be secured, confirm by observing absence of head movement while tugging the tail and moving it in an arc.

3. Drilling the skull: It is imperative to remove the periosteum membrane to completion in and around the area to be drilled. Failure to do so may result in the membrane becoming entangled in the drill bit. In order to ensure absence of the membrane, consider using angled forceps to scrape any remaining tissue from the skull surface.

   Skull thickness can vary by as much as 100% between animals, usually directly correlated with age. When scoring the skull, it is best to err on the side of caution, drilling only deep enough to produce a visible outline. It is important to keep the skull cool and moist with sterile PBS. Selectively dry only the area that will be immediately drilled, and reapply PBS when moving on to another area.

   The skull thickness varies considerably; the skull will be thickest toward the posterior portion and can be quite thin in the anterior region, especially near the midline. It can be challenging to judge whether or not the drill has completely penetrated the skull especially because it is quite transparent once it becomes very thin.

4. Removing skull/reflecting dura: Upon removing the skull cap, be prepared to rapidly address the possibility of bleeding by placing saturated gel foam onto the brain surface. Occasionally, prolonged bleeding may occur. Allowing the gel foam to stay undisrupted on the brain for several minutes may help coagulation, however, caution should be used when removing this gel foam as disruption of the newly formed clot could renew bleeding. When the skull is initially lifted from

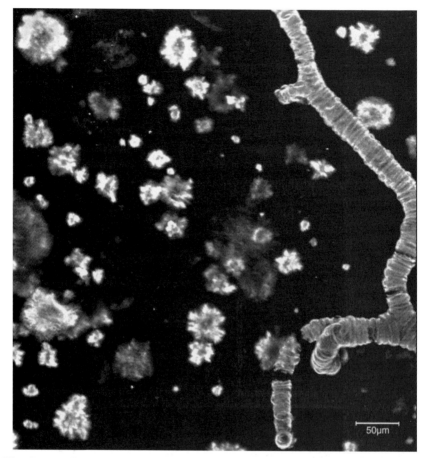

Fig. 7. Z-series maximum intensity projection (325 μm depth) of methoxy-X04 labeled cerebral amyloid angiopathy (right) and amyloid plaques. Scale bar = 50 μm.

    the brain surface, connective tissue may cause the dura mater to tear and be par-
tially excised with the skull tissue.
5. Selection and application of probes: In order to visualize the Aβ protein, first
determine an appropriate biomarker based upon experimental requirements. Thioflavin-S and thiazine red are dyes which selectively bind to proteins with pleated
β-sheet conformations. We use these dyes for the purpose of labeling dense-core
plaques and cerebral amyloid angiopathy in the mouse cortex *(4)*. Both of these
dyes are relatively non-toxic, easy-to-use, and label rapidly with high specificity.
Neither of them, however, readily cross the blood–brain barrier. Additionally,
both only label a certain population of amyloid deposits (diffuse amyloid deposits
lack the required pleated β-sheet conformation). Furthermore, these dyes do not
have tight emission spectra; thioflavin-S is blue/green while thiazine red, although

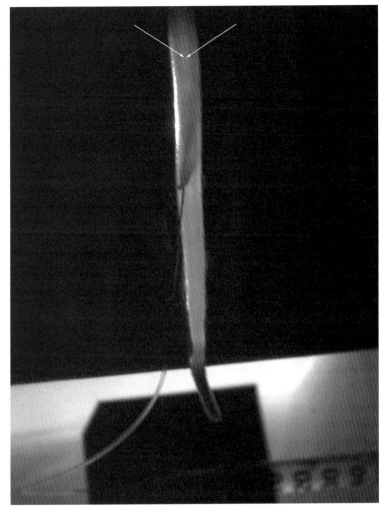

Fig. 8. Trans-illuminated tail; implanted catheter adhered with dental acrylic used to administer drugs or fluorescent compounds. Arrows indicate location of tail vein.

predominantly red, has a blue component detectable with multiphoton excitation. Bear this in mind if co-labeling or performing subsequent fluorescent immuno-histochemistry.

Monoclonal anti-Aβ antibodies can be used to visualize a larger subset of Aβ pathology in the brain (**Fig. 7**). Antibodies can be directly conjugated to a wide variety of fluorophores, making them more suitable for co-labeling experiments. It is important to consider that anti-Aβ antibodies may trigger Aβ clearance *(5,6)*, thereby making subsequent imaging problematic. Much like thioflavin-S and thiazine red, these antibodies are unable to cross the blood–brain barrier, and must be applied topically. This is best done just before cover slipping the exposed brain. For optimal results, carefully reflect the dura mater just before application.

For thioflavin-S and thiazine red, apply approx 20–40 µL at a concentration of 0.01% in sterile PBS for 15 min. It is especially crucial to keep the brain surface moist after the dura has been reflected. Apply copious amounts of moistened gel foam to the brain surface to absorb and wash remaining dye. Touch an absorbent wedge to the top of the saturated gel foam and remove the wedge when approx 50% of the moisture remains. Resaturate the gel foam with sterile PBS and reabsorb as before. Repeat at least three times.

For topical antibody application, follow the same protocol using antibodies at 0.5 to 1.0 mg/mL. The antibodies will label best with a longer incubation period (20–30 min). During this time, the brain should be covered to prevent evaporation. We cover the site by either placing a cover slip gently over the exposed area or by placing a small piece of plastic wrap over the skull. After incubation, wash as previously described.

New compounds such as PIB (Pittsburgh Compound B) *(4,7)* or methoxy-X04 *(8)* readily traverse the blood–brain barrier and offer a slightly less invasive alternative to in vivo detection of Aβ. These compounds can be administered either intravenously (iv) or intraperitoneally (ip). Intravenous injections can be delivered with precision through a tail vein by trans-illuminating the tail and preparing a venous catheter pre-loaded with sterile heparinate PBS (<0.01 mg/mL) (**Fig. 8**). After confirming successful catheter implantation with the heparinate solution, switch to a syringe primed with the treatment compound and inject slowly. For ip injections, we use 12.5 mg/kg and wait 12–24 h before imaging. Injecting methoxy-X04 iv allows for rapid labeling (amyloid deposits will be labeled amidst low parenchymal background) within 25 min.

Intravenous injections of fluorescein (green fluorescence) or Texas Red dextran (red fluorescence) can also be used to create fluorescent angiograms for fiduciary purposes in chronic imaging experiments. For an iv injection of fluorescein, inject a 50-µL bolus at approx 1.5 mg/mL; for an iv injection of Texas Red dextran, inject a 150-µL bolus at approx 20 mg/mL. These concentrations are based upon the assumption that the tail vein has been cannulated successfully. Use higher concentrations if the dye is not directly entering the bloodstream.

6. Cranial window installation: The amount of brain edema is usually proportional to the duration of the surgery and amount of cortical aggravation. Cold, saturated gel foam may help the swelling subside. If the edema is so great that the cover slip does not make contact with the skull, it will be necessary to build-up the dental cement as a platform for the window while being cautious not to allow the acrylic to come into contact with the brain.

7. Microscope information: Our system consists of a BioRad 1024 confocal system mounted on an Olympus BX50WI upright microscope with a custom built three-channel external photomultiplier array. Two-photon excitation is provided by a tunable femtosecond pulsed Ti:Sapphire laser (Mai Tai; Spectra Physics).

8. Tips for chronic imaging: Perhaps the pinnacle of the in vivo imaging technique described here, chronic cortical imaging, requires surgical refinement granted by practice. With the exception of the occasional self-clearance of a hemorrhagic clot, time will invariably degrade the visibility of the brain. Minimally disruptive, efficient surgery, and proper sealing of the cranial window have proven the most reliable combatants of image-site degradation. Whenever possible, avoid any aggravation of the dura mater as the brain tissue below dural lesions is much more susceptible to infection and necrosis and the dura mater will often reform in a thicker, more opaque form.

   Rare occurrences of site infection, hemorrhaging, and dural regrowth may grossly cloud brain visibility. Occasionally, these situations can be rectified by removing the cover slip, and washing the infected area or reflecting the regrown dura, followed by replacing the cover slip. In order to remove the window, it may be necessary to drill through some of the cement to dislodge the glass. Use coarse forceps to pry it free of the cement, being careful to minimize damage to the existing cement platform.

9. Post-procedural: Until the mouse has resumed normal eating behavior, place an easily accessible supply of moisture and nutrients, such as Transgel™, on the cage floor. In the event that the cover slip becomes dislodged, the animal should be sacrificed.

References

1. Christie, R. H., Bacskai, B. J., Zipfel, W. R., et al. (2001) Growth arrest of individual senile plaques in a model of Alzheimer's disease observed by in vivo multiphoton microscopy. *J. Neurosci.* **1,** 858–864.
2. Denk, W., Strickler, J. H., and Webb, W. W. (1990) Two-photon laser scanning fluorescence microscopy. *Science* **248,** 73–76.
3. Hock, B. J. Jr. and Lamb, B. T. (2001) Transgenic mouse models of Alzheimer's disease. *Trends Genet.* **17,** S7–S12.
4. Bacskai, B. J., Hickey, G. A., Skoch, J., et al. (2003) Four-dimensional multiphoton imaging of brain entry, amyloid binding, and clearance of an amyloid-beta ligand in transgenic mice. *PNAS* **100,** 12462–12467.
5. Bacskai, B. J., Kajdasz, S. T., Christie, R. H., et al. (2001) Imaging of amyloid-beta deposits in brains of living mice permits direct observation of clearance of plaques with immunotherapy. *Nat. Med.* **7,** 369–372.
6. Bacskai, B. J., Kajdasz, S. T., McLellan, M. E., et al. (2002) Non-Fc-mediated mechanisms are involved in clearance of amyloid-beta in vivo by immunotherapy. *J. Neurosci.* **22,** 7873–7878.
7. Mathis, C. A., Bacskai, B. J., Kajdasz, S. T., et al. (2002) A lipophilic thioflavin-T derivative for positron emission tomography (PET) imaging of amyloid in brain. *Bioorg. Med. Chem. Lett.* **12,** 295–298.
8. Klunk, W. E., Bacskai, B. J., Mathis, C. A., et al. (2002) Imaging A-beta plaques in living transgenic mice with multiphoton microscopy and methoxy-X04, a systemically administered Congo red derivative. *J. Neuropathol. Exp. Neurol.* **61,** 797–805.

27

Magnetic Resonance Imaging of Amyloid Plaques in Transgenic Mice

Youssef Zaim Wadghiri, Einar M. Sigurdsson, Thomas Wisniewski, and Daniel H. Turnbull

Summary

Transgenic mice are used increasingly to model brain amyloidosis, mimicking the pathogenic processes involved in Alzheimer's disease (AD). In this chapter, a strategy is described that has been successfully used to map amyloid deposits in transgenic mouse models of AD with magnetic resonance imaging (MRI), utilizing molecular targeting vectors labeled with MRI contrast agents to enhance selectively the signal from amyloid plaques. To obtain sufficient spatial resolution for effective and sensitive mouse brain imaging, magnetic fields of 7-Tesla (T) or more are required. These are higher than the 1.5-T field strength routinely used for human brain imaging. The higher magnetic fields affect contrast agent efficiency, and determine the choice of pulse sequence parameters for in vivo MRI, all addressed in this chapter. Ex vivo imaging is also described as an important step to test and optimize protocols prior to in vivo studies. The experimental setup required for mouse brain imaging is explained in detail, including anesthesia, immobilization of the mouse head to reduce motion artifacts, and anatomical landmarks to use for the slice alignment procedure to improve image co-registration during longitudinal studies, and for subsequent matching of MRI with histology.

Key Words: MRI; amyloid deposits; susceptibility; amyloid burden; amyloidosis; molecular imaging; magnetic markers; contrast agent; mouse; transgenic.

1. Introduction

Based on the hypothesis that amyloidosis plays a major role in the pathogenesis of Alzheimer's disease (AD), a number of transgenic mouse lines have been developed to model this aspect of AD. These transgenic mice are currently being used to study the amyloidosis process in vivo, and to test experimental approaches for clearing amyloid. Noninvasive, in vivo imaging methods to map the distribution of central nervous system (CNS) amyloid in these mice would be very valuable for monitoring amyloid plaque formation and progression,

From: *Methods in Molecular Biology, vol. 299: Amyloid Proteins: Methods and Protocols*
Edited by: E. M. Sigurdsson © Humana Press Inc., Totowa, NJ

A   **B**   **C**

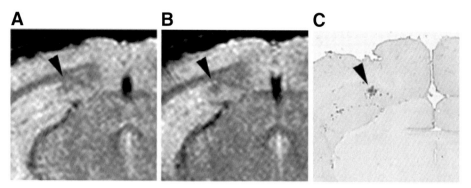

Fig. 1. Large Aβ plaques (arrowhead) can be seen in vivo with T2*-weighted gradient-echo MRI through subtle signal enhancement (**A**). Injection of PUT-Gd-DTPA-Aβ1-40 in the same mouse significantly enhances the Aβ plaque after 6 h (**B**). The matched immunostained histological section confirms these findings (**C**, arrowhead).

and to directly assess the efficacy of plaque clearance therapies. Considerable effort has therefore gone toward developing imaging approaches using such diverse methods as positron emission tomography (PET), multi-photon optical microscopy, and magnetic resonance imaging (MRI). MRI provides some distinct advantages for whole brain assessment of amyloid burden, including higher spatial resolution than PET and much greater penetration than multi-photon microscopy. MRI is also much more widely available for future clinical imaging studies in AD patients, using approaches extended from the mouse imaging methods described in this chapter.

MRI methods to image amyloid plaques in human AD patients are not currently available. It has been suggested that iron concentrated in AD amyloid may enable detection with MR pulse sequences optimized to detect susceptibility-induced contrast from endogenous iron in amyloid plaques in postmortem brain tissue from AD patients *(1)*, and in transgenic mouse models of AD *(2–5)*. However, these results have been controversial and difficult to reproduce *(6)*. Results from our laboratories indicate that although some very large amyloid plaques can be observed in old transgenic mice using susceptibility-induced contrast in T2*-weighted MR images (**Fig. 1**), the majority of plaques in transgenic mice cannot be detected without contrast-enhancement with magnetically labeled ligands targeted to amyloid *(7,8)*.

This chapter describes a method that has been successfully used to map brain amyloid plaques *(7,8)*, involving high field (7-Telsa [T]) MRI of transgenic mice, injected with magnetically labeled peptides for susceptibility-induced contrast-enhanced mapping of amyloid plaques. Details are provided on the peptides and magnetic labeling methods, as well as the experimental setup and protocols required for MRI detection of amyloid in the mouse brain. Although iron-oxide nanoparticles have also been used for magnetic labeling, the protocol provided in this chapter describes labeling with gadolinium (Gd), since this approach resulted in the most sensitive in vivo detection of amyloid *(8)*.

2. Materials

2.1. Labeling of Amyloid-β Peptide

2.1.1. Equipment

1. ABI 430A peptide synthesizer (AME Bioscience, Chicago, IL).
2. Vydac C18 preparative column, 2.5 × 30 cm (Vydac Separations, Hesperia, CA).

2.1.2. Supplies

All reagents available from Sigma unless otherwise noted.

1. Aβ1-40.
2. Diethylenetriaminepentaacetic acid (DTPA).
3. Hydrofluoric acid.
4. Acetonitrile.
5. Trifluoroacetic acid.
6. Gd (III) chloride hexahydrate (Aldrich, Milwaukee, WI).
7. 1 N NaOH.
8. 0.4 M Putrescine in water.
9. EDC coupling agent (Pierce Biotechnology, Rockford, IL).
10. Dialysis membrane (molecular weight cutoff: 2000 g/mol).
11. Mannitol.
12. Phosphate buffered saline (PBS).

2.2. MR Micro-Imaging System (μMRI)

2.2.1. Equipment

1. μMRI scanner: Mouse brain imaging experiments should be performed preferably at a magnetic field strength of at least 7-T (*see* **Note 1**). The experiments described in this chapter were performed with a SMIS console (MRRS, Guildford, UK) interfaced to a 7-T horizontal bore magnet equipped with 250-mT/m actively shielded gradients with 200-μs rise time (Magnex Scientific, Abingdon UK).
2. MRI probe: A radiofrequency (RF) coil fitting closely around the mouse's head should be used for brain imaging. In these experiments, we have developed our own MRI coils, but RF coils can also be purchased from a number of commercial vendors. The results described in this chapter were produced with a custom-made cylindrically shaped Helmholtz coil, sometimes referred to as a saddle coil *(9)*. The coil was developed to resonate at the proton frequency (300 MHz) and to fit

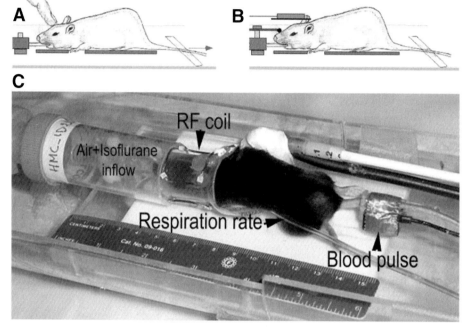

Fig. 2. Overview of mouse positioning/handling in the custom holder. (**A**) Position the mouse head in the bite bar and maintain the head immobilized with hand until taping the tail to stabilize the animal. (**B**) Insert the nosecone for anesthetic delivery (**C**). The mouse holding platform incorporates the RF coil with tooth bar, a nosecone for isoflurane delivery via a vaporizer/scavenger system, and a monitoring system measuring rectal temperature, blood pulse, and respiration rate.

closely around the mouse head (inner diameter, ID = 22 mm; **Fig. 2**). The length along the magnet bore axis (L = 20 mm) was chosen to compromise between high coil-sensitivity and magnetic field homogeneity over the mouse brain (*see* **Note 2**).

3. Mouse holder: The RF coil should be incorporated into a holder that stabilizes the mouse head during MRI, and can be fitted with devices for gas anesthesia delivery and physiological monitoring. MR-compatible mouse holders are becoming available from commercial vendors of small animal MRI systems and RF coils, but most reports to date have used custom holding devices. We have developed our own holder, incorporating the mouse head coil, a nose cone for isoflurane anesthesia, and several physiological monitoring devices (**Fig. 2;** *see* **Note 3**). The main design goal of the mouse holder should be to hold the head in a stationary and reproducible position during the 2–3 h that the animal must be maintained inside the magnet. Predictably, the design closely resembles a stereotaxic injection device, but is fabricated from nonmetallic MRI-compatible materials (*see* **Note 4**). The

head holder should be equipped with a calibrated tooth bar allowing enough vertical and horizontal range (5–10 mm) to center any brain region of interest within the RF coil. Ear bars would be helpful to further stabilize the mouse head, but most RF coil designs are not open structures, and it is difficult to incorporate ear bars within the close-fitting head coil.

4. Gas anesthesia: Isoflurane vaporizer/anesthesia machine (VMS Matrix Medical, Orchard Park, NY).
5. Surgical microscope: (M650, Wild, Heerbrugg, Switzerland).
6. Syringe pump: PHD2000 computer-controlled syringe pump (Harvard Apparatus, Hollison, MA).

2.2.2. Supplies

1. Isoflurane (Aerane, Baxter, Deerfield, IL).
2. Surgical supplies (available from any surgical supplies vendor): small sharp dissection scissors; 2 pairs of no. 5 Dumont forceps; 5-0 silk suture; 30-gauge needle; 70% ethanol for cleaning instruments.
3. Cannulae for infusing magnetically labeled peptides: Polyethylene tubing PE-10 (Intramedic, Becton Dickinson, Parsippany, NJ), Inner diameter ID = 0.28 mm (0.011") and Outer diameter, OD = 0.61 mm (0.024").
4. 10-mL Syringe (Cat no. 309604, Becton Dickinson, Franklin Lakes, NJ).
5. Low-melt agarose: Seaplaque agarose (Cat no. 50100, BioWhittaker, Rockland, ME), or low melt preparative grade agarose (Cat no. 162-0017, Bio-Rad, Hercules, CA).

3. Methods

3.1. Labeling of Amyloid-β Peptide

The peptide-based ligands described in this chapter are experimental probes that can be used for animal MRI after magnetic labeling. Intact Aβ1-40 is unlikely to be used in humans because of its well-documented intrinsic toxicity that may not be noticeable in acute studies. We are currently developing more soluble nontoxic Aβ derivatives that are likely to be more suitable as MRI probes for determining brain amyloid load in patients.

3.1.1. Peptide Synthesis

1. MR imaging ligand based on Aβ1-40 is synthesized on a ABI 430A peptide synthesizer using standard protocols for tBOC (tert-butyloxycarbonyl) chemistry, attaching diethylenetriamine-pentaacetic acid (DTPA) to the amino terminus of the peptide as the final step of synthesis.
2. The peptides are cleaved from the resins using hydrofluoric acid and purification is performed by high-pressure liquid chromatography (HPLC) on a Vydac C18 preparative column, using linear gradients from 0 to 70% of acetonitrile in 0.1% trifluoroacetic acid.

3.1.2. Gadolinium Chelation

Gadolinium (Gd) is chelated to DTPA-Aβ1-40 by incubating the peptide in water or acetonitrile solution at pH 7.0 for 24 h with threefold molar excess of Gd, derived from Gd (III) chloride hexahydrate. For example, if 5 mg of peptide is labeled, the reaction can be performed in 1 mL total volume of 10% acetonitrile (*see* **Note 5**). The pH of the solution can be adjusted with a few microliters of 1 N NaOH and monitored with pH test strips. Mass spectroscopy of the lyophilized end-product, Gd-DTPA-Aβ1-40, can be used to verify the expected molecular weight (4976.6 g/mol).

3.1.3. Putrescine Labeling

For attaching putrescine to Gd-DTPA-Aβ1-40 to increase its blood–brain barrier (BBB) permeability, we have followed the protocol of Poduslo and colleagues *(7)*, with some minor modifications.

1. Gd-DTPA-Aβ1-40 (10 mg) is dissolved in 1 mL of 0.4 M putrescine, pH 4.7, and added to a solution of 1.4 g EDC in 1 mL of 0.4 *M* putrescine, pH 4.7.
2. This solution is mixed at room temperature for 4 h and subsequently dialyzed (molecular weight cut off: 2000 g/mol) at 4°C for 2 d to remove excess putrescine and EDC (*see* **Note 6**).
3. Following dialysis, the peptide is relabeled with Gd, as described earlier for DTPA-Aβ1-40. The resulting Put-Gd-DTPA-Aβ1-40 can be injected intravenously to label amyloid plaques *(7)*.

3.1.4. Mannitol Co-Injection

Gd-DTPA-Aβ1-40 does not cross the BBB alone, and must be either modified with compounds such as putrescine or co-injected into the carotid artery with mannitol. For mannitol co-injection, 400 µg of Gd-DTPA-Aβ1-40 should be suspended in 100 µL of water, and then dissolved in 600 µL of 15% mannitol in PBS immediately before infusion.

3.2. Mouse Preparation

1. Catheter construction:
   a. Heat the PE10 polyethylene tubing using either a heat gun or heated oil and stretch to further reduce its diameter from OD = 0.61 mm to an approximate OD of 0.25 mm.
   b. Cut the tubing to the length required for the syringe pump (*see* **Note 7**). The tapered end of the catheter will be inserted into either the artery or the vein during surgery.
2. Anesthetize the mouse with isoflurane: 5% isoflurane in air for 3 min to induce anesthesia, followed by 1 to 1.5% isoflurane in air to maintain anesthesia.

3. The skin is shaved and cleaned with 70% ethanol.
4. Cut with fine scissors, either on the neck to expose the common carotid artery, or on the inside of the thigh to expose the femoral vein.
5. Under a surgical microscope, the vessel of interest is identified and a small section freed from the overlying muscle tissue with two pairs of fine forceps.
6. For injection into the common carotid artery (CCA), a 5-0 silk suture is tied loosely at the cephalic end of the right common artery and an identical suture is ligated at its central portion. Between the ligations, a puncture is made with a 30-gauge needle. A modified PE-10 tubing, attached to a 1-cc syringe filled with labeled peptide, is introduced into the right CCA through the small puncture. The suture at the cephalic CCA is then tightened around the intraluminal catheter to prevent bleeding. During injection, the left CCA is temporarily clamped with a microvascular clip.
7. For injection into the femoral vein, a small hole is made in the vein with a 30-gauge needle, and the modified PE-10 tubing subsequently inserted.
8. The Gd-DTPA-Aβ1-40/mannitol mixture (600–700 μL) is injected into the carotid artery at a rate of 60-μL per minute, while a similar volume of Put-Gd-DTPA-Aβ1-40 is injected at the same rate into the femoral vein. In either case the injection takes approx 10 min.
9. After injection into the right CCA, the microvascular clip is removed, the catheter is withdrawn from the right CCA, and the cephalic CCA is ligated. The puncture is subsequently sealed with cyanoacrylate glue. The CCA can then be unligated and with the blood flow restored the wound is closed with suture.
10. After injection into the femoral vein, the PE-10 tubing is withdrawn and the puncture is temporarily sealed with a small cotton ball to stop bleeding and the overlying skin then sutured.
11. Mice recover consciousness immediately after removal from anesthesia, and should be kept warm until regaining full mobility.

3.3. In Vivo MR Brain Imaging

1. Mouse setup:
   a. Anesthetize the mouse with isoflurane: 5% isoflurane in air for 3 min to induce anesthesia, followed by 1 to 1.5% isoflurane in air to maintain anesthesia. Care should be taken to properly secure the mouse in the holder before MRI (**Fig. 2**; *see* **Note 8**).
   b. After locking the upper inscissors in the tooth bar, press gently with the index finger just above the nose to avoid unhooking the teeth while pulling the tail taut and immobilizing it with tape (**Fig. 2**; *see* **Note 9**). This simple approach proved to be very efficient for reducing motion artifacts, and provided high-quality MR brain images *(8)*.
2. Slice alignment:
   a. Three orthogonal pilot orientations are needed for accurate image alignment, ensuring accurate matching to histology and reproducibility between imaging sessions during longitudinal studies (**Fig. 3**). Pilot scans are low-resolution MR images, acquired within several minutes at most, and providing anatomical

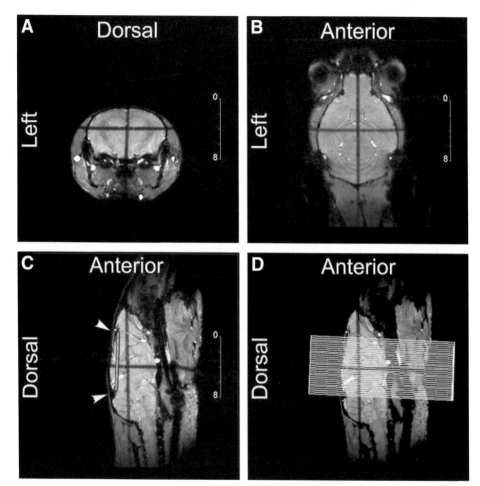

Fig. 3. Orthogonal pilot images (1-mm thick) are acquired corresponding to the coronal (**A**), horizontal (**B**), and sagittal (**C**) orientations, obtained over several minutes using an iterative alignment process. The crossed dark stripes seen in all three images are induced by the saturation effect of the slice inter-crossing, and help in slice positioning. The sagittal pilot orientation (**C**) provides easily identifiable landmarks for the alignment of coronal images. A baseline (white-striped segment) is drawn from the two anatomical notches indicated by the white arrowheads in (**C**). Image (**D**) shows the resulting coronal slice grid, placed orthogonally to the baseline and overlaid on the sagittal pilot.

landmarks adequate to specify the final slice alignment for the high-resolution image acquisition. For amyloid imaging in transgenic mice, we acquire multi-slice MR images in the coronal orientation. The mid-sagittal pilot image is used to align the final slices, checking first to ensure that the pilot is well-aligned by verifying that the midline blood vessels are obvious throughout the image.

b. Two "notches" are then identified, anteriorly between the olfactory bulb and the frontal cortex, and posteriorly at the junction of the cerebellum and midbrain (**Fig. 3**).

c. The final image slices are placed perpendicular to the line between these two notches, checking for symmetric orientation also on the horizontal pilot image.

3. High-resolution MRI acquisition: We determined that amyloid plaques could be better detected with T2*-weighted MRI, 4–6 h after injection of Gd-DTPA-Aβ1-40 and mannitol *(8)*. For each brain, 31 contiguous coronal image slices are acquired from the frontal cortex to the cerebellum, with 78 μm × 78 μm in-plane resolution using a gradient-echo sequence (TE = 15 ms; TR = 1.5 s; flip angle = 55°; slice thickness = 250 μm; total imaging time = 59 min). This sequence provided good anatomical detail and soft tissue contrast, with sufficient susceptibility-induced contrast to detect Gd accumulation in plaques in the brains of the transgenic mice (**Fig. 4**) *(8)*.

3.4. Ex Vivo MR Brain Imaging

1. Brain extraction: Prior to in vivo MRI experiments, ex vivo brain imaging can be used to optimize pulse sequences and to obtain high-resolution amyloid maps in overnight imaging experiments *(7,8)*. For ex vivo MRI, the mice should be intra-cardially or intra-aortically perfused with PBS followed by 4% paraformaldehyde in PBS.

2. Extract brain and submit to fixation by immersion in 4% paraformaldehyde in PBS overnight at 4°C (*see* **Note 10**).

3. After fixation, brain samples can be stored for few days in PBS at 4°C (*see* **Note 11**), but should be embedded in agarose as soon as possible after extraction/fixation.

4. Brain sample preparation: Place the fixed brain gently into a 10-mL syringe (*see* **Note 12**), dorsal side down.

5. Puncture the syringe with a 26-gauge needle, upstream of the expected final position of the syringe plunger. This hole can be used to flush out any air trapped with the agarose inside the syringe.

6. Draw a line along the syringe axis, and align the midline of the brain with this line during the embedding process (*see* **Note 13**).

7. Preparation of 3% low melt agarose: Agarose should be prepared fresh, and can be kept for up to 1 wk as long as it is stored in a sealed container to avoid dehydration and it remains immersed in a water bath at 55°C.

a. Add 3 g of low-melt agarose powder to 100 mL of PBS in a beaker and let it slowly dissolve for a few minutes.

b. Heat the cloudy solution in a microwave oven for 2 min at full power while monitoring the mixture.

c. Stop the microwave oven when the boiling bubbles reach the top of the beaker.

d. Place the hot beaker in a 55°C water bath for at least 30 min to allow the clear agarose solution to cool off and degas.

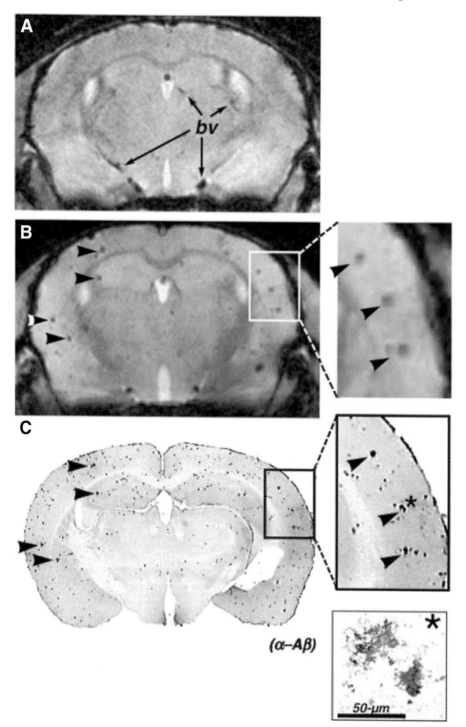

$(\alpha-A\beta)$

50-μm

8. Hold the syringe containing the brain in an oblique orientation (not vertical) and fill the syringe with liquid agarose, using a motorized pipettor to control the flow-rate and avoid introducing air bubbles.

9. Insert the syringe plunger and gently move towards the brain, extracting any trapped air through the puncture hole. The hole on the needle end of the syringe can be used to flush out any final air bubbles trapped in the syringe. It is important to remove all air bubbles before the agarose sets.

10. After the agarose hardens, seal the open end with glue and image the brain immediately or store the sample at 4°C for up to 1 wk before MRI.

11. Ex vivo MRI: Secure the syringe holding the embedded brain sample inside the coil for MRI (*see* **Note 14**). RF coils can be developed specifically for brain sample imaging *(7,8)*, or the same coil used for in vivo MRI can be used in preliminary experiments to optimize the setup. The slice alignment procedure and MR pulse sequences used for ex vivo MRI are the same as described for in vivo experiments (**Fig. 5**).

3.5. Co-Registration of MRI and Histology

Using the slice alignment procedure described before, excellent co-registration of MRI and histology can be obtained by blocking the sample, dorsal side down, and cutting coronal sections vertically across the block in the same orientation as the MR image slices. Take care to align the brain in the same manner as for MRI before sectioning.

4. Notes

1. Static field strength: The diversity of magnet strengths available for MRI studies raises questions about which field strength is optimal. In most centers, MRI studies will be performed on whatever magnet is available for small animal imaging. However, if there is a choice of field strength, the following factors should be considered. MRI sensitivity, generally measured as the signal-to-noise ratio (SNR), is linearly proportional to the magnetic field strength. Therefore, increasing the field strength will enable imaging with increased spatial resolution and/or a reduction in the MRI acquisition time. Currently for mouse brain imaging, magnetic-field strengths $\geq$ 7-T enable in-plane resolution better than 100 µm and slice

---

Fig. 4. (*Opposite page*) Aβ plaques were detected with in vivo µMRI, 6 h after injection of Gd-DTPA-Aβ1-40 and 15% mannitol. In vivo T2*-weighted gradient-echo coronal µMR images show control (**A**) and transgenic (**B**) mouse brains. Many µMRI lesions could be matched to immunostained Aβ plaques (**C**, arrowheads), seen more clearly in the higher-magnification insets. High-power microscopic examination of the immunostained regions revealed that they were indeed parenchymal Aβ plaques, and not blood vessels or vascular amyloid (inset, corresponding to the region marked by an asterisk). bv, blood vessel. (Reprinted with permission from Wadghiri et al., 2003, *Magn. Reson. Med.* **50(2)**, 293–302, Wiley interScience.)

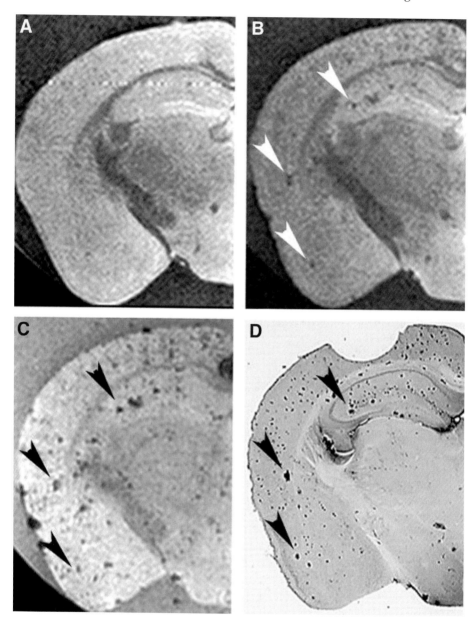

Fig. 5. Ex vivo T2-weighted spin-echo coronal images show control (**A**) and transgenic (**B**) mouse brains. Both brains were extracted and prepared for imaging 6 h after carotid injection of Gd-DTPA-Aβ1-40 with 15% mannitol. (**C**) T2*-weighted gradient-echo (TE = 10 ms, TR = 500 ms, FA = 55°, in-plane resolution = 59 μm × 59 μm, slice thickness = 500 μm, total imaging time = 35 min) of the same brain/level shown in (B). Compared to the control (A), the transgenic mouse (B,C) shows numerous dark spots

thickness lower than 500 µm using acquisition times of 1–2 h. MR image contrast, usually a function of relaxation times T1 and T2, also depends on magnetic field strength. Although a full discussion of MR image contrast is beyond the scope of this chapter, the fact that T1 differences are reduced at higher fields is relevant to amyloid MRI, and as a consequence, T1-weighted MRI is less efficient and requires longer acquisition times. On the other hand, T2- and T2*-weighted sequences are more sensitive to small changes in magnetic properties of tissues at high field, and we and others have demonstrated more sensitive detection of magnetically-tagged amyloid using T2- and T2*-weighted sequences, compared to T1-weighted MRI *(7,8)*.

2. RF coil: Although there is a theoretical advantage in using quadrature RF coils, which are commercially available from a number of commercial suppliers, the relatively simple coil design we employ is efficient at high fields, and the simple impedance matching circuitry facilitates rapid tuning to optimize coil sensitivity for MRI.

3. Physiological monitoring: We have incorporated several devices, available in a computer-controlled system (MP100, Biopac System*s* Inc., Goleta, CA), in our mouse holder, including a rectal temperature probe, an infrared plethysmograph placed around the tail to monitor the blood pulse rate and an air-filled cushion/pressure transducer placed under the mouse body to measure respiration rate. A respiration monitor is required to implement respiration-gated acquisition, although we have found that motion artifacts can be avoided with the mouse setup described in this chapter, without the need for gated acquisition.

4. MR compatible materials: Plastics are generally used to construct animal holders for MRI: acrylic, polycarbonate, nylon, PVC, or delrin are all good choices. Metallic materials should be avoided except in the coil itself, and must be limited to non-ferromagnetic metals: e.g., copper, aluminum, silver, platinum.

5. Performing this reaction in a buffer should be avoided because ions may interfere with the labeling. Removal of the excess Gd by various means is likely to be accompanied by a loss of chelator-bound Gd. Although free Gd can be toxic to mice, the animals seemed to tolerate well Gd-DTPA-Aβ1-40 that was not purified following peptide labeling with threefold molar excess of Gd. The solubility of DTPA-Aβ1-40 depends on the lot of the peptide, and the chelation can be performed in 10–20% acetonitrile in water if needed. Some precipitation can occur when the Gd solution is added, and/or following overnight labeling at 4°C. We have not analyzed the composition of the precipitate but it can be removed by centrifugation prior to lyophilization or injection.

6. In the original procedure by Poduslo and colleagues *(7)*, the pH is maintained at pH 4.7 during the incubation period. This should promote putrescine coupling but we have obtained similar mass spectroscopy profiles when the pH has not been

---

Fig. 5. (*continued*)   throughout the hemisphere confirmed by the matched histology section (**D**) with the arrowheads indicating several major plaques. T2*-GE was more efficient, considering imaging time and lesion detection, a critical consideration for in vivo imaging studies.

kept constant during the reaction. The disadvantage of this method is that numerous peaks of similar molecular weight are obtained, indicating a mixture of DTPA-Aβ1-40 peptides with different numbers of attached putrescine molecules. We have also noticed that with one lot of DTPA-Aβ1-40, substantial precipitate formed during the reaction that reduced its yield. To avoid these issues, it may be better to label amino acids prior to peptide synthesis with putrescine or any other ligand that enhances brain uptake.

7. When cutting the polyethylene tubing, use a sharp scalpel blade or razor blade to avoid distorting/crushing the material. Do not use a pair of scissors.

8. Noticeable image artifacts (blurring) can arise from breathing motion in the neck and caudal head, which will make amyloid detection impossible if not corrected. These artifacts can be greatly reduced if the mouse is properly secured in the holder, as described in the methods (**Fig. 2**).

9. The whole body should remain lightly stretched during MRI, enabling the mouse to breathe freely while greatly reducing head motion from breathing.

10. Immuno-staining with anti-Aβ antibodies such as (4G8) or (6E10) are not affected by overnight fixation at 4°C *(8)* (*see also* Chapter 23).

11. Brain samples should NOT be placed in sucrose or glycerol solution before imaging, because this processing dehydrates the tissue and significantly changes the MRI properties.

12. Brain embedding container: Use any MR-compatible container that can be sealed and fits an adult mouse brain. A 10-mL syringe works well for this purpose, being sealable, disposable, and having a diameter (OD = 16 mm) that closely approximates the size of a mouse head. After imaging, the agarose-embedded brain is easily removed from the syringe by cutting off one end of the plastic cylinder while pushing the plunger from the other side. The agarose can then be peeled away from the brain before histological sectioning and analysis.

13. Careful alignment of the brain sample within the syringe during embedding greatly simplifies the slice alignment procedure during ex vivo MRI.

14. Vibrations can be induced by the gradient coils, particularly when acquiring high spatial resolution MR images. Therefore, the samples must be immobilized inside the coil, either by having a tight fit between the sample holder and coil, or by using tape or other securing measure.

Acknowledgments

The research described in this chapter was supported by grants from the NIH (NS38461 and GM57467 to DHT; AG15408, AG20245, and AG17617 to TW; AG20197 to EMS), the McKnight Endowment Fund for Neuroscience (DHT), and the Alzheimer's Association (DHT, TW, EMS). We thank Yongsheng Li and Jeffrey A. Blind for assistance with the surgical protocols and recent MRI data acquisition. We also thank Dr. Florence Janody for artistic help with **Fig. 2**.

References

1. Benveniste, H., Einstein, G., Kim, K. R., Hulette, C., and Johnson, G. A. (1999) Detection of neuritic plaques in Alzheimer's disease by magnetic resonance microscopy. *Proc. Natl. Acad. Sci. USA* **96,** 14079–14084.
2. Vanhoutte, G., Dewacliter, I., Borghgrae, P., Van Leuven, F., and Van der Linden, A. (2002) In vivo MRI can discern amyloid plaques in the Alzheimer APP[V717I] mouse model. *Proc. Intl. Soc. Mag. Reson. Med.* **10,** 1205.
3. Zhang, J., Yarowsky, P., Gordon, M. N., et al. (2004) Detection of amyloid plaques in mouse models of Alzheimer's disease by magnetic resonance imaging. *Magn. Reson. Med.* **51,** 452–457.
4. Borthakur, A., Uryu, K., Shively, S. B., et al. (2003) *In Vivo* T1r weighted MRI of amyloid in a transgenic mouse model of Alzheimer's disease. *Proc. Intl. Soc. Mag. Reson. Med.* **11,** 2039.
5. Helpern, J. A., Lee, S. P., Falangola, M. F., et al. (2004) MRI assessment of neuropathology in a transgenic mouse model of Alzheimer's disease. *Magn. Reson. Med.* **51,** 794–798.
6. Dhenain, M., Privat, N., Duyckaerts, C., and Jacobs, R. E. (2002) Senile plaques do not induce susceptibility effects in T2*-weighted MR microscopic images. *NMR Biomed.* **15,** 197–203.
7. Poduslo, J. F., Wengenack, T. M., Curran, G. L., et al. (2002) Molecular targeting of Alzheimer's amyloid plaques for contrast-enhanced magnetic resonance imaging. *Neurobiol. Dis.* **11,** 315–29.
8. Zaim Wadghiri, Y., Sigurdsson, E. M., Sadowski, M., et al. (2003) Detection of Alzheimer's amyloid in transgenic mice using magnetic resonance microimaging. *Magn. Reson. Med.* **50,** 293–302.
9. Hoult, D. I. and Richards, R. E. (1976) The signal-to-noise ratio of the nuclear magnetic resonance experiment. *J. Magn. Res.* **24,** 71–85.

Index

A

Aβ standards, 284, 285, 288, 290, 294
Aβ-derived diffusible ligands
    (ADDLS), 19, 28
Amyloid-β (Aβ), 3–55, 59–78, 81–92,
    97–121, 124–142, 145–150,
    153, 154, 157–181, 184–194,
    197–202, 205–228, 231–234,
    237–241, 244–321, 325–362,
    365–378
A/J mice, 240
A30P, 22, 87, 90
A53T, 22, 87, 90, 95
α-Actin, 209, 211
AEC (3-amino-9-ethylcarbazole), 320
AgNO$_3$, 239
AL-amyloid, 106, 237, 238, 245
Alexa Fluor 488, 209
Alexa Fluor 546 goat antimouse, 207
alkaline phosphatase, 317–320
Amide I band, 132, 140, 142
Amino acid side chain vibrations, 143
Amorphous, 44, 78, 130, 222, 263, 264
    Amorphous deposition, 261
Ampholytes, 245–249, 253
Amylin, 103–108, 113, 116, 117, 120,
    126–128, 185
Amyloid burden, 24, 365, 366
Amyloid deposits, 18, 212, 243, 244,
    252, 255, 263, 299–303, 315,
    360, 362, 365
Amyloid enhancing factor (AEF),
    237–241, 331

Amyloid extraction, 243, 255, 268
Amyloid fibril, 19, 21, 26, 31, 47, 62, 67,
    68, 71–73, 78–82, 85–92, 99, 100,
    103, 105, 108, 126, 127, 149,
    153, 183, 184, 194, 218, 221,
    226, 237–245, 252–260, 347
Amyloid P-component, 245, 254, 338
Amyloid plaques, 24, 35, 61, 62, 184,
    268, 278–282, 285, 286, 296,
    303, 307, 310, 321, 347, 360,
    365, 366, 370, 373, 379
Amyloid pores, 26, 31, 32, 100
Amyloid precursor protein (APP), 24,
    32, 103, 185–193, 219, 267–269,
    273–283, 286, 287, 294–297, 379
Amyloid Protein AA, 85, 243–246
    AA, 237–240, 243–247, 250, 253–256
    AA-amyloid, 244, 245, 256
    AA-subtypes, 243
Amylospheroid, 32
Amyotrophic Lateral Sclerosis, 11, 53
Analytical crosslinking, 14, 17
Analytical ultracentrifugation (AU), 12,
    19, 22, 24, 31, 59, 81, 89, 92,
    95, 98
Angstrom scale resolution, 103
Angular dependency, 162, 163
Anisotropic, 129, 148
Annular Protofibrils, 22–26, 31
Annular structures, 19
Antigen retrieval citra solution,
    320, 319
Antipain HCl, 268

Apolipoprotein, 85, 100, 218, 243, 244
Arachidonic acid, 37, 47, 51
Araldite glue, 106
Arctic variant, 19, 20, 26
Asp67His variant lysozyme, 85
Astrocytosis, 310, 311, 315, 316
Ataxia, 310, 312
Atomic force microscopy (AFM), 19,
    22, 28–33, 68, 82,
    101–128, 176
ATR, 133, 138, 139, 142–147
Attenuated total reflection (ATR), 133,
    138, 139, 142–147
ATTR, 237, 238, 258
Avertin, 310, 312, 350, 351, 358, 359

B

β-Sheet, 19, 22, 61, 67, 68, 82,
    131–133, 142–145, 261, 360
β-Sheet content, 129, 132
β-Structure, 36, 37, 124, 145, 177, 182
Band assignment, 141–143
Barium fluoride, 138, 139, 147
Baseline correction, 136
BCA protein assay, 227
Betadyne, 350–353
Biobrene, 225
Black microtiter-plates, 38
Block Ace, 281, 293
Bolton-Hunter, 325, 330, 332, 333, 337,
    344, 346
Bovine lactoperoxidase, 332, 336
bovine spongiform encephalopathy
    (BSE), 53, 54, 60, 63, 65,
    310, 321
Bradford assay, 40, 46, 177
Bragg's law, 78
Brain sectioning, 301

British dementia precursor protein
    BRI, 274
5-Bromo-4-chloro-3-indolyl phosphate/
    nitro blue tetrazolium, 320
Brownian motion, 153, 158, 159, 163

C

C18 ZipTips, 189, 193
C4 ZipTips, 193
Calcite, 74
Calcitonin (CT), 4, 13, 133–138, 146,
    263, 264
Calcium fluorite, 138
(+)-10-Camphorsulfonic acid (CSA),
    135, 147
Cantilever, 104, 105, 124
Capture antibody, 282, 288, 294
Carbon coated formvar film, 83
Carbon coated titanium grids, 95
Carbon-14 (^{14}C), 326, 327, 344, 346
Cataract, 154, 173
Cation exchange chromatography,
    38–40, 45
Cation exchange matrix, 250
CCβ-Met Fibrils, 107, 109
CCA, 137, 371
CCP4 suite, 77, 80
CD1 mice, 240
CDNN, 137
CE/J strain, 240
Cell culture, 146, 195, 197, 200, 210,
    211, 223, 227, 230–233, 302
Cellobiose, 325, 330, 332, 338,
    342, 346
Cerebral amyloid angiopathy (CAA),
    146, 197, 200, 209–212,
    220–221, 358, 360
Cerebral blood vessel, 197, 211

Cerius2, 78
Chloramine T, 325, 329, 332–337,
    343, 344–347
Chloroform-methanol (2:1)
    mixture, 252
Chromatofocusing, 243, 245, 250–253
Chromatography, 3, 4, 12, 16, 19,
    40–40, 45, 95, 135, 165, 187,
    189, 244, 250, 255–259, 335,
    340, 369
Chromium potassium sulfate, 300, 302
Chronic wasting disease (CWD), 310
Chymotrypsin, 199–202, 210
Circular dichroism (CD), 22, 36, 68,
    81, 98, 129–139, 143–151,
    178, 178, 187, 221, 229, 239,
    240, 262, 263, 316, 344, 347
Citrate, 146, 245
CitriSolv, 301, 303
Coherence angle, 161, 162, 166, 172
Collagenase, 199–203, 210, 213, 214
Collagenase/dispase, 213, 214, 217
Colloidal gold assay, 222, 224
Common carotid artery, 371
Conformation, 36, 46–48, 67, 78,
    139–140, 147, 150, 151,
    177–183, 221, 237, 266, 299
Conformational stability, 175, 333
Congo red, 36, 105, 114, 118, 129, 148,
    176, 183, 234, 240, 241,
    299–306, 363
CONTIN, 167
Coomassie blue R-250, 222, 224, 263
Correlation function, 153–168,
    171–173
Correlator, 155, 159, 160, 164, 165
Covalent chromatography, 255–259
Craniotomy, 349, 350

Creutzfeldt-Jakob disease, 53, 54, 62,
    63, 65, 227, 310
Cross-β-structure, 36
Cross-polarized light, 73
Crosslinking, 11–17, 46, 98
Cryo-electron microscopy (CryoEM),
    79, 82, 83, 100
Cryo-loops, 73
Crystal/Mount, 320
Crystallites, 71, 75, 76, 79
CSA solution, 147
Cumulants, 167, 168, 174
Curve fitting, 141–145, 148
Cuvet, 46, 132, 133, 146, 155, 160,
    163–165, 172, 177–181
Cuvets path length, 135
Cyanuric chloride, 340, 346
Cyclic amplification, 53–58, 62–64
Cystatin C, 143–146, 151, 197, 209–212,
    217–226
Cysteine protease inhibitor, 221

D

3D reconstruction, 85–89
DAPI nuclear stain, 209
DC protein assay, 268, 272
DEAE-Sephacel column, 222, 223, 226
Deconvolution, 141–145, 148, 150,
    153, 159
Densitometric analysis, 57
Dental acrylic powder, 350
Denzo, 77
DePex, 303, 306
Deuteration, 143
Deuterium oxide (D_2O), 132,
    140–142, 148
Devcon Epoxy, 106
Dextran, 213–217, 362

Diabetes type II, 53
Dialysis-related amyloidosis, 54
3,3'-Diaminobenzidine
        tetrahydrochloride, 301,306,
        311, 319, 320
Diamond crystals, 133, 138, 147
Diethylamine, 272, 281, 286, 287
Diethylenetriaminepentaacetic acid
        (DTPA), 366–378
Diffraction pattern, 68, 70, 73–80
Diffractogram, 76–78
Diffusion, 153, 155, 158–161,
        168–173
Diffusion coefficient, 153, 155, 158–161,
        166–173
Digital filtration, 165
Dimethyl sulfoxide (DMSO), 4–7, 21,
        105, 117, 138, 186–188, 200,
        205, 281, 284, 337, 344
Direct fit method, 166, 167
Dissection, 267–269, 369
DMF, 337
DOPG, 188
Double labeling, 317
DTT, 13–16, 38, 41–43, 46, 229, 230,
        233, 256–259, 263–266
Dutch type, 145, 197, 209, 295
Dynamic light scattering (DLS), 12, 19,
        28, 89, 98, 153, 173, 174

E

E22Q, 145
EDC coupling agent, 367
Edman degradation, 246
EDTA, 38, 41, 58, 62, 146, 147, 213–217,
        230, 268, 281, 287
Electron diffraction, 79
Electron microscope sample grids, 38

Electron microscopy (EM), 19, 22–28,
        31, 35–38, 41–45, 50, 67–69,
        79–92, 95–108, 113, 161, 176,
        178, 184, 186, 199, 213–215,
        218, 222, 227–230, 311, 314,
        315, 318, 321, 378
Electrophoresis, 11, 12, 59, 223, 243,
        245, 256, 263
Electrospray, 185, 187, 190–194
ELISA sandwich, 279–282, 285–288,
        291–294
Ellipticity, 138, 147
Endothelial cell, 210
Enzyme-linked immunosorbent assay
        (ELISA), 267, 276, 279–293, 297
Epi-fluorescence, 358
Ether, 312
Ethylene glycol, 300, 302, 305
Extraction, 200, 238, 243, 245, 252,
        255–258, 261–269, 272, 279,
        280, 286, 287, 373

F

3F4, 59
Factor VIII-related antigen, 211, 214
Familial amyloid polyneuropathy, 85, 145
Familial amyloidosis, 255
Far-UV, 129, 130, 133–137, 143
Fatty acids, 37, 49
Femoral vein, 371
Fermi resonance, 140
Fibril assembly, 103–127
Fibril elongation rates, 111
Fibril polymorphisms, 107, 108, 116
Fibril structure, 89, 90, 103, 105
Fibrillogenesis, 6, 8, 18, 31–33, 47, 65,
        67, 81–91, 95–101, 128, 153,
        174, 185–194, 222

Filtration, 3–7, 16, 38, 165, 187, 192, 199, 212, 222, 232, 233, 243, 246, 247, 255–259, 265, 340
Fit2d, 77
Fixation, 58, 249, 250, 268–271, 299, 305, 309, 373, 378
Flow-through filtering, 165
Fluorescein, 214, 362
Fluorescence, 11, 12, 35, 37, 41–44, 49, 68, 131, 145, 175, 178–184, 187, 188, 198, 207, 217, 221, 303, 311, 315, 320, 350, 358, 362, 363
Fluorescence quenching, 180, 181
Fluorescence resonance energy transfer (FRET), 12, 175
Fluorophores, 331, 358, 361
Formaldehyde, 98, 318
Formic acid, 186, 266, 267, 272, 279–283, 286–288, 292, 294, 303, 307, 316
Formvar, 69, 70, 83, 95, 96
Fourier self-deconvolution, 141, 142, 145, 150
Fourier transform, 76, 129, 143, 149–151, 158, 221
Fractionation, 22, 27, 30, 50, 165, 267, 268
  Fractionation of membrane proteins, 274
French Press, 38, 40
Frontotemporal dementias (FTPD-17), 37, 50, 176, 184
Fxplor, 77

G

Gadolinium (Gd), 57, 265, 308, 366, 367, 370–377
Gd (III) chloride hexahydrate, 367, 370
Gel filtration, 38, 40, 222, 243, 246, 247, 255–259, 340

Gel permeation chromatography, 38
Gel/Mount, 214, 217, 320, 321
Gelatin, 300–302, 305, 315, 346
GFAP, 309, 316, 317
Globular oligomers, 28–31
GLUTAMAXI, 199
Gluteraldehyde, 98
Goniometer, 68, 74
Growth hormone-releasing factor (GRF), 13
GT-1 cells, 229
Guanidinium hydrochloride, 175, 222, 244, 246, 256–258

H

H & E, 203
Hamster, 55, 59, 241, 316
Hanging drop method, 97
Heart, 200, 218, 226, 253–260, 307, 313
HEK293 cells, 222, 224
α-Helical, 36, 149, 222
α-Helix, 129, 130, 138, 142–145
Helix-inducing agents, 36
Hematoxylin and eosin staining solution, 203
Hemocytometer, 198, 205, 206, 209
Hemolytic anemia, 53
Heparin, 37, 38, 41–46, 49, 50, 177, 178, 287, 300, 302, 305, 362
Hereditary cerebral hemorrhage with amyloidosis (HCHWA), 145, 146, 197, 198, 209, 212, 218, 221, 225, 226
Hereditary cerebral hemorrhage with amyloidosis, Icelandic type (HCHWA-I), 146, 151, 197, 198, 209, 212, 221, 225, 226

Hereditary cystatin C amyloid
angiopathy (HCCAA), 197,
210, 212, 218, 226
1,1,1,3,3,3,-Hexafluoro-2-isopropanol
(HFIP), 21, 105, 110–113, 120,
139, 186–188, 193
Highly oriented pyrolytic graphite
(HOPG), 120, 121
Histology, 175, 299–307, 318, 365,
371, 375, 377
Histopathological staining, 311, 315
Histopathology, 309, 315
Homogenization, 58, 62, 229, 245, 246,
258, 267–272, 275–277, 280
HPLC, 37, 39, 186, 188, 258, 263, 332,
335–346, 369
HRP-coupled detection antibody, 285, 287
Huntingtin, 11
Hydrocortisone, 199
Hydrofluoric acid, 367, 369

I

Imaging, 22, 31, 32, 35, 37, 42, 44, 85,
91, 99, 103–108, 116, 117,
121, 124–128, 307, 349–379
In vivo imaging, 349, 355, 363,
367, 377
Imidazole, 146, 333, 338
Immunoblot, 59, 261, 264
Immunocytochemistry, 213, 216,
307–311, 315, 318
Immunoglobulin lambda light chain
monoclonal antibody, 85
Immunohistochemistry, 60, 261, 264,
268, 271, 299, 303–306, 361
Immunohistopathology, 309
Immunostaining, 197, 207, 209, 307,
309, 315, 316, 319

Infectivity, 54, 61, 64, 65, 149, 230,
234, 322
Inflammation, 237, 238, 244, 358
Infrared (IR), 129–132, 138, 139,
143–151, 221, 377
Inoculation, 233, 309–311, 314, 323
Insulin, 100, 107, 127, 190, 194, 199,
310, 312
Insulin-like growth factor-I (IGF-I), 199
Interferogram, 140
Internal standard, 185–193
Iodination, 335, 341–345
Iodine isotopes ,
^{125}I , 326, 338
^{131}I, 326
^{123}I, 326
Iodo-Beads, 332, 344
Iodo-Gen, 325, 329, 332, 335, 336, 340,
345, 347
Ipdisp, 77
Islet amyloid polypeptide (IAPP), 69,
103, 128, 185–194
Isoelectric focusing, 243, 246–251
Isolectric point (IP), 21, 39, 105, 110–113,
120, 139, 186–188, 193,
243,245, 246, 249–252, 269,
320, 345
Isoflurane, 310, 312, 368–371

K

K$_2$D, 137
KF, 134, 146

L

L-cysteine, 256
L68Q, 145, 146, 209
Lactoperoxidase, 325, 329, 332, 336,
345, 348

Lactoperoxidase method, 325,
    338, 348
Laemmli sample buffer, 229, 230, 233
LALS, 78
Langmuir-Blodgett technique,
    108, 121
Laser, 50, 104, 111, 189, 349, 358,
    362, 363
    Intensity, 154, 155, 161, 166
    Single mode, 159, 161
    Stability, 161
    Polarization, 161
    Overheating, 161
    Focusing, 161, 166
LC-TOF, 189
Leptomeningal tissue, 198, 203
Leptomeninges, 197, 204, 217, 225
Leu60Arg variant of apolipoprotein
    AI, 85
Leupeptin hemisulfate salt, 268
Levamisole, 311, 319
Lewy bodies, 22
Ligand-trapping, 325
Light chains, 265
Light collection, 161
Light scattering, 11, 12, 37, 43, 50, 89,
    135, 153–174
Light-chain amyloidosis, 145
LINCOMB, 137
Lipid bilayers, 105, 106, 116, 125–128
Lipofectamine, 222, 223
Lipofuscin, 358
Low-melt agarose, 369, 373
LSQINT (CCP13), 78
Lysozyme amyloidosis, 145

M

MAC Mode, 104, 125

Magnet, 68, 73, 75, 368, 375
Magnetic alignment, 71, 72
Magnetic markers, 365
Magnetic resonance imaging (MRI),
    365–379
Mannitol, 367, 370–376
Marcvt, 77
MARresearch image plate, 75
Mass spectrometry, 32, 185–189, 192,
    194, 246, 342, 344
Matrix-assisted laser desorption
    ionization (MALDI), 189,
    342, 344
Mayer's hematoxylin, 320
MCT, 133, 146
Meningal tissue, 198, 203, 204
Metabolic labeling, 190, 326
Methoxy-X04, 360–363
Methylamine vanadate, 92, 95, 97
β-2 Microglobulin, 185
Microtubule-associated protein, 35, 49,
    175, 183, 184
Miller indices, 76, 78
Millidegrees, 138, 147
MOM kit, 304
Monomer, 4, 20, 22, 28, 36, 39, 42, 43,
    46, 55, 188, 193
Mosflm, 68, 77, 78
Multiphor II system, 245, 247
Multiphoton microscopy, 349, 363, 366
Multiple scattering, 162, 172

N

N2a cell line, 227
N,N-methylene-bis-acrylamide, 249
N-N-dimethylformamide, 268
NaF, 47, 134, 146, 178, 197
Nanometer, 103, 105, 136

Nanoscope IIIa, 106
Negative staining, 22, 35, 38, 42, 4645,
    47, 81–83, 95–99
Neurofibrillary tangles (NFTs), 24, 35,
    36, 176, 184, 307
Nickel ammonium sulfate, 301, 304
NP-40, 233, 344
Nuclear magnetic resonance (NMR),
    12, 47, 48, 67, 82, 101, 129,
    131, 379
Nucleation-elongation, 37, 113
Nucleation/polymerization model, 55

O

Oligomer, 7, 24, 113, 127, 169
Oligomeric intermediates, 103,
    112, 113
Oligomerization state, 11
OptiMEM, 218, 228, 229, 232
Oxidation, 4, 16, 49, 325, 328, 333–336,
    341, 344, 345

P

Paired helical filament, 35, 183
1-palmitoyl-2-oleoyl-sn-glycero-3-
    phosphocholine (POPC),
    106, 119
Paraformaldehyde, 300–304, 310, 315,
    318, 373
Parkinson's disease (PD), 11, 17–22,
    31, 32, 53, 67, 81, 100, 101
PBE gel 96 or 118, 245
PCMass, 91, 98
Pefabloc, 228, 231, 232
Pepstatin A, 38, 230, 268
Perfusion, 299–302, 305, 313
Periodic acid-Schiff staining solution,
    311, 315, 316

Permount, DPX mount, 311
Peroxidase labeling kit, 281, 285, 294
Pharmalyte, 245, 250
PhastSystem, 247
Phenol red, 21, 25
Phenylmethylsulfonylfluoride (PMSF),
    38, 41, 58, 59, 230, 268, 272, 276
o-Phosphoric acid, 281, 287, 294
Phosphorous-32 (^{32}P), 326, 327, 344
Phosphotungstic acid, 98
Photochemically crosslinked, 7
Photo-Induced Cross-linking of
    Unmodified Proteins (PICUP),
    8, 11–13, 18
Piezo, 104, 120, 125, 133
Piezoelectric scanner, 104
Pittsburgh Compound B, 362
pituitary adenylate cyclase-activating
    peptide (PACAP), 13
Plain transmission, 138
Plasticine, 71, 74
Poly-D-lysine, 213, 216
Polyacrylamide gel electrophoresis
    (PAGE), 7, 12, 13, 16, 28,
    38–41, 59, 177, 223–225, 231,
    232, 243–246, 256, 259,
    264–269, 272–277
Polyanion induced polymerization, 35
Polyanionic cofactors, 37
Polyanions, 37, 46, 176
Polybuffer, 243, 245, 250, 252
Polydispersity, 130, 158, 159, 166, 174
Polyfluorinated alcohols, 4
Polymerization, 35, 37, 41–51, 55, 130,
    173–178, 184, 252
Ponceau S solution, 269, 274
Pore-like, 17, 20–26, 31, 87, 100, 101, 128
Pores, 19, 26, 31, 32, 100

Positron emission tomography (PET), 363, 366
Pre-amyloid deposits, 299, 303
PrepTips, 193
Primary cells, 211
Prion, 11, 17, 53–56, 60–65, 80, 81, 90, 92, 101, 149, 154, 173, 184, 227–233, 237, 238, 241, 274, 316, 321–323, 326, 347
   Clinical scoring, 309–312
   PrP, 11, 54–63, 149, 227, 230–234, 309–318, 321, 322, 326
   PrP^C, 54–59, 62, 227, 230, 233
   PrP^{res}, 54–62, 326
   PrP^{Sc}, 11, 54–61, 227, 230–233, 309–312, 315–318
   Prion strains, 229, 234
      22A, 313, 322
      22-L, 229
      87V, 313
      Chandler/RML, 229
      ME7, 59
   Prion protein monoclonal antibodies
      SAF-32, 228
      SAF-70, 228
Prn-i, 313
Propagation, 54, 60, 63, 211–216, 227–233, 237, 238, 312
Protease, 41, 50, 56, 58, 63–65, 199, 203, 221, 225, 322, 347
   Protease inhibitors, 38, 230, 274, 276
   Protease resistance, 54, 132, 227
Protein
   Assembly, 11, 17, 153, 154, 164, 173, 174
   Conformation, 130, 137, 150, 151, 221, 266

Extraction, 261–265
Protein misfolding cyclic amplification (PMCA), 53–62
Protein sequencing, 261
Proteinase K (PK), 54, 57–59, 62, 228–233, 326
Protofibril, 5, 6, 30, 32, 89, 95, 103
Protofilaments, 75, 79, 85–89, 107, 124
Puralube, 350
Purification, 8, 15, 20, 26, 35–51, 99, 134, 177, 221–226, 243–259, 265, 266, 282, 312, 333–341, 345, 346, 369
Putrescine, 367, 377, 378
   Putrescine labeling, 370

Q

QLS Theory, 154
Quantification, 32, 42, 62, 205, 279, 280, 287, 307, 309, 312
Quantification of Aβ, 42, 287
Quantitation, 18, 35, 42, 47, 65, 190–194, 231, 267, 268, 272, 279–286
Quartz cuvets, 132–134, 146

R

rA(R18H), 113
rA(R18H, L23F), 113
rA(R18H, V26I), 113
Radioiodination of amyloid peptides, 325, 328–330, 333–338, 347
Radioisotopes, 325–328, 331, 346
Radiolabeling, 325, 331–337, 341–348
Raman, 131
Random coil, 22, 36, 37, 143
Regularization, 155, 167–170, 174
Rigaku rotating anode, 75
Ring-shaped, 87

RNA, 37, 46, 49, 50, 55, 63

S

Scanning transmission electron
    microscopy (STEM), 19, 24,
    31, 81, 82, 89–92, 95–101, 176
Scattering Intensity, 168–171
Scattering vector, 155, 156, 159, 162
Scattering volume, 157, 160–166
Scrapie, 9, 53, 55, 59, 63, 64, 80, 149,
    227, 230–234, 309–323
Scrapie strain, 59, 233, 234,
    311–313, 322
    22A, 313, 322
    87V, 313
    ME7, 59
SDS-PAGE, 7, 13, 28, 38–41, 177, 231,
    245, 246, 256, 259, 264–269,
    272–277
Seaplaque agarose, 369
SEC, 4–7, 12, 19, 22–30, 165
Secondary amyloidosis, 243, 254
Secondary structure determination, 129
Sedimentation velocity analytical
    ultracentrifugation (SVAU),
    31, 81, 92, 95
SELCON, 137
Self-assembly, 90, 153, 175
Self-deconvolution, 141–145, 150
Senile systemic amyloidosis, 255, 260
Sephacryl S-200, 256
Sephacryl S-300, 244, 258
Sephadex G25, 251, 256, 257
Sepharose 6B CL, 244, 256
Serpin-deficiency disorders, 53
SH3 domain of phosphatidylinositol-3'-
    kinase, 87
Sickle cell anemia, 154

Single particle analysis, 81–84, 87,
    111, 126
Single particle averaging, 82–87, 90
Size exclusion chromatogram, 22
Size exclusion chromatography (SEC),
    3–7, 12, 19, 22–30, 95, 135,
    165, 187
Slide preparation, 315
Slit bandwidth, 136
Smooth muscle actin, 200, 207, 208,
    214, 216
α-Smooth muscle actin monocolonal
    antibody, 200, 207, 208, 214
Smooth muscle cell, 200
Smoothness parameter, 168
Sodium cyanoborohydride, 332
Sodium deoxycholate, 228, 233
Sodium metabisulfite, 332, 335, 337,
    340, 344
Sodium pentobarbital, 300, 301, 312, 313
Solubilization of Aβ, 25, 285
Spectra acquisition, 135, 138
Spectra/Por membrane, 222, 223
Spectral, 130, 132, 133, 136, 138, 140,
    148, 150, 189, 193, 222
Spectrofluorimeter, 38
Spherical oligomers, 26, 28, 31
SPIDER image processing package, 85,
    90
Steady-state fluorescence, 181, 221
Stereo dissecting microscope, 350
Stereotaxic apparatus, 350
Stern-Volmer,
    Stern-Volmer equation, 181
    Stern-Volmer plot, 180, 181
    Stern-Volmer quenching constant, 181
Straight filaments, 37, 44
Stretch frame alignment, 71

Sucrose homogenate, 267–271, 276, 275
Sulfur-35 (^{35}S), 326, 327, 344
Superdex 75 HR SEC column, 4–6, 21–25, 38
SVAU, 31, 81, 92, 95
SVD, 137
Swedish amyloidotic polyneuropathy, 256
Synchrotron, 75
α-Synuclein, 11, 17–28, 31, 67, 81, 87–92, 95, 100, 101, 149
α-Synuclein fibrils, 28
α-Synuclein protofibrils, 22, 24, 31

T

TappingMode, 104, 105, 111, 126–126
Tau protein, 32, 35–50, 175–177, 181, 184
Teflon, 73, 106, 120, 125, 132, 133, 268, 271, 276
Teflon discs, 106
Template-assisted model, 55
1,3,4,6-Tetrachloro-3α,6α-diphenylglycoluril, 332, 335
Texas Red dextran, 362
TFE, 138, 139
Tg2576 mice, 283
TgCRND8 mice, 283
Thiazine red, 360, 362
Thin film, 72, 132
Thioflavin S, 7, 9, 35, 37, 38, 41–44, 64, 89, 111–113, 134, 135, 139, 141, 148, 150, 156, 159, 178–178, 181, 238, 268, 281, 305, 307, 311, 358, 375
Thioflavin T, 42, 68, 129, 176, 187, 188, 192
Thiopropyl sepharose, 255, 258

Thiopropyl sepharose CB-CL, 258
Thiopropyl-sepharose 6B, 256–260
Time constant, 136
Tissue homogenization, 268–277
Tobacco mosaic virus (TMV), 91, 97, 98
Transgel, 350, 363
transgenic, 24, 54, 211, 212, 219, 267, 268, 273–283, 287, 288, 295–299, 308, 309, 321, 349–351, 363–379
Transmissible spongiform encephalopathies (TSEs), 53, 54, 60, 61, 64, 65, 227, 241, 309, 310
Transmission electron microscope, 38, 44, 70, 97
Transthyretin (TTR), 80, 85, 87, 90, 92, 100, 101, 145, 218, 237, 238, 255–260, 338, 347
2,2,2-Tribromoethanol, 350, 351
Trifluoroacetic acid (TFA), 5, 8, 20, 25, 137, 138, 146, 147, 186, 187, 266, 332, 341, 343, 367, 369
Tris(2,2'-bipyridyl)dichlororuthenium(II) hexahydrate (Ru(bpy)), 12
Tritium (^{3}H), 143, 326, 327
Trypan blue, 200, 205
Trypsin, 199–202, 205, 209, 210, 213, 216, 295
Tryptophan fluorescence, 175, 178, 179, 182
Turbidity assays, 68
β-Turns, 130, 143, 145
Two-photon, 349, 362, 363
Tyramine, 325, 330, 332, 338, 340, 346, 348
Tyramine-cellobiose, 325, 338, 342, 346

Tyrosine and tryptophan fluorescence spectroscopy, 178
Tyrosine residues, 328, 333, 336–338, 344, 345

U

Ultrafree-15 centrifugal filter device (5 kDa), 21, 24, 38, 40, 246, 257
Umbilical cord, 197–202
Uranyl acetate, 38, 44–47, 68, 70, 83, 95–98
Uranyl formate, 83, 95–98
Uranyl salts, 92
Ure2p, 90, 92, 96, 101
Urea, 134, 181, 245–253, 258, 269, 275, 276, 330, 344
UV absorption, 131, 135, 148
UV light, 12, 316

V

V30M, 85, 87, 256
V30M-TTR, 87
Vacuolation, 310, 316
Vandate, 98
Vascular smooth muscle cells (VSMC), 197–209, 218
Vectabond reagent, 315
Vectastain ABC Elite kit, 301
Vectrashield mounting medium, 200

Visible light, 172
von Willebrand factor, 211, 214, 216

W

Wavelength, 135
wax, 68, 71, 72, 318, 350, 357
Western blot, 57–60, 231–233, 277
Western blot analysis, 222–224, 247, 265–269, 273–276
Wet film, 97

X

X-Plor, 77
X-ray, 36, 47, 48, 67–82, 100, 129, 141, 148, 149
X-ray crystallography, 67, 82, 129
X-ray diffraction, 47, 48, 67–80, 148, 149
X-ray fiber diagrams, 36
X-ray fiber diffraction, 67, 70, 79–82, 100, 129
XCONV (CCP13), 77
Xylene, 301, 303, 306, 310, 311, 315, 318, 320
xylocaine/lidocaine, 350

Z

ZipTip, 186, 189, 193
ZnSe, 133, 138, 139, 147